Without Miracles

Without Miracles
Universal Selection Theory and the
Second Darwinian Revolution

Gary Cziko

A Bradford Book
The MIT Press
Cambridge, Massachusetts
London, England

This book was set in Sabon by Greg and Pat Williams, Gravel Switch, Kentucky, and was printed and bound in the United States of America.

Library of Congress Cataloging-in-Publication Data
Cziko, Gary.
   Without miracles: universal selection theory and the second Darwinian revolution / Gary Cziko.
      p.   cm.
   "A Bradford book"
   Includes bibliographical references (p.   ) and index.
   ISBN 0-262-03232-5 (hc)
   1. Natural selection. 2. Learning, Psychology of. 3. Knowledge, Theory of.
I. Title. II. Title: Universal selection theory.
QH375.C95   1995   95-943
575.01´62—dc20   CIP

Copyright Acknowledgments
The author and publisher are grateful to the following for their permission to use excerpts from:
Campbell, D. T. (1974). Evolutionary epistemology. In P. A. Schilpp (Ed.), *The philosophy of Karl Popper* (Vol. 1; pp. 413-463). La Salle, IL: Open Court. Copyright © 1974 by *The Library of Living Philosophers*. Reprinted by permission of Open Court Publishing Company, La Salle, Illinois.
Changeux, J.-P. (1985). *Neuronal man: The biology of mind*. New York: Oxford University Press. Copyright © 1983 by Librairie Arthème Fayard. Reprinted by permission of Georges Borchardt, Inc. and Librairie Arthème Fayard.
Perkinson, H. J. (1984). *Learning from our mistakes: A reinterpretation of twentieth-century educational theory*. Westport, CT: Greenwood Press. Reprinted with permission of Greenwood Publishing Group, Inc., Westport, CT.
Reader, J. (1988). *Man on earth*. Austin: University of Texas Press. Reprinted with permission of the University of Texas Press.

To Carol,
  for having selected me,
and to Anne-Marie and Paul,
  for putting up with her selection

Contents

Preface

One of the most remarkable aspects of the universe in which we find our-selves is that this universe has somehow acquired awareness and knowl-edge of itself. Between 10 and 20 billion years ago, the universe made a dramatic appearance after the Big Bang. As the resulting swarm of lep-tons, bosons, and quarks condensed into atoms, galaxies containing stars and planets slowly took shape, resulting in the dazzling display that the heavens provide on any clear night. And somehow, on at least one of these planets, living organisms emerged, the continued evolution of which even-tually produced a creature who would not only marvel at the stars and planets, but also create theories of their origin, the birth and process of life, and the growth of his own knowledge about the universe and the objects, forces, and life within it.

But it is not only our species that has somehow acquired knowledge of its environment. The fish's streamlined shape suggests functional knowl-edge of the physical properties of water. The design of the eagle's wings reveals sophisticated knowledge of aerodynamics. The deadly effective-ness of the cobra's venom shows useful knowledge of the physiology of its prey. The functioning of the bat's remarkable echolocation system de-pends on knowledge of the transmission, reflection, and speed of sound waves. In all these instances we notice a puzzling fit between one system and another, an adaptation of an organism to some aspect of its environ-ment. Indeed, knowledge itself may be broadly conceived as the fit of some aspect of an organism to some aspect of its environment, whether it be the fit of the butterfly's long siphon of a mouth to the flowers from which it feeds or the fit of the astrophysicist's theories to the structure of the universe.

But how did such remarkable achievements of fit arise? How did the animate world obtain its impressive knowledge of its surroundings? And how do organisms continue to acquire knowledge and thereby increase this fit during their lifetimes?

In this book we will explore various attempts to provide an explanation for the emergence of the knowledge demonstrated by the fit of living organisms to their environments. Chapter 1 provides a quick tour of some striking puzzles of fit, ranging from the biological world of single-celled organisms to the technological world of jet aircraft. The four chapters of part II contrast three approaches to explaining aspects of biological fit and show why natural selection, operating either over the course of many generations or over the course of a single organism's life, is today considered to provide the best explanation for such fit. Part III then extends natural selection by proposing selectionist explanations for other types of fit demonstrated by living organisms, with special emphasis on our own species. Here we will see how, in many different fields from philosophy to technological development, evolutionary solutions involving variation and selection are being increasingly proposed and accepted as explanations for the growth of many different types of knowledge. The two chapters of part IV show how the evolutionary technique of cumulative variation and selection is now being used in the applied sciences to facilitate the function of machines, molecules, and organisms. Finally, part V introduces and argues for *universal selection theory*, the bold conjecture that all knowledge and knowledge growth are due to a process of cumulative blind variation and selection.

Although many and varied disciplines are encountered in this book, they are organized around just two quite simple underlying themes. The first is that in all these fields of inquiry, three major types of explanation have been proposed for the origin and growth of knowledge, that is, for the increase in fit between an organism or product of an organism and its physical or social environment. These three explanations are referred to as *providential*, *instructionist*, and *selectionist* theories. The second theme is that in the many diverse disciplines of knowledge we are about to explore, the first two types of explanations have repeatedly been replaced by the third. Keeping these two themes in mind should help the reader keep his or her bearings as we now commence our excursion through several important and fascinating fields of knowledge.

Acknowledgments

Although I am very pleased to be able to thank the many individuals who contributed in one way or another to the preparation of this book, I must admit to being a bit uneasy about acknowledging them here. My uneasiness has two sources. Given the many different fields of knowledge included in this book, it should not be surprising that I am indebted to a great many individuals for their help. While I cannot guarantee this work to be free of errors of fact, logic, and interpretation, I can assure the reader that there would have been many more such errors had it not been for the expertise, vigilance, and kindness of the individuals who read and commented on all or parts of the manuscript or who took the time to discuss the book's topics with me. However, in acknowledging these many deserving individuals, I fear that I will inadvertently make some omissions. I can only offer my apologies if this turns out to be the case.

The second reason for my uneasiness is that I am quite sure that at least some of the individuals who provided me assistance will not approve of the use to which I have put their help. So I need to emphasize (paraphrasing from the broadcast media) that the views expressed in this book are my own and do not necessarily reflect the views of all or any of the persons acknowledged below.

I wish first to acknowledge those helpful colleagues and students affiliated (now or then) with my home institution, the University of Illinois at Urbana-Champaign. Those who read parts of the manuscript or who responded to my requests for information include Thomas Anastasio, Gary Bradshaw, Richard Burkhardt, the late John Easley, Janet Glaser, Georgia Green, William Greenough, Jeffrey Horn, John Katzenellenbogen, Roderick MacLeod, Jay Mittenthal, William Nagy, Ray Olsen,

Ralph Page, Michelle Perry, Torbert Rocheford, Jenny Singleton, and A. Joshua Wand. The Bureau of Educational Research of the University of Illinois College of Education supported the initial phase of this work with a two-year appointment. The most valuable aspect of this support was the help provided by two graduate research assistants, Joel Judd and Evy Ridwan, both now having attained well-earned doctorates in educational psychology. This book also profited from the comments, criticisms, and lively discussions provided by the graduate students who over the last few years participated in my "Knowledge Processes" course and who were repeatedly subjected to various drafts of this work.

Individuals at other institutions whose assistance I wish to acknowledge are W. Thomas Bourbon of the University of Texas Medical School at Houston, Lee Emerson of Frasca Air Service Inc., Wayne Hershberger of Northern Illinois University, David Hull of Northwestern University, Richard Marken of Aerospace Corporation, Henry Perkinson of New York University, Henry Plotkin of University College London, William Powers of the Control Systems Group, Robert Siegler and G. Richard Tucker of Carnegie-Mellon University, M. Martin Taylor of the Canadian Ministry of Defence, and Diane Schneider of Sanofi Winthrop.

Several people merit special recognition. Greg Williams, who among many other duties has served as publisher and archivist for the Control Systems Group in Gravel Switch, Kentucky, read the entire manuscript and made many valuable comments and suggestions.

Both Paul Bloom of the University of Arizona and Andrew Barto of the University of Massachusetts at Amherst first came to know of my work as anonymous reviewers for the MIT Press. They graciously revealed their identities (and their electronic mail addresses) when they realized that doing so would permit me to take better advantage of their expertise for improving my book. I am grateful to them for going well beyond their expected duties as reviewers. The book also benefited from the opinions of Henry Plotkin of University College in London, four other reviewers who remain anonymous, and from the editorial support and guidance of Fiona Stevens of the MIT Press.

Unfortunately, I have never had the pleasure of meeting Oxford zoologist Richard Dawkins, but I have read much of what he has written. For Charles Darwin, it was the reading of Thomas Malthus's *Essay on Popu-*

lation in 1838 that led him to his natural selection theory of biological evolution. For me, it was my chance encounter with Dawkins's *The Blind Watchmaker* in 1987 that opened my eyes to the awesome power of cumulative blind variation and selection, and suggested to me how selectionist processes could be at the heart of the emergence of all forms of knowledge.

It took me another year to discover that psychologist Donald T. Campbell (now professor emeritus) of Lehigh University had come to the same radical conclusion three decades earlier. Through a series of visits and communication by telephone, letters, and electronic mail, Don served as a constant source of inspiration and encouragement for this book. His influence will be found in many of the following chapters.

For what turned out to be a much longer period than anticipated, my family graciously shared me with the demands of preparing this book, although not without the constantly recurring query, "When will it be finished?" I am grateful to my children, Anne-Marie and Paul, for allowing me the time I needed, and to my wife, Carol Binnington, who provided the emotional support for getting me through the many ups and downs of this project. Paul Cziko is to be acknowledged also for his assistance in preparing the book's figures.

"I THINK YOU SHOULD BE MORE EXPLICIT HERE IN STEP TWO."

I

The Need for Selection

1

Puzzles of Fit

We see beautiful adaptations everywhere and in every part of the organic world.
—Charles Darwin[1]

Are you awed by the exquisite fit between organism and environment, and find in this fit a puzzle needing explanation? Do you marvel at the achievements of modern science, at the fit between scientific theories and the aspects of the world they purport to describe? Is this a puzzling achievement? Do you feel the need for an explanation as to how it could have come about?
—Donald T. Campbell[2]

Much about the world we live in captures our attention and elicits our wonder. Majestic snow-crowned mountains demand our gaze as they rise above the surrounding plain. Raging storms hurl rain, wind, and jagged spears of fire. The normally unyielding ground on which we walk shakes violently as the continents continue their slow drift over the earth's molten core. The rising and setting of the sun, the waxing and waning of the moon, and the orderly march of the seasons and tides provide welcome rhythm and predictability to our lives and activities.

Whereas our ancestors invoked gods, goddesses, and a host of spirits to account for such happenings, science has now provided a naturalistic understanding of these and other physical events as the mechanical consequences of principles of physics and chemistry. The sun is no longer driven across the sky in a chariot, and serpents no longer consume the moon. Myths and religious interpretations for such natural phenomena certainly continue to exist, but few expect miraculous explanations to be offered in our schools and universities.

But there is another side to the world we inhabit, the *organic* side that Darwin mentions in the epigraph above. And here is even more to marvel at. This living world includes both microscopic bacteria and giant redwood trees, the wildly colorful birds and insects of the rain forests, the large mammals of the African savannah, the fishes and bizarre denizens of the ocean's depths, as well as the creature who wrote these words and the one now reading them. These and other living organisms evoke our awe because they appear so marvelously designed to fit the particular environment in which they exist. This fit of organism to environment involves not only the structure and behavior of organisms, but often also the inanimate products of these organisms. Whereas biology now recognizes that living organisms must observe the materialistic laws of the physical world, all forms of life appear somehow to have obtained knowledge of their environment that transcends the blind, uncaring forces of physics, and the ignorant, inevitable principles of chemistry.

The common web-spinning spider provides many examples of wonderful instances of fit. Its eight legs, moving in a coordinated fashion, allow it to travel rapidly across the ground, scale vertical surfaces, and even go with ease upside down across our bedroom ceilings. It can drop quickly and safely from high locations by riding down on its internally produced silk dragline, excreted by special organs called spinnerets, and if it should change its mind, it can hoist itself up again by reeling in the line it has just laid. Being a predator, the spider must catch insects to survive. This is accomplished by constructing a well-designed web whose strands ensnare and quickly entangle any edible insect unfortunate enough to touch them. When this occurs, the spider rushes over and injects a deadly venom into the helpless victim. Then it must decide whether to consume the meal now, drawing out the nutritious body fluids and perhaps spraying it with digesting juices to make it even more palatable to the spider's sucking stomach, or save it for later by wrapping the lifeless prey with silk. Spiders also engage in elaborate courtship behaviors, with the customs of some species requiring the male to present the female with a gift-wrapped fly so that he can mate with her without being eaten himself. Some females protect their eggs until the spiderlings emerge and cast their draglines into the wind to go ballooning off to new locations to establish residence.

The spider thus provides us with many instances of fit—a remarkable meshing of its body, behavior, and products with its environment. Its appendages and their movements fit the demands of locomotion. Thanks to a special oil covering its body, it can move with ease over its sticky web. Its web is designed to fit the space in which it is built and to enmesh unsuspecting insects. Its venom is chemically suited to kill its prey. And we have not even mentioned the intricately designed respiratory, circulatory, digestive, nervous, and reproductive systems that function together to give life to the spider and permit its reproduction.

These and other instances of organic fit are quite unlike anything in the inanimate world. They attract our attention and elicit our wonder because they seem to have a purpose. Legs are for walking, mouths are for eating, and webs are for snaring prey. In biology, the word *adaptation* (from the Latin *adaptare*, "to fit") is used to describe such instances of fit between organism and environment (with the environment understood to include other organisms as well, as in the fit between a male's and a female's sexual organs).

When one looks at the fit and function of these adaptations, it does not seem unreasonable to consider them as forms of knowledge. In an important sense, the spider's legs and feet know something about the terrain that the spider considers home. Its venom has a kind of knowledge of the physiology of its prey. And the male spider knows what it must do to find a female and entice it to mate. Of course, in the case of spiders, we are referring to a type of built-in biological knowledge, not the conscious, reflective knowledge that we associate with human thought. But insofar as biological adaptations are functional and aid in an organism's survival and reproduction in what is usually an uncaring and competitive world, these instances of fit clearly reflect knowledge of the world that the organism inhabits.

But the fit of body parts, behavior, and instinctive products (such as the spider's web) are not the only adaptations in the organic world. More is to be found by looking deeper using the instruments of modern science. The cells themselves and the precisely designed organelles they contain reveal remarkable fit in form and function to their surroundings within the organism. Mitochondria produce energy, ribosomes create new proteins, and DNA of the nucleus is an archive of all the genetic information the organism requires to function and reproduce.

Also at the level of cells is the exquisite fit achieved by the mammalian immune system. Ever since it was discovered late in the nineteenth century that animals are able to produce antibodies that provide protection from disease, the mechanism of this process has been the subject of extensive research, leading to discoveries that have earned 12 Nobel prizes since 1901.

An intriguing puzzle motivated this research. Since many different types of antibodies were found, each effective against a particular antigen (virus, bacterium, chemical toxin), it was thought that information for the construction of each antibody was carried in the individual's genes. However, Austrian-born Karl Landsteiner discovered early in the twentieth century that the immune system is capable of producing antibodies that are effective against *artificially synthesized antigens*, that is, foreign substances with which the animal had no contact either during its own lifetime or during its evolutionary past. The creative ability of the immune system to produce antibodies to novel antigens is even more amazing when it is realized just how well an antibody fits its target antigen, a degree of fit comparable to that a key must have to open a lock.

One also must consider more abstract types of fit, particularly when one reflects on the knowledge that humans require to survive in the world's many different environments. Indeed, the human species lives in a wider range of habitats than any other vertebrate on earth, from the steaming jungles of the tropics to the numbing Arctic tundra, from low-lying coastal areas to the Andean *altiplano* 4000 meters above sea level, from sparsely settled wilderness to teeming cities. We have so far made only short visits away from the surface of our home planet, but it may not be too long before we establish permanent settlements in outer space and on the ocean floor.

The desert nomads of the Sahara wear long, loose-fitting garments to protect them from the sun, dry wind, and blowing sand. The Eskimos of the frozen Arctic wear warm garments made from the skins of the animals they hunt. People living in remote tropical rain forests often wear little or no clothing at all. Astronauts working in space wear pressurized, air-conditioned spacesuits with visors coated with a thin transparent film of gold to reflect the intense, damaging rays of an unattenuated sun. Other aspects of how these remarkably diverse people live—what they eat and how they obtain and prepare it, how many spouses and children they have, how their living quarters are made and maintained—provide clear evidence of fit to features of their particular environment, without which human life would not be possible in its wide range of habitats.

But if human knowledge fits aspects of the world, it must also be the case that our brains, the seat of our knowledge, are adapted to aspects of the world. Arguably the most complex object yet discovered in the universe, the human brain is a rather soggy, irregular sphere, about the size of a large grapefruit. Within its modest volume, however, it has enough "hardware" to make even the most sophisticated supercomputer appear crude by comparison, including about 20 billion electrochemical switching units, with each unit having between 1000 and 10,000 connections to other similar units. The pattern of connections and activation of these neural circuits underlie just about everything we do, from walking, eating, and making love to talking, cooking, and making scientific discoveries. The plasticity of our brain allows us to acquire new skills and knowledge as required by our environment, such as learning to speak another language or learning to ride a bicycle, drive a car, or fly a space ship.

Other instances of fit can be found in the material products of human scientific and technological knowledge. A good example is the modern passenger aircraft, which has helped to make today's world a much smaller place. Able to accommodate hundreds of travelers in relative (if often somewhat cramped) comfort, today's passenger jet is jammed full of sophisticated technology. Its sleek aluminum skin resists corrosion and combines strength with low weight. The configuration of its wing and tail surfaces provides the lift necessary for flight. Its engines deliver amazing thrust for their size and weight. Sophisticated navigational and radar systems permit the jet to avoid both bad weather and other aircraft and keep a steady course for its final destination. Flying 10,000 meters above the ground at 1000 kilometers per hour, passengers are provided with in-flight movies, meals, lavatory facilities, and even telephones to keep in touch with their earthbound families, friends, and business associates.

The sophisticated technology represented by the jet is even more impressive when one takes into account the planning and coordination that is required to construct it. Considering the complexity of these remarkable flying machines, it is hard to imagine how any single person could ever have all the expertise necessary to design one, let alone the time, skill, and strength to build one. But despite these difficulties, air travel in the industrialized countries has become almost as routine as taking the family car on a trip, with accidents much lower on a passenger-distance basis than for any other form of transportation. Clearly, the modern jet airplane is an amazingly fit vehicle for the purpose of transporting passengers safely, quickly, and comfortably over distances that not very long ago would have taken months to cover. And its success in doing so reflects well on the scientific knowledge that is applied to its design, construction, and operation.

As these examples indicate, the world of living organisms and their products provides countless remarkable instances of fit. The spider's body, behavior, and web make it well equipped for the demands of its predatory lifestyle. The mammalian immune system is able to produce antibodies that precisely match the structure of disease-producing antigens. The variety of human cultural knowledge is remarkable for its adaptation to the very different environmental and social conditions throughout the world. The human brain's exceedingly intricate "wiring" permits us to experience and do a remarkable variety of things. Our stubborn insistence on finding ever faster, easier, and more enjoyable ways to satisfy our needs for food, shelter, sex, health, companionship, and entertainment has spurred the growth of science and technology and has led to technological achievements like the jet airplane, gene therapy, and the personal computer.

It could be argued that such fit is a universal characteristic of life and its products. Indeed, if the earth should ever lose the delicate balance that makes life here possible, future extraterrestrial visitors to our then-dead planet would certainly look for artifacts of fit to determine if life had ever existed and the degree to which it might have been accompanied by intelligence.

Fit, as used here, may be difficult to define formally, but as a judge once said in a pornography case, we seem to know it when we see it. What criteria do we use to determine that fit is present? It must appear that it was designed for some purpose and is able to achieve this purpose by functioning in a way that takes into account important, relevant aspects of its environment. A structure or behavior is fit only insofar as it is adapted to its environment and contributes in some useful way to the organism or system that created it or of which it is a part. We recognize such fit when we observe "any system composed of many interacting parts where the details of the parts' structure and arrangement suggest design to fill some function."[3]

Such instances of fit also demonstrate a degree of *complexity* that is highly unlikely to be due to chance. This is certainly the case when we observe any living organism, even one as simple as a common bacterium such as *Escherichia coli*. The probability that billions of different organic molecules would by pure chance just happen to assemble themselves to form the complex arrangement required to produce a cell that is able to take in and metabolize nutrients, eliminate wastes, move about, and reproduce seems (and is) too tiny to consider seriously. It is the astronomical improbability of such functional arrangements of many components that makes them puzzling in the first place. Such achievements of design are instances of what will be called "adapted complexity" throughout this book. When considering examples of adapted complexity in the biological world, such as the spider's prey-catching behavior, it appears as if knowledge had somehow been obtained by an organism about some aspect of its environment.

But there is something even more puzzling about the many instances of adapted complexity among living organisms and their products. In cases such as the mammalian brain and immune system, a continuing process of "fit making" results in the emergence of further achievements of adapted complexity. The ability to fashion novel and more impressive instances of fit is itself an instance of adapted complexity. It is at the same time a very special kind of adapted complexity that can perhaps be best described using the related but distinct term adapt*ive* complexity, with the descriptor *adaptive* intended to indicate a continuing process of fit making in contrast to *adapted*, which describes already achieved fit.

The goal of this book is to explore explanations for both adapted complexity (already achieved fit) and adaptive complexity (ability to achieve new fit) in our world—existing puzzles of fit and the emergence of new ones that are just about everywhere we look, and that we have yet to find anywhere else in the universe. A quite simple and compelling explanation for the puzzles of fit demonstrated by the structure and behavior of living organisms was proposed over 130 years ago and is still regarded as the central unifying principle of biology. No comparable explanation, however, is generally accepted for all other puzzles of fit of the type mentioned above. The purpose of this book is to present a case for just such an explanation—and one that works without recourse to miracles.

II
The Achievements of Selection

2

The Fit of Biological Structures

Providence

Every organised natural body, in the provisions which it contains for its sustenation and propagation, testifies a care, on the part of the Creator, expressly directed to these purposes.

—William Paley[1]

Instruction

All that nature has caused individuals to gain or lose by the influence of the circumstances to which their race has been exposed for a long time, and, consequently, by the influence of a predominant use or disuse of an organ or part, is conserved through generations in the new individuals descending from them, provided that these acquired changes are common to the two sexes or to those which have produced these new individuals.

—Jean-Baptiste Lamarck[2]

Selection

Slow though the process of selection may be, if feeble man can do much by his powers of artificial selection, I can see no limit to the amount of change, to the beauty and infinite complexity of the coadaptations between all organic beings, one with another and with their physical conditions of life, which may be effected in the long course of time by nature's power of selection.

—Charles Darwin[3]

Most of the puzzles of fit mentioned in the preceding chapter have to do with either the structure or behavior of living organisms. The striking diversity and adapted complexity of our planet's life forms have led philosophers and scientists to expend considerable time and energy attempting to account for their existence. This chapter presents three types of explanations for the fit of organisms to their environments.

Fit by Providence

The oldest, most intuitively appealing, and still most widely held explanation for the adapted complexity shown by living organisms is that of a knowledgeable designer who expressly created the remarkably fit forms and behaviors we observe for the very purposes they serve. Thus we find Aristotle (384–322 B.C.) asking how it is that "the front teeth [of humans] come up with an edge, suited to dividing the food, and the back ones flat and good for grinding it," and that "the swallow makes her nest and the spider his web, and that plants make leaves for the sake of the fruit and strike down (and not up) with their roots in order to get their nourishment."[4] Aristotle's answer is that nature created all these things for a purpose, and so they consequently reflect nature's goals and knowledge. This philosophy, which attributes the adapted complexity of living entities to some higher source of purpose and knowledge, became a major component of Judeo-Christian thought that eventually replaced the notion of a purposeful Mother Nature with that of an all-knowing, all-powerful, and personal God.

It is within this tradition of Christian thinking that we find William Paley, an English archdeacon, theologian, and philosopher who lived from 1743 to 1805. In his *Natural Theology,* first published in 1802, Paley used the adapted complexity found in the design of plants and animals as powerful arguments for the existence of God. His thesis is quite simple and, on first hearing, quite appealing. Paley asks us to consider our reaction to finding a watch lying on the ground and being asked to explain its origin. On opening the watch's case, we find an elaborate and complex mechanism consisting of finely interlocking wheels, cogs, and springs. We also notice that when wound, the watch's mechanism functions for a particular purpose, that is, to mark the passage of time accurately. Clearly, it would not occur to us that the watch had always lain at that spot on the ground, or that it had somehow been constructed and deposited there by the blind and random forces of nature in the way that we might explain the presence of a nearby stone. It is absolutely clear to Paley that "the watch must have had a maker: that there must have existed, at some time, and at some place or another, an artificer or artificers who formed it for the purpose which we find it actually to answer: who comprehended its construction, and designed its use."[5]

Paley then goes on to provide numerous examples from plants, insects, and animals that show even more impressive achievements of adapted complexity than that of the watch. These include bones and their joints, muscles, circulatory systems, internal organs, sense organs, and instinctive behaviors. To see that Paley's idea of adapted complexity is consistent with the concept of fit introduced in the previous chapter, we need only to look at the headings listed under chapter 17 in the table of contents of a later edition of his book. There we find "wings of birds—fins of fish—air and water; ear to the air; organs of speech—voice and respiration to air; eye to light; size of animals to external things; of the inhabitants of the earth and sea to their elements; sleep to night." For all of these puzzles of fit there can be, for Paley, only one possible explanation—a grand designer in the form of an all-knowing and benevolent God who fashioned every remarkable contrivance of the living world for the very purpose for which it is seen to be fit.

The argument from design, as Paley's reasoning is now known, has probably been used in some form or another in every human society that ever developed to the point of being able to marvel at the living world of nature and ponder its origin. All human societies, even those without advanced technology, are able to build shelters and fashion clothing for protection against the elements, and produce hunting and farming tools to obtain food. No society, however, not even the most technologically advanced, has developed the technology and tools for creating even the simplest living organism out of nonliving matter. It therefore seems only reasonable to conclude that the existence of life, in all its diversity, must be the creation of an intelligence and power that far surpasses that of humankind. This type of explanation is attractive in that it can reduce many mysteries (the existence of millions of different living plants and animals) to just one (their creation by a supernatural power). If a major purpose of religion is to make sense out of the world, it is hardly surprising that most if not all religions explain the existence of the earth's flora and fauna as being provided by a supernatural being. This can be referred to as a *providential* explanation for the fit of living things to their environments.[6]

The argument from design to divine providence, however, loses its appeal when examined more closely. If God is responsible for the origin of the adapted complexity of all living organisms, what (or who), may we ask, is responsible for the origin of God? A typical answer is that God has no origin, that He always existed and will continue to exist for all eternity. This, however, is tantamount to an admission that *an explanation for the existence of a complex being is not necessary.* If this is indeed the case, the argument from design effectively contradicts itself, for all it does is pass the buck of the origin of adapted complexity back to a being who must be even more complex than any of his creations but for whom no explanation is needed![7] If a being as knowledgeable as God (who, by all accounts far surpasses the complexity of a watch) requires no explanation, then why does the watch? And why do the earth's living creatures, including our own species? We can only conclude, therefore, that this oldest, most intuitively appealing, and probably still most widely held explanation is seriously flawed. That is not to say, however, that we can confidently discount the idea that living organisms are the result of special creation by a supernatural being. Indeed, it could be argued that an explanation *does* exist for a creative deity, but that we are simply unable to grasp it. Nonetheless, when examined critically, the argument from design should fail to convince us of the existence of God in the way that it so compellingly convinced Paley and other natural theologians of his and our time.

Fit by Instruction

Jean-Baptiste Lamarck lived in France from 1744 to 1829, a period of intense interest in natural history in Europe. Lamarck, who first gained fame as a botanist for his practical guide to French plant life, went on to develop a new science to account for the origin, structure, and interrelationships of all living organisms. He called his new science *biologie.*[8]

Lamarck's biological theory had two central tenets. First, although early in his career Lamarck believed that all species had originally come into existence in much the same form as he saw them during his lifetime, he eventually came to accept the mutability of species, that is, that over the course of long periods of time, organisms could change enough to evolve into new species. Second, he saw the complexity of organisms not as the direct work of God, but rather as a natural outcome of the "power of life" and the interaction of organisms with their environment. We can therefore appreciate that Lamarck's theory of evolution was primarily a *constructive* or *creative* one, in contrast with the purely providential perspectives of Aristotle and Paley.[9]

How did Lamarck propose to explain the increasing complexity of organisms and the correlation of their form and behavior to their environments without recourse to God?[10] He invoked two processes. First, he observed that organisms seemed to change in response to their environment. For example, the giraffe, finding little edible ground vegetation, would, on straining upward to find its sustenance from the trees, stretch and thereby permanently elongate its neck and legs:

> In regard to habits, it is interesting to observe a product of them in the peculiar form and the height of the giraffe. . . . This animal, the largest of the mammals, is known to live in the interior of Africa in places where the earth is nearly always arid and without pasturage, obliging it to browse on the leaves of trees and to continually strive to reach up to them. This habit, maintained for a long time by all the members of the race, has resulted in the forelegs becoming longer than the hind legs and the neck being so lengthened that the giraffe, without standing on its hindlegs, with its head raised reaches a height of six meters.[11]

Conversely, a lizard finding it advantageous to remain in a cave completely devoid of light, would eventually lose the use of its eyes. In other words, organs that were used a great deal would be stimulated to continue to develop, whereas those used seldom or not at all would slowly atrophy and perhaps ultimately disappear. This was Lamarck's principle of "use and disuse" by which "the development of the organs and their power of action are always related to the use of these organs."[12]

Second, Lamarck believed that the "characters" acquired by an organism through use and disuse were inherited by its offspring:

> Everything which has been acquired, outlined or changed in the organization of the individuals in the course of their life, is preserved through the reproduction, and is transmitted to the new individuals which spring from those who have undergone these changes.[13]

This principle came to be known as the inheritance of acquired characters, and it is for this theory that Lamarck is still remembered today.

Lamarck was so convinced that changes acquired during an organism's lifetime were passed on to its offspring that he wrote that the "law of nature by which new individuals receive all that has been acquired in organization during the lifetime of their parents is so true, so striking, so much attested by facts, that there is no observer who has been unable to convince himself of its reality."[14] Indeed, the theory was well accepted in Britain and Europe throughout most of the nineteenth century. Yet few such apparent "facts" have caused more grief in biology.

Although Darwin convincingly argued in *The Origin of Species,* published in 1859, that there was more to evolution than the inheritance of acquired characteristics, it was not until the 1880s that the transmission of acquired characteristics from parents to offspring was seriously challenged. German embryologist August Weismann (1834–1914) made an important distinction between cells of an organism that pass genetic information to the offspring (germ cells) and all the other cells that have no direct role in reproduction (somatic cells). According to Weismann, it was simply not possible for changes to the somatic cells (for example, those in the giraffe's neck) to be transmitted to the germ cells so that offspring could inherit the acquired characteristics of its parents. He set out to demonstrate his point by amputating the tails of several generations of mice and showing that successive generations failed to grow shorter tails.

In contrast, Austrian biologist Paul Kammerer appeared to have obtained considerable success in producing what was thought to be clear evidence of the inheritance of acquired characteristics. However, in 1926 it was announced that Kammerer's last remaining specimen of Lamarckian inheritance, the so-called nuptial pads of the midwife toad, showed no evidence of the claimed evolutionary change and had instead been doctored using injections of India ink. Six weeks later on an Austrian mountain path Kammerer put a bullet through his head. Unfortunately, this personal tragedy did nothing to dissuade Russian biologist Trofim Denisovich Lysenko from his doomed attempt to increase the productivity of Soviet agriculture based on Lamarckian principles. Stalin's receptivity to and imposition of Lysenko's Lamarckian ideas were to cripple the development of Soviet biology and genetics until the 1960s.[15]

Weismann's separation of germ from somatic cells remains today in biology an almost universally accepted barrier to Lamarckian inheritance of acquired characteristics.[16] Lamarck's principle of use and disuse also encounters serious difficulties. Let us take the blacksmith's arm as an example.[17] We know that the blacksmith is required to do much of his work with a heavy hammer, and we therefore expect this activity to make him strong and muscular, at least in the arm he uses to swing his hammer. The same expectation leads countless fitness-conscious individuals to subject themselves to various forms of discomfort in gyms and health clubs in an attempt to firm up muscles that would otherwise be flab. This would seem to be a clear example of the acquisition of a characteristic from interacting with the environment. But we must stop and ask ourselves why it is that the increased use of muscles makes them bigger and stronger. Indeed, from our observations of nonliving objects we should expect just the opposite. Our shoes do not grow thicker soles the more we walk in them, nor do they become thin by the disuse of being left unworn in the closet for long periods of time. On the contrary, their soles wear thin from extended use and maintain their original condition only if not used. Automobile engines do not develop more horsepower with increased use, or atrophy and shrink away when left to sit in a garage or museum for extended periods of time. Why then should the blacksmith's arm not also be worn away by its heavy work load? How can the environment instruct the arm's muscles to grow in size, strength, and endurance in response to the work they do? And what is it that tells the retired blacksmith's arm to give up its hard-earned strength? In an attempt to provide an explanation for such adaptive changes, Lamarck made reference to a providential-sounding "plan of nature" and the "power of life" by which organisms could somehow bring about necessary adaptive changes by their own efforts.[18]

Lamarck's theory fails not once but twice for its reliance on environmental instruction. First, according to the principle of use and disuse, the environment must in some way transmit to the organism instructions to make the required adaptive changes. The cold must somehow instruct the beaver to grow thicker fur. The stretching of the giraffe's neck must instruct it to grow longer (and also stronger to support its increased size). And the heavy workload of the blacksmith's arm must instruct its muscles to gain in strength and endurance. Second, since only a single living cell is passed from parent to offspring, the changed body of the parent must somehow transmit instructions to the germ cell (sperm or egg) for these changes to be passed on. Yet biological science has been unable to discover any mechanism that would make either mode of instructive transmission possible.[19]

So today Lamarck's theory of the inheritance of acquired characteristics is rejected by virtually all biologists. It did, however, enjoy wide popularity throughout most of the nineteenth century, and this despite the total lack of any clear experimental support. It even seems likely that it would have retained its popularity much longer if a competing, noninstructionist explanation for the adapted complexity of life forms had not been proposed in the second half of the nineteenth century.

Fit by Selection

Any student preparing for the bachelor of arts degree examination at Cambridge University in 1831 was expected to be thoroughly familiar with the arguments for the existence of God provided by Paley. One such individual who was particularly impressed with the bishop's examples of fit was a divinity student and naturalist named Charles Robert Darwin (1809–1882).

Although Darwin remained impressed until the end of his days by the puzzles of fit shown by nature's living things, he eventually came to reject Paley's providential explanation for their origin. And although he was never able totally to reject Lamarck's theory of the inheritance of acquired characteristics (nor did he understand why such an instructionist theory was genetically problematic), he found it incomplete. Consequently, he developed an alternative theory for the evolution and growth of adapted complexity that did not rely on Lamarckian instruction. Darwin's theory, published in 1859 as *On the Origin of Species by Means of Natural Selection*, was the major scientific achievement of the nineteenth century.

Darwin discovered an explanation for the emergence of adapted complexity in nature that required neither a supernatural provider nor an instructive environment. He proposed a thoroughly natural theory to account for organic fit, the basic principles of which are so easy to grasp that once they are, the discovery itself seems much less remarkable than the fact that so many great minds before him had failed to come up with it first. As Darwin's close friend and defender Thomas Huxley wrote after learning of the theory of natural selection: "How extremely stupid not to have thought of that!"[20]

The theory of evolution has three primary components. First, Darwin observed that all species showed considerable *natural variation* in the forms and behaviors of individual organisms, due primarily to the fact that offspring usually differed, if only slightly, from their parents. Second, he recognized *superfecundity*, that is, species typically produce many more offspring than can be supported by the environment. These two conditions being the case, Darwin's great insight was to realize that the individuals that by their particular variations are better suited to survival and reproduction must leave more offspring (which eventually come to dominate the population) than those that, for whatever reason, are less fit to survive and reproduce. This process Darwin referred to as *natural selection.*

The theory of natural selection seems strikingly simple when so stated, yet it was Darwin's genius to realize that such a process, operating cumulatively and given enough time, could account for the evolution of the wondrous diversity and adapted complexity of all living things, including our own species, starting with the simplest living organism as a common ancestor. For the same reason that the theory became and remains the cornerstone of biology (it relies on neither instruction nor providence), it has evoked more controversy and condemnation than any scientific theory ever proposed.

We will not discuss here the details of the theory, or give examples of evidence supporting it, since many excellent readable accounts are available.[21] However, since a selectionist account of adapted complexity as first proposed by Darwin is the major theme of this book, a few important points have to be made before proceeding.

First, although considerable controversy concerning Darwin's theory of evolution still exists among nonscientists, the basic principles continue to constitute the core of modern biology. The observation of evolution on a small scale both in nature and the laboratory, the patterns of similarity and diversity revealed by the scientific classification of organisms, and fossil records all provide strong support for natural selection.[22] And although biologists still do not agree on all the details of how evolution occurs, no serious rival theories have been proposed to challenge the process.

Second, since Darwin's theory explains evolution as a natural process that does not entail the existence of a supernatural designer, his emphasis was quite different from Paley's. Paley emphasized what he saw to be the perfect design of the earth's living things; Darwin repeatedly made a point of describing how nature's design, although complex and well adapted to its environment, is not necessarily optimal. This is because "ideal arrangement is a lousy argument for evolution, for it mimics the postulated action of an omnipotent creator. Odd arrangements and funny solutions are the proof of evolution—paths that a sensible God would never tread but that a natural process, constrained by history, follows perforce. No one understood this better than Darwin."[23] It is for this reason that Darwin provided so many examples of useless organs and inelegant, although still functional, solutions to the unending challenges of survival and reproduction.

It is crucial to realize that the variation considered by Darwin to be the fuel for natural selection and evolution was not required to be in any way intelligent or foresighted. Confessing ignorance as to the source of variations in nature, he made clear that it was not the presence of variations alone that resulted in adapted complexity, but rather the *cumulative selection* of those relatively few variations that by blind chance gave the organism fortunate enough to possess them even the slightest advantage for survival and reproduction. Thus Darwin concluded chapter 5 of *The Origin* ("Laws of Variation") with the statement that "whatever the cause may be of each slight difference in the offspring from their parents . . . it is the steady accumulation, through natural selection, of such differences, when beneficial to the individual, that gives rise to all the more important modifications of structure, by which the innumerable beings on the face of this earth are enabled to struggle with each other, and the best adapted to survive."[24]

This "steady accumulation" of beneficial differences is an essential part of the theory. Since variations are blindly and ignorantly produced, it is overwhelmingly likely that any given variation will be detrimental to the survival and reproduction of the organism, particularly since the current standards of form and behavior for that species have obviously so far proved successful, or the species would no longer exist. But organic evolution does not abide by the maxim, "if it ain't broke, don't fix it." Through genetic mutation and the sexual reshuffling of genes, new variations are constantly being tried out. And whereas it is a rather rare event that any one of them will prove to be advantageous, given enough variations and enough time, an adaptive change will eventually occur, and to this improvement (as measured by reproductive success) more can be added later. These adaptive changes will tend to be gradual in their evolution since more drastic changes will be even less likely to result in better adaptations. This does not mean that a sudden and dramatic mutation cannot be beneficial to a species, but simply that a very large adaptive change (such as the appearance in one generation of a complex, functioning eye in a species that had been completely blind) is astronomically improbable, and therefore does not provide a reasonable explanation for the puzzles of fit achieved by evolution. The process of adaptive evolution can therefore be summarized as "cumulative blind variation and selection."[25]

Neither is natural selection guided by purpose or planning. Many of Darwin's readers misconstrued just such a purposeful, foresighted, and providential view, imagining nature to be guided by a supernatural intelligence in the production and selection of variations most useful to the organism. It was for this reason that Darwin later adopted Herbert Spencer's term "survival of the fittest," since as Alfred Russel Wallace noted in a letter written to Darwin in 1866, survival of the fittest "is a plain expression of fact; Natural Selection is a metaphorical expression of it, and to a certain degree indirect and incorrect, since, even personifying Nature, she does not so much select special variations as exterminate the most unfavorable ones."[26] It may in some ways be more appealing and certainly kinder to think of nature as selecting out and preserving the most useful variants. However, selection is in fact achieved by a process of elimination, a process that may be kept in mind more easily by thinking of it as Darwin's hammer.[27]

It should not be overlooked, however, that despite the immediate and continuing success of Darwin's theory, his view of evolution suffered from a number of serious defects. Ignorant (as was the rest of the world) of the genetic basis of inheritance as revealed by the work of Austrian monk and pea gardener Gregor Mendel (1822–1884), Darwin had no explanation for how heredity operated or how heritable variation was produced. In addition, he wrote, "I think that there can be little doubt that use in our domestic animals strengthens and enlarges certain parts, and disuse diminishes them; and that such modifications are inherited."[28] Thus like Lamarck, he believed in the inheritance of acquired characteristics. The realization that natural selection alone was the mechanism underlying all biological evolution had to wait until Weismann's work in the 1880s, and it was not until after the tragic Kammerer affair in 1926 that the theory of the inheritance of acquired characteristics became almost universally rejected in biological circles.

Nonetheless, Darwin did make a convincing argument that increases in adapted complexity can arise naturally without instruction, without prior design, and without the miraculous intervention of a supernatural creator or guide. Where Paley's ultimate watchmaker was an all-knowing and benevolent God, Darwin's watchmaker was totally blind. Where Lamarck had seen heritable adaptations as the direct instructive effects of environment on organism, Darwin saw them as the results of the selection of purposeless and accidentally advantageous variations. Darwin dared to see design without a designer and fit without instruction, and the world has yet to appreciate fully the implications of his vision.

Much has happened in biology since Darwin's time. Many more species have been discovered and catalogued, and many others have disappeared forever. Much more is known of the lifestyles of organisms ranging from protozoa to gorillas. Great strides have been made in unlocking the secrets of the molecular foundations of life responsible for the processes of metabolism, reproduction, and heredity. The language of the genetic code has been deciphered. Yet despite these developments, Darwin's theory of natural selection remains as the central pillar of modern biology.

This is not to say that the theory has remained unchanged since 1859. Indeed, many biologists today are more likely to refer to "neo-Darwinism" or the "synthetic theory of evolution" than to "Darwinism" or "Darwinian theory." But despite these newer terms, which reflect updatings based on recent advances in evolutionary theory, population genetics, and molecular biology, the fundamental features of Darwinian theory remain essentially unchanged. An exception is the almost universal rejection today of the inheritance of acquired characteristics, a belief that Darwin himself was never able to reject completely. Darwin's theory has been challenged, but as yet it has no serious rivals. Not "neutralism,"[29] "mutationism,"[30] or "molecular drive"[31] can account for the adaptive evolution of organisms.[32] (Other challenges to natural selection will be considered in chapter 16.) Furthermore, since current evolutionary theory rejects all forms of Lamarckian instruction and considers natural selection to be the sole mechanism for adaptive evolutionary change, neo-Darwinian theory can be considered in this respect to be more Darwinian than Darwin was himself.

The theory of evolution by natural selection remains central to the field of modern biology, but it has not met with similar success among the general population, at least not in the United States. In a poll conducted by the Gallup Organization in 1991, 47% of Americans expressed a belief that "God created man pretty much in his present form at one time within the last 10,000 years," and only 9% agreed that "man has developed over millions of years from less advanced forms of life. God had no part in this process."[33] Perhaps even more troubling, a survey of 387 United States high school biology teachers conducted in 1987 found that about one-third of them held beliefs incompatible with Darwinian evolution, such as that "Adam and Eve were the first human beings and were created by God" and that "the Bible's account of creation should be taught in public schools as an explanation of origins."[34]

It is hard not to come to the conclusion that, for the majority of Americans, the argument from design, whether explicitly considered or not, plays a major role in the widespread rejection of Darwinian evolution. We see a world full of fit plants and animals, but very little of the blind and therefore overwhelmingly unfit variation that the theory says must be generated for natural selection to work. It therefore seems reasonable that the design we see was planned in advance. But the folly of such a view is made clear in a simple thought experiment described by philosopher Daniel Dennett:

You obtain a mailing list of serious gamblers, divide it in half, and send one half the prediction that team A will win the championship next week, and the other half the prediction that team A will lose. A week later, half your mailing list has received a true prediction from you—free of charge. Discard the other half of the mailing list; divide the remainder in half again, and send them a second brace of complementary predictions; this cuts down your pool of suckers, but now they have two "proofs" of your clairvoyance. After a few more "successes," you announce that the free trial period is over; for your next prediction they will have to pay.[35]

The same sort of phenomenon occurs in biological evolution: since we only see those organisms which survive, we are like one of Dennett's "suckers." Because a sucker does not know about all the wrong predictions which are sent out, he thinks the con man knows in advance which team will win. Similarly, we think that evolution is purposeful, but it is not—nature is only guessing. Only when he looks at the larger picture can a sucker see that he is being conned; only when we look at the larger picture can we see how random variation is producing organisms which fit.[36]

Of course, it must be admitted that no argument can ever be produced that will offer definite proof against a supernatural designer who is responsible for planning and implementing biological adapted complexity. If such an entity were indeed able to create all the earth's living organisms, it could also no doubt be successful in hiding its presence from us, if that were also part of its plans. Indeed, such an omnipotent being could have fabricated the fossil evidence for evolution as a test of spiritual faith, making the believers of natural selection the real (and damned) suckers. But admitting these possibilities does nothing to weaken the argument that Darwin's theory is immeasurably more plausible than Paley's conclusion, for the simple fact that "Darwin's theory does not require positing things for which we have no evidence. The argument from design involves a being of unimaginable powers, while the theory of natural selection involves forces and mechanisms that we observe today and that are easily explained."[37] In short, natural selection provides a nonmiraculous explanation of puzzles of fit.

We should now be better able to appreciate the power and appeal of Darwin's account of adaptive biological evolution. It alone provides a non-providential and noninstructionist account of the emergence of adapted complexity. Indeed, its power and appeal are such that Oxford evolutionary zoologist Richard Dawkins proposed "universal Darwinism." He believes that if life were ever discovered elsewhere in the universe, it too would almost certainly be the product of a long series of gradual increases in adapted complexity brought about by cumulative variation and natural selection.[38] This, he argues, is because natural selection turns out to be the only known theory that, *in principle,* is able to offer a natural explanation for the many puzzles of fit we observe in living things. It might be difficult to find a better test of the intelligence of any advanced extraterrestrial beings we might someday encounter than to inquire as to whether they have yet discovered the cumulative process of blind variation and selection that is responsible for their own existence.

3

The Emergence of Instinct

Providence

Be it that those actions of animals which we refer to as instinct are not gone about with any view to their consequences, but that they are attended in the animal with a present gratification, and are pursued for the sake of that gratification alone; what does all this prove, but that the prospection, which must be somewhere, is not in the animal, but in the Creator?
—William Paley[1]

Instruction

As the penchants that animals have acquired through the habits they have been forced to contract have little by little modified their internal organizations, thus rendering the exercise of [these penchants] very easy, the modifications acquired in the organization of each race are then propagated to new individuals through generation. Indeed, it is known that generation transmits to these new individuals the state of organization of the individuals that produced them. It results from this that the same penchants already exist in the new individuals of each race even before they have the chance to exercise them, so that their actions can only be of this one kind.
—Jean-Baptiste Lamarck[2]

Selection

To my imagination it is far more satisfactory to look at such instincts as the young cuckoo ejecting its foster-brothers,—ants making slaves,—the larvae of echneumonidae [wasps] feeding within the live bodies of caterpillars,—not as specially endowed or created instincts, but as small consequences of all organic beings, namely, multiply, vary, let the strongest live and the weakest die.
—Charles Darwin[3]

Among the most marvelous puzzles of fit are those observed between the behavior of animals and their environments, environments that include not only a world of objects and forces but also other organisms. Indeed, many of the highly specialized structures of animals would be quite useless if they were not coupled to correspondingly specialized behaviors. The marvelously designed wings of a hawk would be of little use without the ability to move them in such a way as to generate lift for flight. The rattlesnake's rattle would be a frivolous accessory without the snake's ability to shake it vigorously to sound its warning. And the adroit human hand with its opposable thumb would be a worthless appendage if it were not coupled with the coordinated movements that allow it to plant crops, throw spears, construct cathedrals, and perform piano sonatas.

When we consider animal behavior, we are first struck by what appear to be two quite separate categories of actions. One category consists of complex behaviors that all the individuals of a given species are somehow able to perform without first experiencing the behaviors performed by others and without being in any way guided or instructed in them. Thus a mother rat will build a nest and groom her young even if she is raised in total isolation from other female rats.[4] Usually, the evolutionary significance of such behavior is quite clear. In the case of mother rat, it is not difficult to see how building a nest to keep her pups warm and secure and keeping them clean increases their chances of survival, thereby enhancing her own reproductive success and the survival of her genes. The behaviors involved in the spider's spinning a web, the beaver's constructing dams, and the honeybee's sculpting a honeycomb are additional examples.

The other type of behavior consists of those acts that appear to result from an animal's *particular experiences,* and it is here that we notice striking differences across individuals of the same species. The circus shows us what dogs, bears, horses, lions, tigers, and elephants can do given the special environment provided by the animal trainer. Dogs do not normally walk upright on their two hind legs, and bears are not to be seen riding motorcycles through the woods or seals balancing beach balls on their noses in the Arctic. Yet these creatures *can* perform these and other unnatural acts if provided with a special type of environment. Similarly, whereas all normal children manage to walk and talk without formal instruction, this is not the case for reading, writing, and mathematics skills, the development of which normally requires many years of formal schooling.

Of these two types of behavior, the first is typically referred to as instinctive, innate, or inherited, and the second learned or acquired. Both provide countless instances of puzzles of fit. We will examine the evolution of explanations for the fit of instinctive behavior in this chapter, and address learned behavior in chapter 7. These two chapters will also show how traditional theories of behavior fail to account for the full extent of the observed fit.

Instinct Through Providence

Our prehistoric ancestors' survival depended on their understanding the ways and habits of the animals that shared their surroundings. They hunted the animals useful as food, and avoided and repelled those that would have humans as *their* food. This delicate relationship led to an appreciation of the ways in which animal behavior is adapted to survival. The bird's ability to fly, the frog's skill in catching insects with its long, sticky tongue, the beaver's construction of dams and canal systems—these and countless other examples provide evidence of apparently built-in abilities that emerge without long study or labor, and that all normal members of the species share. This adapted nature of behavior led many traditional societies to venerate certain animals as embodiments of divine spirits.

Two interrelated questions must be considered in attempting to understand instinctive behavior. The first deals with the origin of the behavior itself and the second with the propagation of the behavior to new generations. It is important to address both of these questions separately, but the most satisfactory answer to each turns out to be very much the same.

The providential view of instinctive animal behavior came to us in Western philosophy primarily through the influence of Aristotle, Thomas Aquinas, and Descartes, and it remained popular and virtually unchallenged through the eighteenth century. As Thomas Aquinas reasoned:

Although dumb animals do not know the future, yet an animal is moved by its natural instinct to something future, as though it foresaw the future. Because this instinct is planted in them by the Divine Intellect that foresees the future.[5]

The views of the Aristotelians and Cartesians of the eighteenth century differed in many respects concerning animal behavior. Nonetheless, they agreed that "complex animal behavior (e.g., birds building their nests and bees their cells) should be explained by appeal to instincts, which they understood as blind, innate urges instilled by the Creator for the welfare of his creatures."[6]

Reverend Paley was also interested in using the instinctive behavior of animals as evidence for the existence, goodness, and wisdom of God. He continued his argument from design by emphasizing those behaviors that could not possibly have been the result of any instruction provided during the lifetime of the organism. Thus he described how moths and butterflies

deposit their eggs in the precise substance, that of a cabbage for example, from which, not the butterfly herself, but the caterpillar which is to issue from her egg, draws its appropriate food. The butterfly cannot taste the cabbage—cabbage is no food for her; yet in the cabbage, not by chance, but studiously and electively, she lays her eggs. . . . This choice, as appears to me, cannot in the butterfly proceed from instruction. She had not teacher in her caterpillar state. She never knew her parent. I do not see, therefore, how knowledge acquired by experience, if it ever were such, could be transmitted from one generation to another. There is no opportunity either for instruction or imitation. The parent race is gone before the new brood is hatched.[7]

From this passage we see that Paley put great emphasis on the "unlearn-ability" of complex behaviors that are essential to the survival and continuation of a species, since if the animal has no way to learn such important behaviors, the origin of the behaviors must lie in God. From this providential perspective, the question of transmission of behaviors to the next generation simply does not arise, since the behavior is seen as an integral part of the organism as designed by the Creator.

The Instruction of Instinct

It is in the work of Charles Darwin's grandfather, Erasmus Darwin (1731–1802), that we find an early alternative to the providential view of instinct. Erasmus Darwin's annoyance with this view can be seen in his observation that from this perspective, instinct "has been explained to be a *divine something*, a kind of inspiration; whilst the poor animal, that possesses it, has been thought little better than a machine!"[8] He and other "sensationalists" of the time emphasized the role of *sensory* experience in the development of behavior. They believed that all behavior was based on the experience and intelligence of the individual organism and described ways in which apparently instinctive behavior could be explained as such. But this explanation fared less well with behaviors demonstrated immediately after hatching or birth. Here Lamarck's concept of the inheritance of acquired characteristics seems at first consideration to do much better.

Lamarck's understanding of instinctive behavior was intimately tied to his theory of evolution. Whereas today biological evolution is used to explain the emergence of instinctive behavior, Lamarck saw behavior as the motor of evolution.[9] For him, changing environmental conditions forced organisms to change their habits to survive, and changed habits involved the increased use of certain body parts and systems accompanied by the decreased use of others. According to his theory, such organic changes would be passed on to succeeding generations. Since behavior is clearly influenced by biological organs, including appendages, the inheritance of such modified organs would result in the accompanying instinctive behavior dependent on the organs in succeeding generations. In this way Lamarck attempted to provide explanations both for the origin and transmission of new instinctive behaviors.

This theory depends crucially on mechanisms of instruction in at least three ways. The first two we have already considered and rejected; namely, the transmission of instructions from behavior to biological structure without previous selection, and the instructive transmission from the somatic (body) cells to the germ (sperm and egg) cells. We now have to ask how it is that a changing environment results in the animal assuming just those habits that are adapted to the new environment. If a particular source of food is no longer available, what is it in the environment that instructs the animal to adopt a new effective behavior to feed itself? Certainly, a type of instruction is implied, although Lamarck and others of his day did not consider the problems inherent in understanding *how* new adapted behaviors could initially arise. Particularly problematic in this regard are behaviors that cannot be imagined as the result of individual learning, as in the egg-laying behavior of the moth and butterfly (as Paley describes in the first epigraph to this chapter) and the egg-laying behavior of the wasp (mentioned below). So even if Lamarck's theory could account for the transformation of learnable habits into instinctive behavior (which it does not), it still offers no explanation for the origin of those instincts that could not have been acquired as habits during the lifetime of any individual organism in the first place.

The Selection of Instinct

Charles Darwin's initial attempt to explain instinct had much in common with Lamarck's theory of the acquisition of acquired characteristics. Darwin "supposed that habits an animal might adopt to cope with a shifting environment would, during the course of generations, slowly become instincts, that is, innately determined patterns of behavior. Instincts in turn would gradually modify the anatomy of an organism, adapting the creature to its surroundings."[10] Thus, habits that persisted over many generations, such as eating berries from a certain bush, and were beneficial to the survival and reproduction of the species would "inscribe themselves in the heritable structures of the brain."[11]

Gradually, however, Darwin became dissatisfied with the idea of inherited habits as the sole explanation for instinctive behaviors, particularly when he realized (as Lamarck apparently had not) that many of these behaviors could not have originated as habits. Another example was provided by British natural theologian Henry Lord Brougham, writing in 1839 about the female wasp who provides grubs as food for the larvae ("worms") that will hatch from its eggs "and yet this wasp never saw an egg produce a worm—not ever saw a worm—nay, is to be dead long before the worm can be in existence—and moreover she never has in any way tasted or used these grubs, or used the hole she made, except for the prospective benefit of the unknown worm she is never to see." We know that Darwin was intrigued by this observation, since he added the comment "extremely hard to account by habit" to his copy of Brougham's work.[12] It was, in fact, more than "extremely hard" since "an act performed once in a lifetime, without relevant experience, and having a goal of which the animal must be ignorant—this kind of behavior could not possibly have arisen from intelligently acquired habit."[13]

It was clear to Darwin that habit could not be the explanation for such instinctive behaviors. Therefore he considered them to be not the result of inherited useful habits, but rather the *selection of individuals with useful habits*, although he never completely abandoned the former idea. Thus natural selection provided an explanation for instinctive behaviors that never could have originated as habits, such as the wasp's egg-laying behavior; however, one particularly thorny problem remained, that of the evolution and behavior of the neuter insects.[14]

The *Hymenoptera* order of insects includes bees and ants together with some wasps and flies. Many of these insects live in a well-structured society where their survival depends on a specialized division of labor among its members, reflected in different castes such as the queen, drones, and workers in a beehive. Particularly intriguing and troublesome for Darwin's theory of natural selection was the fact that the worker caste in these colonies often comprise sterile insects *that therefore cannot genetically pass on their instinctive behaviors to the next generation of workers.* This posed no small threat. Indeed, it may well be that "the case of the neuter insects presented the most serious obstacle to his general theory of evolution."[15]

Darwin's answer came to him after he learned how cattle were selected for breeding to produce meat with desirable characteristics. As described in a book by William Youatt published in 1834 and read by Darwin in 1840, animals from several different families would be slaughtered and their meat compared. When a particularly desirable type of meat was found, it was, of course, impossible to breed from the slaughtered animal. But it *was* possible selectively to breed cattle most closely related to those slaughtered to produce the desired meat. In like manner, a colony of insects that produced neuters that helped the survival of the community (by taking care of young, providing food, or defending the colony against enemies) would be naturally selected to continue to produce such neuter insects. As Darwin noted, "this principle of selection, namely not of the individual which cannot breed, but of the family which produced such individual, has I believe been followed by nature in regard to the neuters amongst social insects."[16] The concept of selection by community or kin rather than the individual later became a powerful idea in the understanding of the evolution of altruistic behavior,[17] and it provided Darwin with an explanation for complex and useful instinctive behaviors that could not be explained by Lamarckian inheritance of habits. But where the transmission of inherited habits seemed conceivable, particularly where Darwin could see no selective advantage for the behavior, he made use of Lamarckian principles. And since he was unable or unwilling to see any survival or reproductive advantages accruing from the expression of emotions, he explained these as inherited habits.[18]

The mix of natural selection and instruction may seem like an odd combination to us today, but when seen from Darwin's own perspective it makes rather good sense. Certain instinctive behaviors (as well as the existence of neuter insects) could not be explained by the Lamarckian theory simply because he had no way to conceive how they could have developed *originally* as habits. For these behaviors, natural selection of individuals possessing behaviors leading to reproductive success was the only way to explain their appearance. However, other instinctive behaviors, such as the expression of emotions through facial or postural means, appeared to have no functional value. Since natural explanation is no explanation at all for the development of traits that provide no survival and reproductive benefits, Darwin reasonably concluded that they must have resulted from the inheritance of useless habits accompanying more useful ones.

The Ultra-Darwinians

Others, however, both during and after Darwin's lifetime, completely rejected the Lamarckian idea that acquired characteristics could be inherited. Consequently, they dared to explain all of evolution, including that of complex instinctive behaviors, solely through natural selection. The most prominent of these ultra-Darwinians, as Darwin's young disciple and defender George J. Romanes called them, was fellow British naturalist and codiscoverer of the theory of natural selection, Alfred Russel Wallace (1823–1913). And since they were ultra-Darwinian only in the sense that they discounted all mechanisms of evolution other than natural selection, they are probably more accurately referred to as ultraselectionists.

But Wallace's view of evolution was more than just a rejection of all Lamarckian instruction. He and his fellow ultraselectionists

viewed each bit of morphology, each function of an organ, each behavior as an adaptation, a product of selection leading to "better" organisms. They held a deep belief in nature's "rightness," in the exquisite fit of all creatures to their environments. In a curious sense, they almost reintroduced the creationist notion of natural harmony by substituting an omnipotent force of natural selection for a benevolent deity.[19]

Wallace's own words express this view quite clearly:

None of the definite facts of organic selection, no special organ, no characteristic form or marking, no peculiarities of instinct or of habit, no relations between species or between groups of species, can exist but which must now be, or once have been, useful to the individuals or races which possess them.[20]

But this concept posed a special problem when Wallace considered the human intellect. This is because he believed, in contrast to other European intellectuals of his day, that the brains of "savages" were not intrinsically inferior to those of civilized Europeans. He knew of non-European military bands that very capably performed Western music and was much impressed by the very fine singing of Africans who had learned this music in England.[21] So he reasoned that if natural selection can select only behaviors and abilities that are advantageous to the species, natural selection could *not* be responsible for such "latent" intellectual and musical abilities since they served no purpose in their native environment. Therefore, the human brain and the intellectual and moral qualities that it confers could only be the work of a divine provider.

So although both Darwin and Wallace were concerned with the same problem, they arrived at very different conclusions. Darwin saw certain behaviors that to him served no useful function and explained them by inheritance of acquired characteristics. Wallace saw abilities that served no useful function and explained them as the work of God. It was only much later that Darwin's selectionism and rejection of providentialism would be combined with Wallace's selectionism and rejection of Lamarckian instruction to give birth to the modern neo-Darwinian view of biological evolution that we embrace today.

Despite the enormous impact that Darwin's work had on the life sciences during his own lifetime, it had relatively little immediate impact on the comparative study of animal behavior. Two of the reasons for this are the methodological difficulties of both naturalistic and experimental research on animal behavior, and the heavy use of anecdotal evidence and anthropomorphic[22] interpretation practiced by Romanes, who wrote extensively about animal behavior and mind from a Darwinian perspective while maintaining a belief in the inheritance of acquired habits.[23]

The Beginning of an Evolutionary Ethology

It was not until the 1930s that a serious attempt to conduct research on animal behavior from evolutionary and selectionist perspectives was begun. Konrad Lorenz (1903–1989) grew up sharing his family's estate near Vienna with dogs, cats, chickens, ducks, and geese. In this setting his observations of animal behavior led to the founding of the field of *ethology,* which he defined as "the comparative study of behaviour . . . which applies to the behaviour of animals and humans all those questions asked and methodologies used as a matter of course in all other branches of biology since Charles Darwin's time."[24]

As suggested by his definition of ethology, Lorenz was primarily interested in finding an evolutionary explanation for the instinctive behavioral patterns characteristic of a species. For example, it was brought to his attention that greylag geese that were reared by humans would follow them in much the same way that naturally hatched goslings waddled after their mother. Lorenz confirmed these findings for the greylag goose and extended them to a number of other birds as well. This pattern of behavior, resulting from a type of bonding with the first moving object seen by the bird, he called "imprinting," and it is for this finding that Lorenz is still best known today.

By extending Darwin's theory of natural selection to behavior, Lorenz posited a genetic basis for specific behaviors that was subject to the same principles of cumulative blind variation and selection that underlie the adapted complexity of biological structures. In the case of the greylag goose, goslings that maintained close contact with the first large moving object they saw (usually the mother goose) would be in a better position to enjoy her protection and nurturance. Consequently, they would be more likely to survive and to have offspring that would similarly show behavioral imprinting. Those goslings that lacked this behavioral characteristic would be less likely to survive to maturity and reproduce. In much the same way that we understand how a tree frog can become so well camouflaged over evolutionary time through the elimination by predators of those individuals who are less well camouflaged, we can also understand how instinctive behavior can be shaped through the elimination of individuals whose behaviors are less fit to their environment.

Another example of Lorenz's conception of instinctive behavior is the egg-rolling behavior of the greylag goose.[25] When the goose sees that an egg has rolled out of her nest, she stands up, moves to the edge of the nest, stretches out her neck, and rolls the egg back into the nest between her legs, pushing it with the underside of her bill. This behavior depends on what Lorenz called a "fixed motor pattern," that is, a pattern of activity in the central nervous system of the goose that is released or triggered by the sight of the egg outside the nest. In other words it is a specific fixed response released by a specific type of stimulus. The purpose of this instinctive act is clearly to return the egg to the security of the nest, and it is easy to appreciate its value for the survival of the species.

But a problem with Lorenz's conceptualization of instinctive behavior becomes apparent when one considers that exactly the same pattern of behavior will *not* be successful in returning the wayward egg to the nest unless all environmental conditions are exactly the same for each egg-rolling episode. Instead, for the goose to be successful in keeping her eggs nestbound, she must be able to modify her behavior not only from episode to episode but also *within* each episode to compensate for the variability in conditions and disturbances that she inevitably encounters, such as differences in the distance between her and the egg at the beginning of the behavior, and irregularities in the terrain between the egg and the nest. As American philosopher and psychologist William James noted over 100 years ago, the behavior of all living organisms is characterized by the attainment of *consistent ends using variable means.*[26] And it is for this reason that Lorenz's analysis fails to explain the typical success (and therefore the fit to environment) of instinctive behavior. This essential aspect of adapted behavior will be considered in more detail in chapter 8.

But despite this shortcoming, Lorenz must be acknowledged as being the first to attempt to provide a truly Darwinian (or, more accurately, a purely selectionist) account of species-specific behavior patterns, and he was recognized for his achievement in 1973 when he shared a Nobel prize with fellow ethologists Nikolaas Tinbergen and Karl von Frisch. In the same way that biologists constructed evolutionary trees (or phylogenies) by comparing the anatomical similarities and differences among living animals and fossils, Lorenz proposed patterns of instinctive behavior for the same purpose.[27] Indeed, he based his comparative study of animal behavior "on the fact that *there are mechanisms of behavior which evolve in phylogeny exactly as organs do.*"[28] His evolutionary perspective also led him to emphasize that understanding animal behavior required one to appreciate its purposefulness in preserving the species, its role in the entire repertoire of the animal's activities, and its evolutionary history.[29] For these reasons, among others, Lorenz regarded with suspicion research on animal behavior being conducted by experimental psychologists using domesticated animals in artificially contrived environments on the other side of the Atlantic. But to understand how animals are able to modify their behavior during their own lifetime, we must turn to the work of these American animal psychologists later in chapter 7.

4

The Immune System

Selection by the Enemy

From Providence to Instruction

The extraordinary number of specific antibodies, including those against artifi-cial antigens, defied the genetic origin originally propounded by Paul Ehrlich. A somatic, custom-made template mechanism seemed more logical.
—Debra Jan Bibel[1]

Selection

It follows that an animal cannot be stimulated to make specific antibodies, unless it has already made antibodies of this specificity before the antigen arrives. It can thus be concluded that antibody formation is a selective process and that instructive theories of antibody formation are wrong.
—Niels Jerne[2]

The previous two chapters described how Darwin's theory of cumulative variation and selection provides a naturalistic explanation for the fit of bio-logical form and behavior observed in living organisms. But although the fit thus achieved is striking, natural selection is a very wasteful process in the sense that it depends on eliminating entire organisms, those less fit, from the evolutionary process. Working on large populations of organisms over long periods of time involving many generations, Darwinian natural selection is what is referred to as a *phylogenetic* process, since it results in new *phyla*, or new branches of the evolutionary tree of life. It would thus seem to be in an organism's own interest if it could somehow increase its fit to its envi-ronment during its lifetime, that is, be capable of *ontogenetic*[3] adaptation.

Indeed, individual organisms undergo ontogenetic adaptation in many ways. Muscles grow stronger with increased use, and animal and human behaviors change over time in functional, adaptive ways. These changes are readily observable, but the best understood ontogenetic adaptation involves the mammalian immune system.

Probably the best example of a puzzle of fit at the microscopic level is demonstrated by the antibodies produced by the mammalian immune system. For the immune system to be able to rid the body of antigens—foreign invaders in the form of chemical toxins, viruses, bacteria, cells, and tissues—there must be a precise fit between antibody and antigen. Attempts to explain this fit go back over 100 years and resulted in one of the most striking triumphs of the relatively new field of molecular biology.

A (Genetically) Providential Theory of Antibody Production

After the general acceptance of Louis Pasteur's germ theory of disease late in the nineteenth century, a number of scientists became interested in understanding the mechanisms responsible for the appearance of antibodies in an animal's blood after it had been infected with disease-producing bacteria. Antibodies were of special interest since they were known to protect the animal from subsequent infection by the same pathogenic bacteria.

Paul Ehrlich (1854–1915) was a German who in the 1890s developed a technique for estimating the quantity of antibodies in blood. He was intrigued by the explosive increase in antibody production after exposure to an antigen and attempted to account for this phenomenon by formulating his side-chain theory. According to this theory, the surface of white blood cells is covered with many side chains, or receptors, that form chemical links with the antigens they encounter. For any given antigen, at least one of these receptors would bind, stimulating the cell to produce more of the same type of receptor, which would then be shed into the blood stream as antibodies. According to Ehrlich's theory, an antibody could be considered an irregularly shaped, microscopic, three-dimensional label that would bind to a specific antigen but not to the other cells of the organism.[4] This analogy of antibodies as labels is used here since the antibodies themselves usually do not destroy the antigen, but label it, providing a molecular kiss of death for destruction of the antigen by complement proteins, macrophages, or other agents that either perforate the antigen's cell membrane or completely engulf the antigen.[5]

The major assumption of Ehrlich's theory, and one that makes it a type of providential theory, was that white blood cells possessed numerous genetically specified side chains, at least one of which would bind to any encountered antigen. That is, the information essential for the production of all possibly necessary antibodies was *provided* by the animal's genes. For this reason, this is known as a germ-line theory, the germ line referring to the entire set of genes (or *genome*) that is passed from an organism or pair of organisms to its offspring.

But the germ-line theory soon encountered a major difficulty. During the early 1900s, Karl Landsteiner (1868–1943) clearly demonstrated that there was apparently no discernible limit to the range of antibodies that an animal could produce. His finding that antibodies could even be produced in response to completely novel artificial substances revealed that the animal could not possibly possess in its finite genome the information required to produce an infinite number of all possibly necessary antigens. This led to the rejection of Ehrlich's theory and to the consideration of constructivist theories of antibody production. Such constructivist theories had to account for the fact that the immune system was not only adapted to the task of producing fit antibodies, but that it was adapt*ive* as well, able to create new puzzles of fit in response to completely unpredictable and novel antigens.

An Instructionist Theory of Antibody Production

The first well-known theory that attempted to account for the immune system's ability to produce antibodies in response to novel antigens appeared in 1930 with Breinl and Haurowitz's introduction of the template instruction theory,[6] which was further developed and advocated by Nobel prize-winning chemist Linus Pauling.[7] The template theory attempted to explain that antibodies could be made to bind with any novel antigen because they were produced by direct contact with antigens. By proposing that the antigens themselves served as models for antibody production, the template theory was able to account for the seemingly limitless range of antibodies. To pursue the label analogy, if Ehrlich's innatist side-chain theory could be likened to an innately determined set of labels designed so that at least one would stick to any given antigen, the template theory saw the immune system as supplied with the knowledge of a *procedure* whereby labels would be custom-made for each antigen in the same way a tailor makes a suit of clothes using the customer as a template. The template theory could therefore be considered *instructionist* since it did not require innate, germ-line information for the production of all antibodies, but rather required only a general label-construction procedure for building antibodies using the antigens themselves as the required source of instructions.

Although one version or another of the instructionist template theory remained influential for over 20 years, it too encountered a number of serious difficulties. In the 1950s Danish immunologist Niels Jerne noted several immunological findings that the theory could not explain.[8] First of all, it could not account for the increasing rate of antibody production during the initial immune response. If the antigen itself was necessary for the production of each antibody, how could the antibodies so quickly outnumber the templates? Second, it could not explain the memory of the immune system by which a second exposure to a given antigen results in a much more rapid production of antibodies than does the initial contact. If the antigen itself served as a template, its total elimination by the immune system would also necessarily entail destruction of templates, so a second exposure to the same or similar antigen should be accompanied by an immune response no different from the first. Third, since it was thought that antibody cells were quite short lived, the template theory could not account for the fact that antibodies continued to be present long after the antigen had been eliminated from the body. Finally, the theory did not explain the fact that the antibodies produced during the latter stages of an immune response are usually more effective in binding with the antigens (are better-fitting labels) than the antigens initially produced.

It was therefore clear that in attempting to account for the formation of new antibodies, the instruction-based template theory ran into a number of rather serious problems that motivated scientists to pursue other explanations.

A Selectionist Theory of Antibody Production

Jerne's paper provided not only arguments against the template theory but also an alternative theory that in some respects resembled Ehrlich's original side-chain theory. Jerne's natural-selection theory of antibody production stated that a mammal initially possesses a relatively small number of antibodies. The successful binding of an antibody to an antigen triggers the antibody to produce a large number of copies of itself. In this way, a pre-existing antibody is effectively *selected* by the antigen, which stimulates the chosen antibody to produce a multitude of clones. Australian virologist and immunologist Sir Frank Macfarlane Burnet (1899–1985) further developed the theory using the term clonal selection to describe it.[9]

Much research still has to be done to understand fully the complex dynamics of the immune system, but the Jerne-Burnet clonal-selection theory of antibody production is generally accepted. Although its major characteristics are not difficult to understand, we have to go into a bit more detail concerning the immune response to appreciate its selectionist functioning.

First, instead of thinking of an antibody as either binding or not binding to an antigen, we must appreciate that antibodies have a very wide range of affinity (that is, attraction or binding power) to a given antigen.[10] Thus whereas a well-fitted antibody will almost always bind to an encountered antigen for which it is well fit, another less well-fitted antigen may also bind with the same encountered antigen, although its rate of binding may be considerably less than 100%. In contrast, an antibody that is completely unfit for binding with a given antigen will seldom, if ever, bind to the antigen.

Second, instead of every antigen having a single "handle" (called a *determinant* or *epitope* by immunologists) for the antibody to grasp, they all have many such determinants, each of which is *different* and thus can be bound by a different type of antibody. These determinants correspond to small patterns of molecular structure on the surface of the antigen.[11]

The large number of determinants on each antigen effectively increases the likelihood that the immune system will be able to produce an antibody that will bind to any antigen introduced into the body. It still has to create a staggering number of different antibodies to ensure its effectiveness, however, a diversity that includes up to 10 billion B lymphocyte cells, each able to produce more than 100 million different antibody proteins. And since a person has only about 100,000 genes, there is simply no way our genes could specify each and every one of these proteins.[12]

The answer to this enigma required a radical reconceptualization of genes and how they function. Until the mid-1970s it was a generally accepted principle of biology that one gene always leads to the synthesis of one and only one protein, and that the genetic makeup of an organism remains constant throughout its life. (Relatively rare and usually harmful mutations do occur due to genetic copying errors during cell division or exposure to irradiation or other environmental mutagens.) But in 1976 at the Basel Institute for Immunology, Susumu Tonegawa discovered that antibody genes are not inherited complete, but rather as fragments that are shuffled together to form a complete gene that specifies the structure of a given B lymphocyte and the antibodies it produces (Tonegawa received a Nobel prize in 1987 for his discovery). In addition, as the DNA segments are combined to form the complete B lymphocyte gene, new DNA sequences are added at random to the ends of the fragments, ensuring even more antibody diversity.[13] In this manner:

The receptors used in the adaptive immune response are formed by piecing together gene segments, like a patchwork quilt. Each cell uses the available pieces differently to make a unique receptor, enabling the cells collectively to recognize the infectious organisms confronted during a lifetime.[14]

This active, random reshuffling of immunoglobulin genes, together with the insertion of random DNA sequences during the recombination process, is responsible for the diversity of antibody receptors attached to each B cell. Such diversity virtually ensures that at least one antibody, although perhaps not fitting perfectly, will be able to bind with at least one of the many determinants presented by a new antigen.

Once an antibody is selected by an antigen by binding, it stimulates the B lymphocyte to which it is attached to divide and make exact copies of itself. Some of the selected clones remain as circulating B lymphocytes and as such serve as the immune system's memory. Increased numbers of these cells provide for a faster immune response to subsequent infections and establish the immunity that follows many infections and vaccinations. Other selected clones stop dividing, grow larger, and turn into plasma cells whose sole function is to produce large numbers of free antibodies to fight the current infection.

The clonal-selection theory explains the great diversity of antibodies and the ability of the immune system to bind with completely novel antigens. It also provides an account for Jerne's first three findings, noted above, in his criticism of the template theory. The theory as described so far, however, still fails to account for the finding that the antibodies produced during the latter stages of the immune response are more effective in binding with the antigens than the antibodies initially produced. This fine-tuning of antibodies is accomplished by another mechanism that also changes the genetic makeup of the antibodies—the random mutation of the genes within the B cell clones. "By altering individual nucleotide bases the mutations fine-tune the immune response, creating immunoglobulin genes whose products better match the antigen."[15]

We therefore see that the clonal-selection process of antibody production has a number of noteworthy characteristics. First, it is constructive in that the actual structure of the antibodies is not explicitly included in the genome. Thus, antibodies are quite unlike other physiological structures (such as livers, eyes, and noses) whose basic design depends on what is believed to be a fixed set of genes.[16]

Second, the structure of antibodies is determined by what appears to be an essentially blind process. This blindness shows itself at three levels: the random recombination of immunoglobulin genes as B lymphocyte cells are formed, the random insertion of DNA segments into the recombined gene, and the consequent blind hypermutation of the clone B cells to fine-tune them to the antigen. Thus, the immune system does not attempt to predict the antibody structure that will bind with an antigen, but rather uses a type of "shotgun" approach that sends in a diverse army to meet the invaders. Almost all of these produced antibodies will turn out to be quite ineffective in binding with the antigens, but the diversity of this army virtually ensures that at least one of them will be effective. Indeed, studies have shown that if antibodies are produced blindly, the probability that a novel antigen will be recognized is virtually assured if very many antibody types are present.[17]

Finally, the immune system is designed so that only those antibodies that are able to bind with the antigen are reproduced and remembered the next time the same or similar antigen invades the animal. Antibodies that are not successful leave no offspring and therefore soon become extinct, to be replaced by the estimated one million new B lymphocytes produced in the bone marrow every second.[18]

But we still have to account for another important ability of the immune system—that it does not attack the cells and products of its host body. Although it was first thought that the immune system was provided this information in the germ line, research has now demonstrated that the immune system in fact *learns* to distinguish self from nonself. The process by which it does so will not be described in detail here, but it is virtually a mirror image (with an important reversal) of how it produces antibodies to recognize invaders. That is, mature lymphocytes are triggered to reproduce and mutate when they encounter a foreign antibody; however, immature lymphocytes that form while the animal is still in utero or shortly after birth "go through a stage when binding of their receptors causes them to die. Self-reactive cells are killed before they have a chance to proliferate and damage their host."[19] This is the clonal-deletion theory of how the immune system learns tolerance to self, and was first proposed by Joshua Lederberg in 1959.

To summarize, the clonal-selection theory states that a very large number of unique B lymphocytes to which are attached antibody receptors are always circulating throughout the body. Their great diversity results from the random recombination of immunoglobulin gene fragments and random insertion of DNA sequences as the B cells develop. This blind diversity of B cells virtually ensures that at least one will produce an antibody that will bind with any antigen that makes its way into the organism. The binding of a B cell's antibody with an antigen stimulates the cell to divide and produce clones, with successive generations of reproducing clones resulting in an exponential rise over time in the number of circulating antibodies of the selected type. Some of the B cell clones remain in circulation to form the immune system's memory of the antigen. Others terminally differentiate, forming plasma cells that produce large numbers of antibodies that fight the current infection. Finally, as the B cell clones reproduce they undergo a high rate of somatic mutation that, when combined with the continued selection pressure exerted by the antigen, fine-tunes the fit of the antibodies to the antigen. A similar process of variation and selection of immature B cells (although now with selected cells eliminated rather than reproduced) accounts for the immune system's ability to tolerate the cells and products of its host body.

Antibody Production as a Microcosm of Darwinian Evolution

Even this much abbreviated and simplified account of the clonal-selection functioning of the immune system reveals it to be a remarkable microcosm of Darwinian evolution with the three major principles of superfecundity, variation, and natural selection each playing an essential role. Super-fecundity is evident in that the immune system produces far more anti-bodies than will be effective in binding with an antigen. In fact, it appears that the majority of produced antibodies do not play any active role what-soever in the response of the immune system. Natural (and blind) variation is provided by the variable gene regions responsible for the production of a highly diverse population of antibodies. And selection occurs, as only anti-bodies able to bind with an antigen reproduce.

The similarity between adaptive biological evolution and the production of antibodies is even more striking when one considers that the two central processes involved in the production of antibodies, genetic recombination and mutation, are the same ones responsible for the biological evolution of sexually reproducing species. We have seen that the recombination of im-munoglobulin genes underlies the large diversity of the antibody popula-tion, and the mutation of these genes serves as a fine-tuning mechanism.[20] In sexually reproducing species, the same two processes are involved in pro-viding the variations on which natural selection can work to fit the organ-ism to the environment. Thus cumulative blind variation and natural se-lection, which over many millions of years resulted in the emergence of mammalian species, remain crucial in the day-to-day survival of these species in their ceaseless battle against microscopic foreign invaders.

A final similarity between the functioning of the immune system and bio-logical evolution is worth noting—the evolution of our knowledge of how each operates. Our understanding of the fit of organisms to their environ-ment has progressed from a providential explanation to an instruction-based (Lamarckian) one to a purely selectionist (neo-Darwinian) account. The same stages of thought can also be seen in biology's attempt to account for the puzzle of fit of antibody to antigen.

Ehrlich's side-chain theory can be considered providential in that the organism's genes were believed to provide all the knowledge necessary to construct antibodies that would be able to fit all the antigens the organism would ever encounter. Since this knowledge was considered to be the result of past evolutionary selection, the genetically providential theory does not lead to the same problem of ultimate origins that the supernaturally provi-dential theory of the origin of species encounters. However, it did run into difficulties when it was found that the immune system could produce antibodies that fit novel antigens never encountered before, either in the organism's own lifetime or in that of its ancestors.

The appreciation of the immune system's ability to adapt to a changing environment of novel antigens led to the instruction-dependent template theory of antibody formation in which the environment (in the form of antigens) somehow transmitted instructions for the formation of close-fitting antibodies to the B lymphocytes. This theory, however, did not account for many of the characteristics of the immune system that were observed subsequently. In addition, no transmission of information from antigen to antibody was ever observed.

Given the problem of accounting for adaptive change without recourse to instructionist theories, it should perhaps not be surprising that the field of immunology would eventually hit on the same solution that Darwin had discovered, even if it took an additional century. By combining the basic principles of superfecundity, blind variation, and selection that explain the adaptation of organism to environment, the clonal-selection theory provides an understanding of how the immune system can produce antibodies adapted to its environment of novel antigens. This it does without recourse to providential or instructionist explanations. But whereas adaptive biological evolution proceeds by cumulative natural selection *among* organisms, research on the immune system has now provided the first clear evidence that ontogenetic adaptive change can be achieved by cumulative blind variation and selection *within* organisms.

5

Brain Evolution and Development
The Selection of Neurons and Synapses

Instruction versus Selection

The 10,000 or so synapses per cortical neuron are not established immediately. On the contrary, they proliferate in successive waves from birth to puberty in man. . . . One has the impression that the system becomes more and more ordered as it receives "instructions" from the environment. If the theory proposed here is correct, spontaneous or evoked activity is effective only if neurons and their connections already exist before interaction with the outside world takes place. *Epigenetic selection acts on preformed synaptic substrates. To learn is to stabilize preestablished synaptic combinations, and to* eliminate *the surplus.*

—Jean-Pierre Changeux[1]

The most complex object yet discovered anywhere in the universe is the organ that fills the space between our ears. Although weighing only about 1300 to 1500 grams (three to four pounds), the human brain contains over 11 billion specialized nerve cells, or *neurons*, capable of receiving, processing, and relaying the electrochemical pulses on which all our sensations, actions, thoughts, and emotions depend.[2] But it is not the sheer number of neurons alone that is most striking about the brain, but how they are organized and interconnected. And to understand how neurons communicate with each other we first must consider their typical structure.

Although there are many different types of neurons, almost all of them share certain common features as portrayed in figure 5.1. The cell body, or soma, contains the nucleus of the neuron, which in turn houses a complete set of the organism's genes. The nucleus is surrounded by cytoplasm, the chemical "soup" of the cell that contains the organelles essential to the neuron's functioning and metabolism. In these respects, neurons are similar to other cells throughout the body, except for the fact that unlike most other cells they rarely divide to reproduce new neurons.

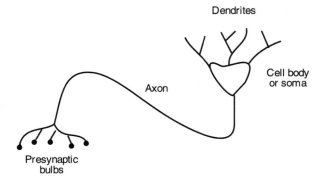

Figure 5.1
A typical neuron (after Churchland, 1990).

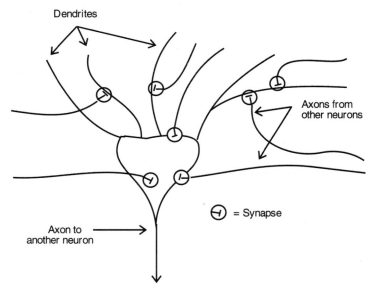

Figure 5.2
Neural synapses (after Churchland, 1990).

The ways in which neurons are specialized to carry out their communicative function is made evident by closer examination of the appendages they sport, that is, their dendrites and axons. The dendrites can be likened to a bushy antenna system that receives signals from other neurons. When a dendrite is stimulated in a certain way, the neuron to which it is attached suddenly changes its electrical polarity and may fire, sending a signal out along its single axon where it may be picked up by the dendrites of other neurons.[3] Considering the small size of the neuron's body, the length of an axon can be considerable, up to several meters in the neck of the giraffe. Thus the firing of one neuron can influence the firing of another one a considerable distance away.

For one neuron to influence another, the two must be connected, and this is accomplished by junctions called synapses (figure 5.2). These synaptic junctions usually connect the axon of one neuron with the dendrites of another, a typical neuron in the cortex of the human brain having about 10,000 synapses. The synapses therefore constitute an exceedingly complex wiring system that surpasses by many orders of magnitude the complexity of even the most advanced supercomputers. It is this organization of connections both within the skull and to more distant sense organs and muscles that gives the brain its amazing abilities. Indeed, it is widely believed today by neuroscientists, psychologists, and even philosophers that all of the knowledge the human brain contains—from being able to walk to the ability to perform abstract scientific and mathematical reasoning—is a function of the connections existing among the neurons.

How this unfathomably complex organization allows us to perceive, behave, think, feel, and control our environment presents us with what may be the most striking puzzle of fit we have yet encountered. The puzzle actually has three aspects. First, we must consider how over millions of years the primitive nervous system of our early ancestors evolved into an organ that has made it possible for the human species to become the most adaptable and powerful organism on the planet—living, thriving, and modifying the environment (both intentionally and unintentionally) from the tropics to the polar regions, and perhaps soon in outer space and on other planets.[4] Second, we must understand how it is possible for the intricate structure of the brain to develop from a single fertilized egg cell. Finally, we must try to comprehend how the mature brain is able to continue to modify its own structure so that it can acquire new skills and information to continue surviving and reproducing in an unpredictable, ever-changing world. In this chapter we will consider the research and theories that are beginning to provide answers to these questions. Indeed, the 1990s has been referred to as "the decade of the brain," as scholars and scientists in fields from philosophy to molecular neurobiology focus their energies on understanding humankind's ultimate inner frontier.

The Evolution of the Brain

Neurons are quite distinct from other body cells in ways that make them suited to their specialized role of signal processing and communication, but it is not too difficult to see how they could have evolved from less specialized cells. All living cells are surrounded by a cell membrane that separates the special chemical composition of its interior from that of the external world. This difference in chemical composition results in a small electrical potential between the inside and outside of the cell, in much the same way that a voltage exists between the two sides of a battery. When a part of a cell's membrane is disturbed in a certain way, it loses its electrical potential, becoming depolarized at the site of the disturbance. This sudden change in electrical potential can itself be a disturbance, causing additional depolarizations along the membrane. In most cells, such depolarization would not spread far, certainly not to neighboring cells. But a few changes in the shape and arrangement of cells (in just the way that neurons are fashioned) permits depolarization to propagate quickly from one neuron to the next, and allows it to travel quickly as an electrochemical signal from one end of an animal to the other.

An example of a simple nervous system is provided by the jellyfish (or *Medusa*). The jellyfish's nervous system forms an undifferentiated network and serves primarily to coordinate the animal's swimming motions. Since the jellyfish's skirt must open and contract in a coordinated manner for the animal to move through the water, its nervous system serves as a simple communications network making it possible for all parts of the skirt to open repeatedly and then contract at the same time.

Worms are the simplest organisms to have a central nervous system, which includes a distinct brain that is connected to groups of neurons orga-

nized as nerve cords running along the length of its body. This more complicated nervous system allows worms to exhibit more complex forms of
behavior. An anterior brain connected to a nerve cord is the basic design for
all organisms with central nervous systems, from the earthworm on the
hook to the human on the other end of the fishing rod. But although we can
discern a separate brain in worms, it is not the case that the brain is the sole
"commander" of the animal that the rest of the nervous system and body
obeys. Indeed, even with its brain removed, worms are able to perform
many types of behaviors, including locomotion, mating, burrowing, feeding, and even maze learning.[5]

As we move to insects we find increased complexity in all aspects of the
brain and nervous system. So-called giant fiber systems (also found to some
extent in worms and jellyfish) that allow rapid conduction of nerve impulses connect parts of the brain to specific muscles in legs or wings. Such
connections permit the cockroach to dart away as soon as it senses the
moving air preceding a quickly descending human foot. The brain itself is
typically divided into three specialized segments, the protocerebrum, the
deutocerebrum, and the tritocerebrum. In addition, insects possess a greater
variety of sensory receptors than any other group of organisms, including
vertebrates, that are sensitive to the odors, sounds, light patterns, texture,
pressure, humidity, temperature, and chemical composition of their surroundings. The concentration of these sensory organs on the insect's head
provides for rapid communication with the tiny yet capable brain located
within.

Although minuscule by human standards, the range of abilities made
possible by insect brains is impressive. These creatures show a remarkable
variety of behaviors for locomotion, obtaining food, mating, and aiding
the survival of their offspring. They can crawl, hop, swim, fly, burrow, and
even walk on water. The female wasp hunts down a caterpillar, paralyzes
it with her venom, and then lays its egg on the motionless prey so that her
offspring will have a fresh and wholesome meal immediately after hatching.
Leafcutter ants harvest leaves and bring them into their nest where they
use them to cultivate indoor gardens of edible fungus. Honeybees live in
social communities where there is a strict division of labor, and where food-
gathering worker bees perform a special dance to communicate the location
and richness of food sources to their hivemates. It is the evolution of their

brains, together with the complementary evolution of their other body parts, that make insects the most abundant multicellular organisms on our planet.

The brain becomes both much larger and still more complex as we move to vertebrates such as fish, amphibians, and reptiles. The spinal cord, now protected within the vertebrae of the backbone, has become primarily a servant of the brain, a busy two-way highway of communication with fibers segregated into descending motor pathways and ascending sensory ones. The brain itself is now composed of a series of swellings of the anterior end of the spinal cord (the brain stem), the three major ones making up the three major parts of the vertebrate brain: the hindbrain, midbrain, and forebrain. From the hindbrain sprouts a distinctive structure, the "cerebellum" (Latin for "little brain").

Among mammals, the brain keeps its three major components, but with two new structures. The neocerebellum ("new cerebellum") is added to the cerebellum, looking much like a fungal growth at the base of the brain, and the neocortex ("new cortex") grows out of the front of the forebrain. In most mammals, these new additions are not particularly large relative to the brain stem. In primates they are much larger, and in the human they are so large that the original brain stem is almost completely hidden by this large convoluted mass of grey neural matter. In keeping with this remarkable increase of neocerebellar and neocortical tissue, humans enjoy the largest ratio of brain weight to body weight of any of earth's creatures.

It is not possible to know exactly why the human brain evolved as it did, but consideration of the structural evolution of the brain and results of comparative research on human and nonhuman brains provides some useful clues. It is now believed that during the long evolution of our brain, nervous systems changed in four principal ways. First, they became increasingly *centralized* in architecture, evolving from a loose network of nerve cells (as in the jellyfish) to a spinal column and complex brain with impressive swellings at the hindbrain and forebrain. This increasingly centralized structure also became increasingly *hierarchical*. It appears that newer additions to the human brain took over control from the previous additions and in effect became their new masters. Accordingly, the initiation of voluntary behavior as well as the ability to plan, engage in conscious thought, and use language depend on neocortical structures. Indeed, the human neocortex can actually destroy itself if it wishes, as when a severely depressed individual uses a gun to put a bullet through his or her skull.

Second, there was a trend toward *encephalization*, that is, a concentration of neurons and sense organs at one end of the organism. By concentrating neural and sensory equipment in one general location, transmission time from sense organs to brain was minimized. Third, the size, number, and variety of elements of the brain increased. Finally, there was an increase in *plasticity*, that is, the brain's ability to modify itself as a result of experience to make memory and the learning of new perceptual and motor abilities possible.

One way of understanding the evolution of the human brain is to see it as the addition of higher and higher levels of control. We will see in chapter 8 that the function of animal and human behavior can be understood as the control of perceptions, with perceptions corresponding to important aspects of the environment. For a sexually reproducing organism to survive and leave progeny, it must be able to control many different types of perceptions, that is, sensed aspects of its environment. At a minimum, it must be able to find food, avoid enemies, and mate. But as life evolved, the environment of our ancestors became more complex due to increasing numbers of competing organisms. So it would have been of considerable advantage to be able to perceive and control increasingly complex aspects of this environment. The bacterium *E. coli* can control its sensing of food and toxins only in a primitive way; organisms with more complex brains are able to sense and control much more complex aspects of their surroundings.

This capacity for increased environmental control is nowhere more striking than in our species. Using the advanced perceptual-behavioral capacities of our brain together with our culturally evolved knowledge of science and technology, we can visit ocean floors, scale the highest peaks, and set foot on other worlds. (The role that language is believed to have had in the evolution of the human brain will be considered separately in chapter 11.) But can the most complex human abilities and mental capacities be explained by natural selection? Our brain has certainly not changed appreciably over the last couple of hundred years, and yet we can solve mathematical, scientific, technological, and artistic problems that did not even exist a hundred years ago. So how could natural selection be responsible for the striking abilities of today's scientists, engineers, and artists?

This is actually the same problem that troubled Alfred Russel Wallace, as mentioned in chapter 3. It will be recalled that Wallace, despite being an independent codiscoverer of natural selection, could not, for example, imagine how natural selection could account for Africans' ability to sing and perform European music, since nothing in their native environment could have selected for such an ability. Consequently, for him the brain could only be a creation provided to us by God. We now know that in his embrace of this providential explanation, Wallace failed to realize that natural selection can lead to new abilities unrelated to those that were originally selected.

To use an example from technological evolution, the first personal computers were used to perform financial calculations in the form of electronic spreadsheets. However, these same machines with the proper software could also be used for word processing, telecommunications, computer games, and many other purposes, even though they were not originally designed with these functions in mind. A classic example of this phenomenon of functional shift in biological evolution is the transformation of stubby appendages for thermoregulation in insects and birds into wings for flight.[6] In the same way, selection pressure was undoubtedly exerted on early hominids to become better hunters. The ability to understand the behavior of other animals and organize hunting expeditions must have been very important in the evolution of our species. And the increasingly complex and adapted brain thus selected would have made other skills possible, such as making tools and using language, traits that in turn could become targets for continued natural selection. This transformation of biological structures and behaviors from one use to another was given the unfortunate name of preadaptation by Darwin, unfortunate since it can too easily be misunderstood to imply that somehow evolution "knows" what structures will be useful for future descendants of the current organisms.

American evolutionary paleontologist Stephen Jay Gould provided a better term for this phenomenon—exaptation. He made a major contribution to our understanding of evolution by insisting that we distinguish *adaptation*, the evolutionary process through which adaptedly complex structures and behaviors are progressively fine-tuned by natural selection with no marked change in the structure's or behavior's function, from *exaptation*, through which structures and behaviors originally selected for one function become involved in another, possibly quite unrelated, function. Exaptation makes it difficult if not impossible to understand why our brain evolved as it did. Although the brain allows us to speak, sing, dance, laugh, design computers, and solve differential equations, these and other abilities may well be accidental side effects of its evolution. As Gould and his associate Vrba cautioned:

> . . . current utility carries no automatic implication about historical origin. Most of what the brain now does to enhance our survival lies in the domain of exaptation— and does not allow us to make hypotheses about the selective paths of human history. How much of the evolutionary literature on human behavior would collapse if we incorporated the principle of exaptation into the core of our evolutionary thinking?[7]

But although we may never know the actual events and specific selection pressures responsible for our brain power, we have no scientific reason to believe that evolution could not have fashioned our brain through natural selection. The fact that living organisms today have nervous systems and brains ranging from quite simple to amazingly complex is compelling evidence that our brain evolved through forgotten ancestors in progressive stages from simple to complex. And somehow, as a part of this evolutionary process, that most remarkable and mystifying of all natural phenomena came into being—human consciousness.

The Development of the Brain

From this evolutionary perspective, one might be led to conclude that our brain in all its striking adapted complexity is an inherited legacy of biological evolution. That once evolved it is thereafter *provided* to each individual by good old natural selection, specified in all its fine detail in the genome and transmitted through the generations from parent to offspring.

This type of genetically providential thinking of course is selectionist from the viewpoint of biological evolution, but nonetheless providential at the level of the individual organism. It can be seen in the pioneering work on brain development and function of Roger Sperry for which he shared a Nobel prize in 1981. This research in the 1950s involved disturbing the normal location of nerve fibers in the developing brains of fish and rats. For example, nerve fibers that normally connect the top part of the fish's retina with the bottom part of the brain, called the optic tectum, were surgically removed and reconnected to the top part of the optic tectum. Despite this modification, the nerve grew back to its normal position in the brain. Similar experiments carried out by other researchers on rats indicated that fibers that innervate muscles also "knew" to which muscle they should be attached and made their proper connections despite surgical disturbances. This led Sperry to conclude that the connections of the nervous system are completely specified in the organism's genes. As his former student Michael Gazzaniga explains:

In the original Sperry view of the nervous system, brain and body developed under tight genetic control. The specificity was accomplished by the genes' setting up chemical gradients, which allowed for the point-to-point connections of the nervous system.[8]

But there is a vexing problem with the notion that the genome provides complete information for the construction of the nervous system of humans and other mammals. It is estimated that just the human neocortex alone has about 10^{15} (one followed by 15 zeros, or one thousand million million) synapses.[9] Since the human genome has only about 3.5 billion (3.5×10^9) bits of information (nucleotide base pairs), with 30% to 70% of these appearing silent,[10] some neural and molecular scientists have concluded that our genes simply do not have enough storage capacity to specify all of these connections, in addition to including information on the location and type of each neuron plus similar information for the rest of the body. The problem is not unlike trying to save a document made up of 100 million characters on a computer disk that can hold only 1.4 million characters. As Changeux noted:

Once a nerve cell has become differentiated it does not divide anymore. A single nucleus, with the same DNA, must serve an entire lifetime for the formation and maintenance of tens of thousands of synapses. It seems difficult to imagine a differential distribution of genetic material from a single nucleus to each of these tens of thousands of synapses unless we conjure up a mysterious "demon" who selectively channels this material to each synapse according to a preestablished code! The differential expression of genes cannot alone explain the extreme diversity and specificity of connections between neurons.[11]

Additional understanding of the relation between the genome and the nervous system can be gained by considering *Daphnia magna*. Commonly referred to as the water flea or daphnid, this small fresh-water crustacean is familiar to many aquarium owners since it is relished by tropical fish. But what makes the daphnid interesting for our current purposes is that when the female is isolated from males, she can most conveniently reproduce by the asexual process of parthenogenesis, giving birth to genetically identical clones. In addition, the daphnid has a relatively simple nervous system that facilitates its study. If its genome completely controlled the development of its nervous system, it should be the case that genetically identical daphnids should have structurally identical nervous systems. However, examination of daphnid eyes using the electron microscope reveals that although genetically identical clones all have the same number of neurons, considerable variation exists in the exact number of synapses and in the configurations of connections leading to and away from the cell body of each neuron, that is, the dendritic and axonal branches. As we move to more complex organisms, the variability of their nervous systems increases. This provides clear evidence that the structure and wiring of the nervous system are not the result of following a detailed construction program provided by the genes.

How then is the brain able to achieve the very specific and adapted wiring required to function in so many remarkable ways? For example, how does a motor neuron know to which particular muscle fiber it should connect? How is a sensory neuron in the visual system able to join itself to the correct cell in the visual cortex located in the occipital lobe of the brain? If this detailed neuron-to-neuron connection information is not provided by the genes, whence does it come?

The first clues to solving this puzzle go back to 1906 when it was observed that in embryonic nerve tissue, some neurons did not stain well and appeared to be degenerating and dying.[12] Since it had been assumed that in a developing embryo, nerve cells should be *increasing* in number and not dying off, this finding was somewhat surprising. But nerve cell death in the developing nervous system has since been observed repeatedly. The extent to which it occurs was dramatically demonstrated by Viktor Hamburger. He found that in a certain area of the spinal cord of the chicken embryo over 20,000 neurons were present, but that in the adult chicken only about 12,000, or 60%, of these cells remained.[13] Much of this neuronal death occurs during the early days of the embryo's existence. Nerve cells continue to expire thereafter, albeit at a slower pace.

A particularly striking example of neuronal elimination in development involves the death of an entire group of brain cells:

Most frequently, neuron death affects only some of the neurons in a given category. However, in one case . . . a whole category of cells dies. These particular neurons of layer I, the most superficial layer of the cerebral cortex, characteristically have axons and dendrites oriented parallel to the cortical surface rather than perpendicular to it, like the pyramidal cells. These cells were first observed in the human fetus but have since been found in other mammals. Purely and simply, they disappear in the adult.[14]

But the death of obviously useless brain cells cannot account for the specific connections that are achieved by the remaining neurons. For example, the visual cortex of cats and monkeys has what are called *ocular dominance columns* within a specific region known as cortical layer 4. In any one column of this brain area in the adult animal we find only axons that are connected to the right eye, while in the neighboring column are located only axons with signals originating from the left eye. So not only must the axons find their way to a specific region of the brain, which can be quite far from where their cell bodies are located, they must also find a specific address within a certain neighborhood.

The ability of axons to connect to the appropriate regions of the brain during development has been studied in careful detail since the beginning of this century. Axons grow in the brain like the stem of a plant. At the end of the growing axon is found a growth cone which was described by Spanish neuroscientist Ramón y Cajal in 1909 as "a sort of club or battering ram, possessing an exquisite chemical sensitivity, rapid amœboid movements, and a certain driving force that permits it to push aside, or cross, obstacles in its way . . . until it reaches its destination."[15] Although the exact mechanisms by which this is accomplished are still unknown, it appears that the growth cone is sensitive to certain chemicals along its path that are released by its target region. In this way visual system axons originating in the lateral geniculate nucleus find their way to cortical layer 4 in the occipital lobe of the brain in much the same way that a police bloodhound is able to sniff out the escaped prisoner hiding in an Illinois cornfield.

But although these growth cones lead their axons to the proper region of the brain (or muscle in the case of motor neurons), they cannot lead them to the precise target addresses. For a particular growth cone, it appears that any cell of a particular type will serve as a target. Indeed, in the newborn cat, ocular columns receive axons from both eyes, not just from one or the other, as in the adult brain. For this final and important fine-tuning to be

achieved (on which stereoscopic vision depends), many of the original ter-
minal connections of the axon must be eliminated. In the case of vision, all
axonal connections from the wrong eye are eliminated, and those from the
correct eye are retained. In the case of motor systems that initially have
many-to-many connections between motor neurons of the spinal column
and muscle fibers (that is, many motor neuron axons connected to same
muscle fiber, and many muscle fibers connected to the same axon), the
mature animal possesses a much more finely ordered system with each muscle fiber enervated by one and only one motor neuron. The mammalian
nervous system changes from birth to maturity from a degenerate system
having many redundant and diffuse connections, to a much more finely
tuned system that makes both adaptedly complex behaviors and perceptions (such as stereoscopic vision) possible.

So now the question naturally arises, how does the nervous system know
which connections to retain and which to eliminate? The work of David
Hubel and Torsten Wiesel in the 1970s (both of whom shared a Nobel prize
with Sperry in 1981) provided the first clue. They conducted their ground-
breaking experiments by closing the lid of one eye of newborn cats, and
found that even one week without sight altered the connections of the eyes
to layer 4 of the occipital cortex. Axons carrying nervous signals from the
closed eye made fewer connections with the cortex, whereas axons from the
open eye made many more connections than was normal. This suggested
that visual system axons *compete* for space in the visual cortex, with the
result of the competition dependent on the amount and type of sensory
stimulation carried by the axons. Subsequent research by others using drugs
to block the firing of visual system neurons, as well as artificial stimulation
of these neurons, showed that it is not neural activity per se that results in
the selective elimination of synapses, but rather that only certain types of
neural activity result in the retention of certain synapses, while all others are
eventually eliminated.

In a sense, then, cells that fire together wire together. The timing of the action-poten-
tial activity is critical in determining which synaptic connections are strengthened
and which are weakened and eliminated. Under normal circumstances, vision itself
acts to correlate the activity of neighboring retinal ganglion cells, because the cells
receive inputs from the same parts of the visual world.[16]

The dependence of the development of the visual system on sensory
stimulation would seem to indicate that the fine-tuning of its connections

would have to wait until the birth of the animal when it is delivered from the comforting warm darkness of the womb to the cold light of day. However, recent evidence suggests that this fine-tuning actually begins to take place in utero. Prenatal development appears to depend on spontaneous firing of retinal cells that do not depend on light stimulation from the external world. Similar endogenous patterns of activity may also exist in the spinal cord, and may refine the synaptic connections of motor systems as well.[17]

Nonetheless, interactive postnatal experience of the external world *is* required for normal development of senses and nervous systems in mammals. Cats who have one eye sewn shut at birth lose all ability to see with this eye when it is opened several months later. The same applies to humans. Before the widespread use of antibiotics, eye infections left many newborn infants with cloudy lenses and corneas that caused functional blindness, even though their retinas and visual nervous systems were normal at the time of birth. Years later a number of these individuals underwent operations to replace their cloudy lenses and corneas with clear ones, but it was too late. Contrary to initial expectations, none of these people was able to see after the transplant.[18] It was simply not known at the time that early visual experience was essential to the normal maturation of the brain's visual circuitry. Similarly, some children are born with a wandering eye that does not fixate the same part of the visual field as the normal eye, and other children have one eye that is seriously nearsighted or farsighted; in both cases, the retina of the abnormal eye must be provided with clear visual stimulation, usually by age four years, or it will become functionally blind since its connections to the brain's vision centers will be eliminated in favor of the normal eye.

We thus see that the normal development of the brain depends on a critical interaction between genetic inheritance and environmental experience. The genome provides the general structure of the central nervous system, and nervous system activity and sensory stimulation provide the means by which the system is fine-tuned and made operational. But this fine-tuning does not depend on adding new components and connections in the way that a radio is assembled in a factory, but rather it is achieved by *eliminating* much of what was originally present. It is as if the radio arrived on the assembly line with twice as many electrical components and connections as necessary to work. If such an overconnected radio were plugged in and turned on, nothing but silence, static, or a hum would be heard from its speaker. However, careful removal of unnecessary components and judicious snipping of redundant wires would leave just those components and connections that result in a functioning radio. This snipping is analogous to the elimination of synapses in the human brain as part of its normal development.

The process by which brain connections change over time as maturing animals interact with their environments has been studied in detail by psychologist William Greenough of the University of Illinois at Urbana-Champaign. Using sophisticated techniques for determining the numbers and densities of neurons and synapses in specific regions of the rat's brain, he and his associates found that during the first months of the rat's life a rapid spurt in the growth of synapses occurs regardless of the amount or type of sensory experience.[19] This period of synaptic "blooming" is followed by a sharp decline in the number of synapses. That is, an elimination or "pruning" of synapses then takes place based on the activity and sensory stimulation of the brain, and ultimately results in the configuration of connections characteristic of the mature rat's brain. Greenough refers to this initial blooming and pruning of synapses as "experience-expectant" learning, since the initial synaptic overproduction appears to be relatively independent of the animal's experiences. It is as though the brain is expecting important things to be happening during the first weeks and months of life, and is prepared for these experiences with an overabundance of synapses, only a fraction of which, however, will be selectively retained.

The work of Greenough and his associates is limited to rats and monkeys, but their findings have much in common with those of Peter Huttenlocher of the University of Chicago who counted the synapses in specific regions of the brains of humans who died at various ages. Huttenlocher found that:

The increase in synaptic density plus expansion of total cortical volume leave no doubt that the postnatal period is one of very rapid synaptogenesis in human frontal cortex. By age 2 years, synaptic density is at its maximum, at about the same time when other components of cerebral cortex also cease growing and when total brain weight approaches that of the adult. Synaptic density declines subsequently, reaching by adolescence an adult value that is only about 60% of the maximum.[20]

This wealth of synapses is thought to be responsible for the striking plasticity of the immature brain that permits the learning of skills that can be learned only with much greater difficulty or not at all by the already pruned adult brain. We already saw how immature animals and children are unable to develop normal vision if they are not exposed to a sharply focused visual world during this period of brain development. It has also been repeatedly observed that although many adults initially may make quite rapid progress in learning a foreign language, young children appear to have an important advantage over adults in being able to master the sounds of languages. Canadian child language researchers Janet Werker and Richard Tees observed that children younger than one year appear able to distinguish between the speech sounds used by any human language. By age 12 months, however, they begin to lose the ability to discriminate between sound contrasts that are not used in the language they hear every day. So whereas all normal infants can distinguish between the two related but distinct sounds represented by the letter *t* in Hindi, those who hear only English quickly lose this ability, and Hindi-speaking children retain it.[21] The work of Werker and Tees therefore provides important human behavioral evidence that is consistent with the view that normal brain development involves the loss of synaptic connections, which results in the loss of certain skills as the brain approaches its adult form.

A sensitive period for the acquisition of a first language was demonstrated by the plight of Genie, an American girl who was brutally isolated from all normal human interaction until she was found at age 13 years, and who never subsequently developed normal language abilities.[22] There is striking evidence that the immature, overconnected brain is also better suited than a mature one to acquiring second languages and sign languages.[23]

Taken together, these findings paint a picture of the developing brain that contrasts sharply from the genetic providentialism favored by Sperry. Instead of the brain unfolding according to a genetically specified blueprint, we see instead a process of selection by which overly abundant neuronal connections are eliminated through a weeding-out process, leaving only those connections that permit the animal to interact successfully with its environment.

Learning and Memory: Rewiring the Brain

The mammalian brain appears most adaptive during the early postnatal period, and continues to adapt and learn from new experiences throughout its adult life. During the 1960s and 1970s a series of studies offered impressive evidence that rats grew thicker brains and new synapses when they were placed in complex and challenging environments. These findings were consistent with the then-popular belief that learning and memory in mature mammals (as opposed to the brain development of immature animals) were *additive* processes involving the formation of new synaptic connections or the strengthening of already existing ones. The influential Canadian psychologist Donald Hebb assumed that "the changed facilitation that constitutes learning" was the result of "the growth of synaptic knobs."[24] Similarly, Sir John C. Eccles, who shared a Nobel prize in 1963 for his research on the transmission of nerve impulses, believed that memory and learning involved "the growth . . . of bigger and better synapses."[25]

However, it was also suggested that more than just adding synapses was involved in learning. One of the first to propose that *subtractive* brain changes could be involved in adult learning and memory was J. Z. Young, who in 1964 posited that such learning could be the result of the *elimination* of neuronal connections.[26] Several years later J. S. Albus hypothesized that "pattern storage must be accomplished principally by weakening synaptic weights rather than by strengthening them,"[27] and Richard Dawkins speculated that the selective death of neurons could underlie the storage of memories.[28]

But how could a subtractive process of neuron elimination be involved in learning and memory? It is particularly difficult to understand how the learning of a new skill, such as riding a bicycle or speaking a foreign language, or acquisition of new memories, such as learning the words to a poem or song, could be made possible by loss of synapses. We saw in the development and maturation of the brain that synaptic connections that are rarely used are weakened or eliminated, whereas those in active neural pathways are retained and perhaps strengthened. This subtractive process makes sense when dealing with an overwired, immature brain that may have close to twice as many synapses as it will have as an adult. But how can it work for a mature adult brain that has already been substantially whittled down by synaptic pruning?

To illustrate this problem, imagine an adult Spaniard learning English. To do this, the Spaniard will have to learn to hear and produce certain sound distinctions that are not used in Spanish, such as the contrasts involved in *ship* versus *sheep*, *sue* versus *zoo*, and *watch* versus *wash*. The research of Werker and Tees would lead us to predict that the Spaniard would not initially be able to make these distinctions since they are not made in the language he has heard and spoken all his life. The synaptic connections necessary for making these discriminations were present when he was born, but we would expect them to have been promptly pruned away since they were not used in the language of his environment. It is therefore not clear how any further pruning of synapses would permit him to learn this aspect of the English language.

Instead, it seems more likely that a process involving the *addition* of new synapses, or at least reorganizing current ones, would be necessary for this learning to take place. But then we run into the equally thorny problem of understanding how the brain could ever know which new synapses to add or modify! Surely, some combination of synaptic changes should allow the Spaniard to learn English, since many adults learn English and other languages, and such learning must be the result of changes in the synaptic connections of the brain. But just which new combination of synapses will do the trick? At the very least it would appear that the brain would somehow have to try out a number of new combinations and select the best ones. But to select the best ones, a source of variation is necessary, perhaps not unlike the initial variation of synaptic connections present in the immature, over-connected brain.

A possible solution to this riddle was offered by French neurobiologist Jean-Pierre Changeux in 1983. In his book *L'Homme Neuronal* (published in English in 1985 as *Neuronal Man*), Changeux proposed a "Darwinism of the synapses"[29] to account for the development of the brain and the learning it undergoes within its cultural environment.

> According to this scheme, culture makes its impression progressively. The 10,000 or so synapses per cortical neuron are not established immediately. On the contrary, they proliferate in successive waves from birth to puberty in man. With each wave, there is transient redundancy and selective stabilization. This causes a series of critical periods when activity exercises its regulatory effect.[30]

In effect, he was suggesting that all adaptive brain changes, or at least those occurring between birth and puberty in humans, involve the elimination of preexisting synapses, but that these preexisting synapses were not necessarily all present at the same time. From birth to puberty, Changeux hypothesized that waves of synaptic growth would occur, with subsequent experience serving to retain the useful ones and eliminate the useless and redundant ones. These waves of synaptic overproduction would provide the source of variation on which synaptic selection could operate. Such learning resulted in an absolute increase in synaptic growth and numbers over time. This growth was not constant, but was rather envisioned as analogous to repeatedly taking two steps forward—randomly adding new synapses—followed by one step backward—eliminating the useless ones just added.

Changeux provided no hard evidence for his hypothesis that synaptic variation in the form of overproduction would precede the elimination of synapses as part of the brain's restructuring to permit the learning of new skills and acquisition of new knowledge. But such evidence was found a few years after the publication of his book. William Greenough and his associates, whose work on the maturational development of the rat's brain was noted earlier, also conducted research on changes in the brain induced by placing adult rats in special, enriched environments. In one study this resulted in a 20% increase (roughly 2000) in the number of synapses per neuron in the upper layers of the visual cortex.[31] Later research showed that such dramatic increases in synapses were not restricted to the rat's visual cortex.[32]

These and other similar findings led Greenough's group to propose that the waves of synapse proliferation first described by Changeux could be elicited by the complex demands placed on the adult brain in a new, challenging environment. These researchers referred to this process as "experience-dependent" development since it depends on the environment triggering the formation of new synaptic growth on which the selective process can act.[33]

Greenough's conception of how the adult brain is able to learn new skills and form new memories offers an appealing solution to the problem concerning the additive and subtractive processes underlying the adult's brain adaptation to new environments. According to this theory, experience-dependent learning combines both additive and subtractive processes. The additive component involves the blooming of new synapses in response to the animal's attempt to control aspects of a new, complex environment. Although the brain does appear to know what part of itself has to be involved in this new synapse-construction project, it need not (indeed, could not) know which particular connections to make. By forming a large variety and number of new connections, the brain can select the combinations that work best, in the same way that the immature, developing brain retains useful connections from its initial oversupply of synapses. The long-term result is an overall addition to the number of synapses. But the actual selection process that fine-tunes the connections is a subtractive one in which the useful connections are selectively retained and less useful ones eliminated. Although clear evidence exists for synaptic increase in learning, as I write this we still have no such evidence in mature learning for an over-production of synapses that are then pruned away. However, recent research has found evidence for an overproduction of dendrites in mature rats during readaptation of the brain after brain injury, which at least suggests that synaptic overproduction may be involved as well.[34] These findings fit very nicely with the subtractive synapse findings on brain maturation and provide a solution to the mystery of how the brain could know exactly which new synaptic connections to establish to enable it to acquire new knowledge, skills, and memories.

Although only a relatively small number of neuroscientists have opted for a selectionist approach to their research and theorizing, Changeux and Greenough and their associates are not the only ones whose research suggests that the adult brain develops and learns through a process of cumulative neural variation and selection. This theory has now been embraced and given additional support by several other leading neuro-scientists. William Calvin refers to the brain as a "Darwin machine" that follows the plan "make lots of random variants by brute bashing about, then select the good ones."[35] Gerald Edelman, who shared a Nobel prize in 1972 for his research on the chemical structure of antibodies in the immune system, has contributed a remarkable outpouring of books describing aspects of his "neuronal group selection theory" of brain development and learning through a selectionist process he refers to as "neural Darwinism."[36] And noted psychologist and neuroscientist Michael Gazzaniga, best known for his ground-breaking research on humans with split brains, recently embraced a selectionist account of brain functioning and development.[37]

Current research is under way to determine whether unambiguous physical evidence can be found for the overproduction and elimination of newly formed synapses in the adult brain in response to environmental changes. Such a finding would place the brain alongside the immune system as another striking example of how cumulative variation and selection processes during the lifetime of an organism make it possible to adapt to complex, changing environments.

We have now seen that understanding both the adapted and adaptive complexity of the human brain involves finding answers to three questions: how did the brain originate as a biological organ?; how does it develop from a fertilized egg into a mature brain?; and how is it able when mature to rewire itself to learn from and adapt to changes in its environment?

Much more work must be done before we have detailed answers to these questions. But substantial progress has already been made as we move midway into the "decade of the brain." To a large extent this progress has consisted of rejecting providential and instructionist explanations for these puzzles of fit, and finding considerable evidence and reason in favor of selectionist explanations. The powerful process of cumulative blind variation and selection working over millions of years is not only the only reasonable theory for the biological evolution of the brain, but we find that it has surfaced again in a different but still recognizable form as an explanation for the brain's embryonic growth and continued development during its relatively brief lifetime.

It is here, as Changeux remarked, that "the Darwinism of synapses replaces the Darwinism of genes."[38] To close the circle, it should be noted that a striking consequence of the joint effects of among-organism genetic and within-organism synaptic selection is the brain's understanding both itself and the process of selection that is responsible for its extraordinary abilities.

III

The Promise of Selection

6

The Origin and Growth
of Human Knowledge

Providence

*Now, if the truth of things is always in our soul, the soul is immortal. So it is right
to try boldly to inquire into and recollect what you do not happen to know at
present—that is, what you do not remember.*

—Plato[1]

Instruction

*Let us then suppose the mind to be, as we say, white paper, void of all characters,
without any ideas:—How comes it to be furnished? Whence comes it by that vast
store which the busy and boundless fancy of man has painted on it with an almost
endless variety? Whence has it all the materials of reason and knowledge? To this I
answer, in one word, from experience.*

—John Locke[2]

Selection

*Our categories and forms of perception, fixed prior to individual experience, are
adapted to the external world for exactly the same reasons as the hoof of the horse
is already adapted to the ground of the steppe before the horse is born and the fin of
the fish is adapted to the water before the fish hatches.*

—Konrad Lorenz[3]

Since the time of Socrates, the study of the origin, nature, validity, and
limits of human knowledge has been of much interest to philosophers. This
field of inquiry, referred to as epistemology,[4] continues to be a major focus
of philosophy today. Here we will examine the major approaches that
philosophers have taken to attempt to account for the puzzle of fit between
knowledge and our universe, and explore the extent to which these approaches can be understood as providential, instructionist, and selectionist
explanations for human knowledge.

Knowledge by Recollection

Plato, who lived in Athens during the fourth and fifth centuries B.C., provides us with the first written discussions concerning the origin of knowledge. His philosophy was very much influenced by his belief that knowledge seems to go well beyond what we can learn about the world through our senses. For example, we can imagine (and therefore must possess knowledge about) a perfect circle despite the fact that we have never seen one, since any particular circle will, on close examination, reveal imperfections. Plato thus reasoned that the world we see, hear, smell, and touch cannot be the sole source of our knowledge, since although we can imagine perfect circularity, goodness, beauty, or justice, no particular object or event in our experience can ever be a perfect instance of such a quality. The problem of understanding how we can know so much, given our limited sensory experiences of particular objects and events, has been referred to as "Plato's problem" by the influential American linguist Noam Chomsky.[5]

A related problem concerning the source of knowledge appears in Plato's dialogue between Socrates and Meno. In their conversation concerning the nature of virtue, Socrates' friend begins to doubt the utility of their inquiry and presents an intriguing dilemma:

And how will you enquire, Socrates, into that which you do not know? What will you put forth as the subject of enquiry? And if you find what you want, how will you ever know that this is the thing which you did not know?[6]

Meno is in effect asking, if you don't already possess the knowledge you are looking for, how will you know when you have found it? And if you *do* know what knowledge you are seeking, then mustn't it be the case that you already possess the knowledge in question and therefore have no need to look for it? Socrates recognizes the essence and importance of Meno's question and paraphrases it thus:

I know, Meno, what you mean; but just see what a tiresome dispute you are introducing. You argue that a man cannot enquire either about that which he knows, or about that which he does not know; for if he knows, he has no need to enquire; and if not, he cannot; for he does not know the very subject about which he is to enquire.[7]

To appreciate fully Meno's dilemma, we have to make a distinction between two kinds of knowledge. Clearly, one can be ignorant of a certain fact or piece of information, make an attempt to find it out, and, if successful, be quite certain that what was found out was what one had wanted to know in the first place. For example, it is easy to imagine not knowing the telephone number of an acquaintance, looking it up in the telephone directory, and then knowing that the number so found is indeed what one wanted to know. In this case what we seem to be doing is simply filling in a specific piece of *factual* knowledge. However, since Meno's question arose in the context of discussing the nature of virtue, it is obviously not this kind of factual knowledge he and Socrates found troublesome. That problem has to do with *conceptual* knowledge.[8] If you do not already know what virtue is, how will you ever discover its meaning? Now, if virtue is simply some combination of other concepts already well understood, say a mix of one-third goodness, one-third fairness, and one-third moral strength, it may be the case that you could simply be told what virtue is with respect to these other concepts and thereby acquire a new concept. However if, as is usually the case for concepts, there is no generally agreed upon way to define a new concept as some easily understood combination of old ones, the full force of Meno's dilemma becomes apparent.

Let us look at another example in what is perhaps the more familiar context of a child learning to add.

Children in the preschool years have been found to make a transition, without any instruction, from a crude addition algorithm to a more efficient one. . . . According to the earlier algorithm, the problem of 4 + 3 = ? is solved by a procedure analogous to counting out four blocks, then three blocks, and then counting the combined set. A more advanced algorithm consists of starting with four and counting three more. A key step in this transition is eliminating the counting out of the first addend. To attribute this step to some kind of insight—to "realizing" or "seeing" that counting is unnecessary because the resulting number is already given—is, of course, to tumble right into the learning paradox [that is, Meno's dilemma]. *For such an insight presupposes an understanding of the more sophisticated procedure in advance of discovering it.*[9]

This pattern of behavior was demonstrated by my son when he was about four years old in his interaction with a computer program designed to teach addition. To teach 4 + 3 = 7, the computer program would display images of four objects (say, puppies) in a box on the left side of the screen with three like objects in a box on the right side. Above the images was the equation "4 + 3 = ?" with the number "4" directly above the box containing the four puppies and the number "3" directly above the box containing the three puppies. Although my son had no difficulty counting all the puppies and coming up with the right answer, he initially used the first, inefficient procedure described above of counting all the puppies instead of just looking at the first "4" of the equation and then continuing to count the three objects on the right as "5, 6, 7." After a few such solutions, I could not resist pointing out to him that there was no need to count the first set of objects since he could just use the number provided. I was initially pleased to see that my son was able to make use of this more efficient strategy, but then quickly dismayed to discover that he didn't seem to realize that the answer obtained using this new method was the same as that obtained by counting all the objects! So left on his own he returned to the less efficient method he understood, only to move back spontaneously within a few days to the more efficient method, this time with an understanding of the correspondence between ordinality and cardinality that he somehow managed to acquire on his own. Just how he was able to acquire this new knowledge constitutes the crux of Meno's dilemma.

Bereiter's dismissal of *insight* in the quotation above as a solution to the problem is of particular interest here, and we will consider in chapter 9 how this word raises important issues concerning the origin and growth of knowledge. Let us for now simply note that making an appeal to insight as a type of foresight, foreknowledge, prescience ("to know before") or clairvoyance does nothing to solve Meno's dilemma. It would imply that the knowledge in question was somehow already available to the child, and therefore what seemed to be *new* knowledge was instead actually *old* knowledge.

But this is exactly the unsatisfactory answer that Plato (through the character of Socrates) offers. Since Socrates was devoted to the pursuit of knowledge, he could not leave Meno unanswered. He provides not only a response, but first a demonstration of how Meno's slave boy, having never been educated in geometry, is nonetheless able to solve a problem concerning the area of a square by answering a series of questions. Socrates takes the results of this demonstration as evidence that the slave boy *must already have known* the fundamentals of geometry, and that since these were not learned during his present life as an uneducated slave, they must have been acquired during some previous existence of the boy's soul. Thus Plato, through Socrates' demonstration and response, argues that all knowledge is in essence remembered or *recollected*, a view that is known in philosophy as the doctrine of recollection, or to use the Greek word, *anamnesis*. As Plato explains:

> Now, if the truth of things is always in our soul, the soul is immortal. So it is right to try boldly to inquire into and recollect what you do not happen to know at present—that is, what you do not remember.[10]

Plato's view that all knowledge is simply recollected may have seemed reasonable at the time. We see today, however, that it is seriously deficient in a number of important respects, of which we will consider just two. First, the doctrine of recollection does not adequately address the core of Meno's original question, since substituting the words *recollect* and *recollection* for *inquire* and *inquiry* in the original shows that the essence of the original dilemma remains. In other words, if you don't already know what you are trying to remember, how will you know when you have indeed remembered it? And if you *do* know what you are trying to remember, then it must be the case that you already have access to the knowledge in question and therefore have no need to remember it!

Second, and perhaps of greater importance, the doctrine of recollection does not address the issue of the *origin* of knowledge. If Meno had wished to continue his discussion with Socrates, he might have asked where the soul obtained its knowledge in the first place. From Plato's other writings, it is quite clear that Socrates would have answered that the soul is immortal and therefore its knowledge simply has no beginning and no end. So we see that Plato's proposed solution to the question of the origin of knowledge is a providential one, with no beginning and no end in much the same way that religions view the existence of God.

Knowledge Provided by a Benevolent God

The scientific revolution that began in Europe at the end of the Middle Ages was accompanied by increasing interest in the problem of knowledge, particularly the problem of the reliability of our perceptions of our surroundings. Is the universe actually as it appears to be? Can we trust our senses that fire is hot, that water is wet, and that rocks are hard? Or are these impressions merely illusions, perhaps in the way that our dreams appear to be? Although science was beginning to make remarkable progress at this time, its achievements were accompanied by a fair amount of philosophical skepticism concerning the truth and accuracy of the laws of nature being discovered. Are these scientific laws to be completely trusted, or are they illusions or at best only *likely* to be true or only *approximately* true? The first person in Renaissance Europe to consider these issues seriously was René Descartes, who lived from 1596 to 1650.

Descartes began with a skeptical view of human knowledge and hoped to prove by his "method of doubt" that we can indeed trust our impressions of the physical world. His thinking on this matter can be summarized as follows. First, he realized that all his knowledge was subject to doubt, *except* the knowledge that, to doubt in the first place, there must certainly be a doubter. From this reasoning arose his famous dictum, *cogito, ergo sum* ("I think, therefore I am"). Second, Descartes convinced himself that something external to himself exists. He reasoned that the cause of anything must have as much perfection as what it has caused. Therefore, since he was able to imagine an entity that is the perfect instantiation of all that is good, such an entity (that is, a perfectly good God) must truly exist. The third and final step in Descartes's method was to conclude that a benevolent God would not deceive. Therefore we can be confident that our impressions of the world do indeed correspond to what is really there.

Descartes consequently believed that God played an indispensable role in human knowledge. Not only does God provide assurance that the world we experience corresponds to the world that truly exists, He also provides us with certain innate ideas that are not in any way based on experience, for example, the very idea of God itself, and other abstract concepts such as beauty, goodness, justice, and virtue. Because Descartes's epistemology depends crucially on the knowledge and assurance provided by an all-knowing and benevolent God, it is, like Plato's, a providential view. But to the extent that our knowledge also depends on information gathered by our senses, it is also an instructionist view in which the reliability and validity of sensory instruction are certified and guaranteed by God Himself.[11]

Descartes's arguments have since been much criticized, especially his proof of the existence of a benevolent and all-knowing God, but they are widely regarded as marking the beginning of modern philosophical thought. The belief that the basis of human knowledge cannot reside solely in sensory experience is still considered a key insight. And the quest to provide a justifiable and rational basis for knowledge is very much alive in philosophy today.

Knowledge Instructed by the Senses

Three British philosophers who lived during the seventeenth and eighteenth centuries developed theories that are in striking contrast to the providential epistemologies of Plato and Descartes. Since they all emphasized the role of sensory experience in the acquisition of knowledge, they are referred to as the British empiricists.[12]

The first of these, Englishman John Locke (1632–1704), completely rejected the concept of preexisting innate ideas and argued that all knowledge has its origin in sensory experience. To him the mind at birth was like a blank slate, a *tabula rasa*. To repeat part of Locke's epigraph used at the beginning of this chapter:

Let us then suppose the mind to be, as we say, white paper, void of all characters, without any ideas:—How comes it to be furnished? . . . Whence has it all the materials of reason and knowledge? To this I answer, in one word, from experience.

It was therefore the sensory experiences provided by vision, hearing, smell, and touch that wrote on the mind, leaving impressions that would then be our knowledge of the world. Since Locke saw the role of sensory experience as that of transmitting knowledge to the mind from the outside world, his epistemology is essentially nonprovidential and instructionist.

But although he believed that the senses were the origin of all human knowledge, he did not assume that they provide accurate knowledge concerning all aspects of the external world. He made a distinction between what he called the *primary* and *secondary* qualities of things, the primary including shape, weight, number, and movement, and the secondary including color, taste, smell, texture, and temperature. He believed that our senses provide accurate knowledge about the primary qualities so that, for example, if we see or feel a round object, it is the case that the object actually is round. But this is not the case for the secondary qualities (such as the taste or color of a lemon) since these are sensations produced by an object and do not reflect the properties of the object itself.

Locke's idea of secondary qualities was an important recognition that not all our perceptions of the external world necessarily indicate the actual state of the world. But in making this admission, he opened the way for others to doubt that our perceptions correspond to anything in a real, material world at all. One such philosopher was the Irish-born Anglican Bishop George Berkeley, who lived from 1685 to 1753. Like Locke, he believed that all knowledge was based on sensory experience. However, whereas Locke insisted that our perceptions of primary qualities provide knowledge about what the external, physical world is really like, Berkeley logically concluded that it is not possible to check whether any of our perceptions accurately correspond to a real world. If all we can ever know about the world are our perceptions of it, how can we ever know whether any of our perceptions actually do correspond to physical objects and events?

Berkeley's conclusion was simply to deny that an independently existing, physical world exists. This was in some respects a quite logical continuation of the train of thought begun by Locke, and this immaterialist, idealist philosophy makes even more sense when it is realized that the primary purpose of the bishop's writings was to argue against "skepticism, atheism, and irreligion." According to Berkeley, what we experience through our senses is due to the direct action of God, and so it is through the senses that God communicates with us and informs us as to what things are good for us and what are harmful. As one contemporary critic of Berkeley put it, "Roughly speaking, [Berkeley's] immaterialism is what you get if you start off with Locke's picture and replace matter by God."[13] We therefore see that Berkeley's epistemology is both providential, in that all knowledge is provided by God, and instructionist, in that all knowledge is transmitted to us through the senses.

Scottish philosopher David Hume (1711–1776) was the last of the three British empiricists and arguably the most interesting, troubling, and influential. He was very much impressed by Newton's success in discovering laws of physics, and attempted to apply Newton's experimental method to understand the content and abilities of the human mind and to create a method for discovering truth.

However, Hume's attempt to be scientific and empirical in his search for truth led him to an inescapable and distressing conclusion. Locke believed that from our experience of the world we can know that an external world exists and know at least some of its actual characteristics. Berkeley concluded that we can know only that our ideas and God exist. Hume, almost in spite of himself, reasoned that we cannot claim to know anything about an external world (or even if it exists), about God, or about our own mind.

Like Berkeley, Hume recognized that we can never know the external world directly since all that we know of such a world are our sensory perceptions of it and the ideas that these perceptions generate. But unlike

Berkeley who consequently rejected belief in a material world, Hume insisted that by our very nature we are *compelled* to believe that external objects exist despite the fact that such belief is clearly irrational. His conclusion is therefore essentially that humans are at the core irrational in their beliefs that sensory experiences can reveal to them anything of an external, material world.[14]

Let us be a bit less skeptical for at least a moment and accept for the sake of argument our intuitions that a world of real objects exists independent of our experience of them, and that our perceptions provide generally accurate information about the world. If we grant this much, can we not establish an empirical basis for knowledge? Unfortunately not, as Hume demonstrated in pointing out the problem of induction. Induction refers to the process by which we derive general knowledge based on observations of a limited number of instances. For example, after having eaten bread a few times, one might be expected to believe that bread is an edible and nourishing substance, and not hesitate to eat it in the future. That is, if the bread that one has eaten so far has been nourishing, one might well conclude that *all* bread is nourishing. This seems reasonable enough until one realizes that no matter how much nourishing bread one eats, this can in no way guarantee (or even make it probable) that some bread somewhere is *not* nourishing. Indeed, the bread baked by the local baker tomorrow could possibly be contaminated with the ergot fungus, resulting in bread that could cause illness.

The problem of induction as raised by Hume had, and continues to have, a significant impact on epistemology, particularly on the philosophy of science. In essence, Hume realized the impossibility of an instruction-dependent explanation of knowledge in that no amount or kind of sensory experience (which, even if trustworthy is always limited to a particular time, place, and context) could ever result in certain, justifiable knowledge in the form of universal generalizations or laws of nature. In essence, he could not escape from the conclusion that all human knowledge must be *fallible*, and that no kind or amount of experience, logic, or reasoning could be trusted to eliminate the possibility of error. In this sense, he effectively destroyed empiricism by revealing the irrationality of human belief based on sensory experience and reflections on this experience. As Bertrand Russell, British philosopher, mathematician, social reformer and Nobel laureate in literature noted, "the growth of unreason throughout the nineteenth century and what has passed of the twentieth century is a natural sequel to Hume's destruction of empiricism."[15]

Although this chapter is meant to be philosophical, it should be noted before leaving this section that much psychological research has revealed the inadequacy of the empiricist view that our senses provide us with trustworthy information about the world. Perceptions of the same object can vary from person to person, and even within the same person. The chemical phenylthiocarbamide (PTC) has an unpleasant taste for some people and no taste at all for others. Plunge your right hand into a container of hot water and your left hand into one of cold water. After 15 seconds or so, place them both into the same container of warm water and the very same water will feel simultaneously cold (by the right hand) and warm (by the left hand). Observe an oar that has been placed at an angle in water and it will appear bent, although you know that it is straight. The three visual illusions in figure 6.1 show objects that appear to be of different lengths and sizes although a ruler indicates they are actually the same. Many other examples of how we can be fooled by our senses could be given, and of course a magician's livelihood depends on such misperceptions. We will also see in chapter 11 how our understanding of spoken and printed words may depend as much on our expectations and prior knowledge as on the actual words

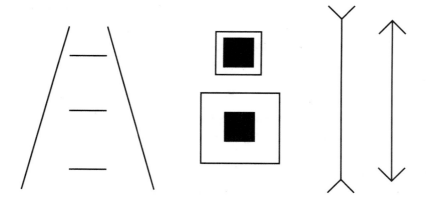

Figure 6.1
Three visual illusions. The three horizontal lines appear to be of different lengths, although they are all the same length. The top solid square seems larger than the bottom one, but they are both of identical size. The left arrow shaft looks longer than the right one, but they are of equal length.

themselves. Add to this the problem of induction raised by Hume, and it should not be surprising that few philosophers consider sensory experience to be an absolutely reliable source of knowledge of the external world.

The Importance of Prior Knowledge

Prussian philosopher Immanuel Kant (1724–1804) was also concerned with the problem of human knowledge and attempted to show how human reason could lead to objectively valid knowledge despite the problems raised by Hume. Kant admitted that it was Hume who caused him to awaken from his "dogmatic slumber."[16] He came to the conclusion that neither coherent experience nor knowledge of the world would be possible without prior ("a priori") possession of certain types of knowledge.

This preexperiential knowledge includes concepts of space, time, and causality. Kant concluded that we cannot possibly gain knowledge about the world without at least some guiding prior knowledge about what to expect. Accordingly, we naturally expect that events take time, that objects take up space, and that if event A always precedes event B (and A and B are irreversible), A is the cause of B. Because these are concepts that we cannot gain from experience, since our very experience of the world depends on them in the first place, and since Kant knew nothing about how a selectionist process can generate new knowledge, he could not and so did not attempt to explain the origin of such a priori knowledge. In this respect, his view of knowledge can also be considered both providential and instructionist since knowledge of the world results from the interaction of mysteriously provided a priori knowledge with instructionist sensory experiences.[17]

In our pursuit of a naturalistic, nonmiraculous account of the knowledge that appears to be prior to sensory experience, we once again meet Konrad Lorenz, who, in addition to his important contribution to the understanding of the evolution of animal behavior, made an important contribution to epistemology. In a paper first published in 1941, he argued that the necessary a priori knowledge, including concepts of space, time, and causation, is actually the product of the biological evolution of the human nervous system.[18] As such, this knowledge does not result from the limited experience of an individual but rather is the hard-won product of the long and arduous evolution of our species. Thus Lorenz states that "all laws of 'pure reason' are based on highly physical or mechanical structures of the human central nervous system which have developed through many eons like any other organ."[19]

This use of biological evolution as an explanation for human knowledge has three important consequences. First, Lorenz is able to account for the fit of human knowledge to its environment without falling back on the providential and instructionist explanations of the philosophers who preceded him. Second, an evolutionary perspective responds to the challenge of Plato's problem, that is, how is it that we are able to know so much despite our limited personal experiences of the world. According to Lorenz, this is possible since biological evolution endows us with a central nervous system that reflects past knowledge obtained through the natural selection of our ancestors, and that is consequently not limited to the experience of the individual. Finally, Lorenz concludes that since human knowledge evolves from an interaction of the species with its environment, it must, at least in some important respects, reflect the environment in which it evolved. In this way he stands opposed to those who assert that what we seem to know of the world may in fact be simply an illusion bearing no resemblance to reality. To repeat another chapter epigraph:

Our categories and forms of perception, fixed prior to individual experience, are adapted to the external world for exactly the same reasons as the hoof of the horse is already adapted to the ground of the steppe before the horse is born and the fin of the fish is adapted to the water before the fish hatches.[20]

Lorenz's evolutionary account of the fit of human knowledge to its environment relies on Darwinian selection operating between organisms, and thus leads to a primarily innatist view. But others have extended Darwinian cumulative variation and selection to knowledge processes occurring *within* humans. It is in this framework that the evolutionary epistemologies of Sir Karl Popper and Donald T. Campbell are found. Discussion of their ideas will be saved for chapters 9 and 10 where thoroughly selectionist views of knowledge, thought, and science are presented.

This admittedly cursory discussion of how philosophy has dealt with the puzzle of fit of human knowledge to its environment cannot pretend to do justice to the vast amount of thought devoted to this issue. Nevertheless, we have touched on the major themes and approaches since Plato's time, and detected a trend from providential to instructionist to selectionist epistemologies. Before Darwin, philosophy had essentially two ways of accounting for human knowledge—providence and instruction. Plato's doctrine of recollection is a rather pure providential epistemology. Locke's and Hume's empiricism emphasizes sensory-based instruction of knowledge from the environment to the individual. Hume recognized the essential irrationality of such knowledge. And the epistemologies of Descartes, Berkeley, and Kant are based on a mixture of both providentialism and sensory instruction.

It was only after Darwin's revolutionary theory became known that a third perspective was imaginable; namely, that human knowledge owed its origin and development to something other than providence or instructive sensory experience. Although certainly no philosopher himself, Darwin made possible a reconceptualization of knowledge as a type of adaptation of the brain to its environment, an adaptation resulting from the same processes of cumulative blind variation and selection that underlie the adaptation of other biological structures and behaviors. Both Lorenz and Popper present such an evolutionary epistemology, although this Darwinian perspective is embraced by only a small minority of philosophers today. As evolutionary biologist Ernst Mayr observed:

No one resented Darwin's independence of thought more than the philosophers. How could anyone dare to change our concept of the universe and man's position in it without arguing for or against Plato, for or against Descartes, for or against Kant? Darwin had violated all the rules of the game by placing his argument entirely outside the traditional framework of classical philosophical concepts and terminologies. . . . No other work advertised to the world the emancipation of science from philosophy as blatantly as did Darwin's *Origin*. For this he has not been forgiven to this day by some traditional schools of philosophy. To them, Darwin is still incomprehensible, "unphilosophical," and a bête noire.[21]

Mayr makes an interesting point here, but he goes too far in implying that many philosophers find Darwin's selectionist theory of evolution incomprehensible. Instead, the philosophers who take the time to reject explicitly an evolutionary epistemology are invariably well acquainted with Darwinian selectionism. However, they advance reasons why they believe that the processes underlying the growth of human knowledge are very different from the those underlying adaptive organic evolution (some of their reasons will be considered in the last two chapters).

It does appear that an evolution-inspired epistemology is resisted by many philosophers because it is inconsistent with their attempts to establish an infallible, justifiable foundation for human knowledge. In this sense, the continually reappearing themes of providentialist rationalism and instructionist empiricism can be seen as attempts to find some bedrock, some firm base on which to base our knowledge, whether it be infallible prior knowledge, God, or completely trustworthy sensory experience. An evolutionary, selectionist epistemology cannot provide such a foundation since selectionist processes are not foresighted and give no guarantee of errorless fit, especially not with future environments not yet encountered.

Yet just such a selectionist perspective is the basis for an alternative epistemology that avoids the problems of providential and instructionist epistemologies while at the same time accounting for the increasingly better fit of our knowledge to the world. Because of this, a selectionist, Darwin-inspired epistemology has gained many proponents over the last century.[22] This trend toward a selectionist account of knowledge growth will likely continue as philosophers become more interested in and familiar with the evolution and development of the human brain and the selection of synapses that, as discussed in the previous chapter, is now believed to underlie all memory and learning.[23] Such complementary selectionist perspectives on learning and thinking are presented in the next three chapters.

7

The Adaptive Modification of Behavior

The environment made its first great contribution during the evolution of the species, but it exerts a different kind of effect during the lifetime of the individual, and the combination of the two effects is the behavior we observe at any given time. Any available information about either contribution helps in the prediction and control of human behavior and in its interpretation in daily life. To the extent that either can be changed, behavior can be changed.[1]

In certain respects operant reinforcement resembles the natural selection of evolutionary theory. Just as genetic characteristics which arise as mutations are selected or discarded by their consequences, so novel forms of behavior are selected or discarded through reinforcement.[2]

—B. F. Skinner

In addition to the inherited instinctive behaviors demonstrated by animals as discussed in chapter 3, we cannot fail also to notice behaviors that are modified to fit the circumstances of each animal during its own lifetime. Biological evolution can account for the emergence of adapted instincts through the natural selection of organisms having useful behaviors. It is simply too slow, however, to generate new behaviors adapted to rapidly occurring changes in the environment. To keep pace with these environmental changes, organisms must be able to learn or acquire new behaviors during their lifetimes.[3]

The ability to modify behavior adaptively is most impressive among the more complex animals such as birds and mammals. In temperate forests these creatures must search for new sources of food as the seasons change and learn to avoid enemies and physical dangers. We have all seen how dogs, cats, and birds learn new behaviors that allow them to adapt better to the artificial world of their owners. The ability of humans to modify their behavior to acquire new job-related and leisure abilities, from computer programming and speaking foreign languages to bicycle riding and piano playing, is striking.

One of the major tasks that the field of psychology has set for itself is to discover the ways in which an animal's experiences lead to the acquisition of new behaviors. Countless worms, snails, rats, pigeons, monkeys, humans, and other animals have been subjected to a wide variety of experimental treatments to help us understand under what conditions and how this adaptive modification takes place. And although relatively little consensus exists today in the field of psychology concerning the mechanisms of learning, especially for the most intelligent animals such as dolphins, apes, and humans, it will be informative to consider a short history of psychological research and theory that have attempted to analyze learning into its basic components.

Pavlovian Conditioning

The first scientific attempts to study changes in behavior began during the 1890s at the Institute of Experimental Medicine in St. Petersburg. There, Ivan Pavlov (1849–1936) was director of what was at the time the world's best equipped physiology laboratory, with facilities to support a large number of dogs. Working with dogs into which stomach tubes had been inserted to collect gastric juices, Pavlov and his assistants observed that the animals would secrete gastric juices not only when food was placed in their mouths but also at the mere sight of food and even at the sight of anyone who regularly fed them. This was followed by the observation of Stefan Wolfsohn, a student in Pavlov's laboratory, that a dog that had repeatedly had sand injected into its mouth (causing salivation to remove the sand) began to salivate at the mere sight of sand. Anton Snarsky, another of Pavlov's students, then demonstrated that a dog could learn to salivate in response to completely arbitrary stimuli. One of Snarsky's experiments involved coloring an acid black and allowing the dog to see it before introducing it into the dog's mouth. After a few such repetitions, the dog would salivate profusely at the sight of any black liquid in a jar.

In this way research began on the type of learning that is still referred to today as Pavlovian conditioning.[4] It is said to have occurred when a neutral stimulus, for example a sound or a light that at first elicits no strong behavioral response, is paired with an unconditional stimulus that normally always results in a specific response, that is, the unconditional response. In Snarsky's experiment, the black acid placed into the mouth would be considered the unconditional stimulus and the secretion of saliva the unconditional response. The original unconditional response to the unconditional stimulus was not considered to be the result of any previous learning experiences (hence the term unconditional), but due instead to an inherited prewired reflex arc connecting the perception of the particular stimulus to a specific behavioral response. By repeatedly presenting a neutral stimulus such as the sounding of a bell immediately before the unconditional stimulus such as placing meat powder in the mouth, Pavlov's dogs soon learned to produce the response (in this case salivation) at the presentation of the previously neutral stimulus. In this way, the dogs would learn to salivate at the sound of a bell if the sound had regularly preceded the placing of food in their mouths.

It is interesting to note that although Snarsky attempted to explain this change in behavior by appealing to the dog's higher mental processes involving feelings, expectations, and thoughts, this was resisted by his professor.[5] Pavlov wished to remain "in the role of a pure physiologist, that is, an objective observer and experimenter."[6] This led him to reject any such mentalistic interpretations, preferring to consider the observed change in behavior as the result of the modification of a simple reflex. After Pavlov received a Nobel prize for his work on the physiology of the digestive system in 1904 (the first Russian and the first physiologist to be so honored), he shifted his attention away from digestion and focused his research efforts almost exclusively on the learning phenomenon discovered in his laboratory.

While Pavlov restricted his research to dogs, American psychologist John B. Watson (1878–1958) applied Pavlov's theory of conditioning to understanding the emotional development of human infants. Watson's observation and experiments led him to believe that during the first month of life babies showed only three emotions—fear, rage, and love—and that these emotions could be elicited only by specific unconditional stimuli, such as a loud sound to evoke a fear response. To demonstrate how an initially neutral stimulus could elicit emotional reactions, he performed what remains one of the best-known psychological experiments with an 11-month-old boy referred to ever since as Little Albert.[7]

Little Albert was presented with a number of live animals and showed no fear as he reached out to touch them. The conditioning procedure then began as a white rat was presented to him. Albert reached out to touch the rat, but as he did so, Watson produced a very loud sound by striking a steel bar behind Albert's head. This pairing of animal and sound was repeated once more. One week later when brought back to Watson's laboratory, Albert was more cautious toward the rat. After five more pairings of the rat with the loud sound, Albert would cry and attempt to move away when he saw the rat. Watson reports that Albert also showed some transfer of his fear to other furry objects, such as a rabbit, dog, and fur coat.

Watson used results such as these to argue that our emotions are largely habits acquired as a result of various experiences, and believed that his findings had important implications for psychological therapy. As he explained, "if we do possess, as is usually supposed, many hundreds of emotions, all of which are instinctively grounded, we might very well despair of attempting to regulate or control them and to eradicate wrong ones. But according to the view I have advanced it is due to environmental causes, that is, habit formation, that so many objects come to call out emotional reactions."[8]

Pavlovian conditioning and habit formation had a great impact on psychology, and continue to influence the practice of clinical psychology in treating individuals suffering from various psychological disorders. But since the theory deals only with the bonding of new stimuli to old responses, it cannot account for the development of new behaviors. For this a different theory of learning was required.

Operant Conditioning

It is a noteworthy coincidence that the same year (1898) in which Wolfsohn submitted his dissertation in St. Petersburg on the Pavlovian conditioning of the dog's salivary response, Edward Thorndike (1874–1949) deposited his dissertation at New York City's Columbia University on learning in cats and dogs. Like Pavlov, Thorndike's experimental studies of animal behavior were to convince him that all learning was dependent on establishing connections between environmental stimuli and specific behaviors. However, the learning task he investigated was very different from that studied in Pavlov's laboratory.

Thorndike was interested in the ability of animals to learn and remember new behaviors. To this end, he constructed a number of "puzzle boxes" into which he would place a hungry dog or cat. The animal could open the door of the puzzle box only by performing some special action such as turning a catch or pulling on a loop of string. Since a dish containing a small amount of food was placed in the animal's view just outside the box, the famished animal was quite eager to escape to obtain a morsel.

Thorndike found that a dog or cat made what appeared to be many random movements when first placed in a puzzle box, but would eventually stumble across the behavior that would allow it to escape. When placed repeatedly in the same box, the animal would generally take less and less time to escape until it was able to perform the specific action required to open the door with no hesitation. Thorndike was also surprised to discover that contrary to previous reports of animal learning, his dogs and cats were not able to learn by observing the successful actions of other dogs or cats, nor were they able to learn from being guided passively by Thorndike through the motions that would free them.

Based on this and other animal research, Thorndike boldly concluded that all learning in all animals (including humans) followed certain laws, the most important being his law of effect:

Of several responses to the same situation, those which are accompanied or closely followed by satisfaction to the animal will, other things being equal, be more firmly connected with the situation, so that, when it recurs, they will be more likely to recur; those which are accompanied or closely followed by discomfort to the animal will, other things being equal, have their connections with that situation weakened, so that, when it recurs, they will be less likely to occur.[9]

He saw his animals acting with no knowledge whatsoever of the consequences of their actions, the sole purpose of the reward being to stamp in the connection between their perception of the situation and a behavioral response. So like both Pavlov and Watson, he came to the conclusion that the formation of connections between stimuli and responses were responsible for learning. But unlike Pavlovian conditioning, which deals with the formation of connections between new stimuli and old responses, Thorndike's animals demonstrated the gradual "wearing smooth of a path in the brain of connections between old stimuli and new responses."[10]

Thorndike was the first psychologist to propose that all new learned behavior results from the combination of random responses and reinforcement. His fellow American B. F. Skinner (1904–1990) did most to popularize this type of learning. Skinner called it "operant conditioning" since it dealt with how animals could learn new ways of *operating* on their environment. In addition to his extensive, detailed research on animal learning, particularly rats and pigeons, he wrote a number of popular books about behaviorism and its applications for solving social and educational problems.[11] For these reasons, he remains among the best-known psychologists of all time. His name remains most firmly connected to the theory of radical behaviorism, a theoretical perspective that ignores the role of internal mental states, purposes, and thought processes in behavior, and instead sees all changes in learned behavior resulting from contingencies of environmental reinforcement.

Limitations of Conditioning Theory

The combination of the conditioning of Pavlov and Watson with the operant conditioning of Thorndike and Skinner might appear to go a long way toward accounting for the adaptive changes in behavior occurring during the lifetime of an organism as a result of experience. Since the description of Watson's research with Little Albert might leave the impression that Pavlovian conditioning can lead only to maladapted behaviors, it should be pointed out how such learning can be adaptive. To return to Pavlov's setting, learning to salivate at the sight of food, or at the sound of a bell signaling its arrival, readies the mouth with the moisture and enzymes necessary for digestion. For a more striking example, consider the flight reaction of most wild animals such as deer to a loud, sudden sound. If this sound is repeatedly preceded by the sight of men holding rifles, the survival value of fleeing at the mere sight of hunters becomes obvious. Pavlovian conditioning can therefore be understood as a type of *stimulus substitution*, or the attaching of old meanings, such as danger, to new experiences, such as hunters. Previously the loud sound of a shotgun elicited an automatic fleeing response; now the mere sight of a hunter will do the same. Thus the development of conditional responses to previously neutral stimuli that regularly precede an unconditional stimulus would allow an animal to anticipate and thereby react more quickly to avoid danger, locate food, and win mates. One might invoke Pavlovian conditioning to explain how any previously meaningless stimuli could come to acquire a meaning for an individual, even in the case of learning a language or learning to read.

As already noted, however, Pavlovian conditioning cannot account for the emergence of new patterns of behavior, and it is here that the theory of operant conditioning is relevant. Operant conditioning can be seen as a way for animals to find and retain creative solutions to problems, as did Thorndike's dogs and cats in learning how to open the door of their puzzle boxes to escape, and Skinner's rats and pigeons as they discovered how to obtain food by pressing a lever, pecking at a key, or walking in a figure-eight pattern. The situations of these animals were contrived and controlled, but it is not hard to imagine natural settings in which such learning would be very valuable, as when an animal discovers a new source of food or finds a new location in which to find shelter. In natural settings, monkeys have learned to wash sand from their food, and birds have learned to get their breakfast by sipping from milk bottles left on doorsteps. Our own species would appear to be the most adept at this type of learning, as we constantly find new and creative ways of feeding, clothing, sheltering, and entertaining ourselves, and providing for our families.

Both Pavlovian and operant conditioning theories of learning gained great popularity during the first half of the twentieth century, particularly in the United States. Seeing behavior as responses to stimuli, and explaining learning as the formation of new stimulus-response connections, made up the core of a behaviorist movement that attempted to make psychology scientific by focusing on publicly observable stimuli and responses. The behaviorist approach reacted against and contrasted sharply with that of the so-called structuralists, who saw psychology as the study of consciousness and used research methods that relied on the subjective verbal reports of subjects. Behaviorists such as Watson, Thorndike, and Skinner began their research with animals and eventually extended their theories to include all human actions. In so doing, they intentionally disregarded any role that thought and other higher mental processes might have in the adaptive change of animal or human behavior.

This neglect of the role of cognitive processes in human learning led to a number of serious difficulties in the application of theories to human behavior. In 1974 cognitive psychologist William Brewer published a review of a large number of studies that were designed to determine whether the change in behavior demonstrated by adult humans in conditioning experiments could be explained by unconscious, automatic stimulus-response connections or if higher mental processes were necessary.[12]

For example, in several of the studies reviewed, a Pavlovian conditioning procedure was employed that paired an initially neutral light or sound stimulus with an electric shock. This pairing, as predicted, resulted in a response[13] to the light or sound (now a conditional stimulus) that was like the original unconditional response to the shock. However, the conditional response to the previously neutral stimulus often quickly disappeared when subjects were informed that the shock would no longer be administered, and whether the conditional response disappeared quickly depended on whether a particular subject actually believed it.

Other studies reviewed by Brewer showed that immediate Pavlovian conditioning often occurred when adults were informed of the purpose of the experiment, and that it did not occur when subjects were prevented from discovering the relationship between the conditional and unconditional stimuli. These and a large number of other studies led Brewer to conclude that "all the results of the traditional conditioning literature are due to the operation of higher mental processes, as assumed in cognitive theory, and that there is not and never has been any convincing evidence for unconscious, automatic mechanisms in the conditioning of adult human beings."[14]

Skinner's theory of operant conditioning has also been criticized by cognitive scientists. Perhaps the most important assessment was provided by American linguist Noam Chomsky who in 1959 reviewed Skinner's attempt to explain language behavior using operant conditioning theory.[15] Chomsky's review will be considered in chapters 9 and 11.

Let us now examine in a bit more detail how both Pavlovian and operant theories, although formulated to account for different types of learning, are both stimulus-response views of learning. By this is meant that a particular stimulus causes activity in some sensory system that is connected by the central nervous system (spinal cord and brain) to motor neurons, and causes a reaction in some muscles that results in an observable response. The view of learning described by Pavlovian conditioning can be seen as the development of new connections between new stimuli and old responses. In other words, if the organism is innately wired so that stimulus A (gunshot) is connected to response Z (fleeing), the pairing of a new, neutral stimulus B (hunters) with A (gunshot) will cause a new connection to form between stimulus B (hunters) and response Z (fleeing). Since the pairing of the unconditional and conditional stimuli (sight of hunters and sound of gunfire) is provided by the environment, and since no trial and error or selection of responses is apparent in such learning, Pavlovian conditioning seems to be a form of instruction by the environment. As British psychologist Henry Plotkin remarked:

When respondent [Pavlovian] behavior enters into a learning relationship it is explained by a process of "instruction." That is, some stimulus or stimulus configuration becomes associated with a reinforcing stimulus and comes to elicit (in some way cause) a response similar to that previously elicited by the reinforcing stimulus.[16]

The apparent instructionist nature of an animal's ability to make new, useful Pavlovian connections between stimuli and responses suggests that not all processes resulting in adapted complexity have to be selectionist in their operation. But although ostensibly instructionist in its basic operation, Pavlovian conditioning may nonetheless have certain important selectionist aspects. First, the ability to learn in this way must itself have been the product of biological evolution, and hence has its roots ultimately in selection, that is, the natural selection of organisms who could learn adaptively to associate new stimuli with old behaviors. Second, as we will see in chapter 9, the perception of stimuli, without which no learning could take place, may be best explained as a selectionist process. Third, it does not appear unreasonable to suspect that the synaptic changes in the nervous system that underlie Pavlovian conditioning may depend on a blind variation and selection of neurons, as discussed in chapter 5. And it should be kept in mind that the instructionist nature of Pavlovian conditioning imposes severe limits on what can be learned, in contrast to the more creative process of operant conditioning. To continue Plotkin's quotation:

What is learned [in Pavlovian conditioning] is an absolutely determined association between stimuli and reflexive responses—the learning does not, cannot, go beyond these explicit events, and the temporal parameters that relate them. This is what I mean by "instruction."[17]

In marked contrast to Pavlovian conditioning, operant conditioning involves a stimulus that initially does not elicit any particular response.

Instead, the organism responds with a series of varied, random behaviors; or to use Skinner's term, the organism "emits" behaviors. Eventually, one of these creatively fashioned behaviors leads to a reward for the animal (for example, food, water, or reduced discomfort), and as a consequence, the behavior is more likely to occur in the same or similar situation. Thus stimulus A (the sounding of a tone after the rat is placed in a Skinner box) originally does not evoke any particular response, but rather the organism emits behaviors X (sniffing the floor), Y (scratching itself), and Z (pushing the lever) after A has sounded. If behavior Z (pushing the lever) is followed by a reward (the appearance of a food pellet), a connection between stimulus A (sound) and response Z (pushing the lever) will be first established and then strengthened by additional reinforcement so that response Z will be likely to occur in the future when stimulus A is again encountered. As described by Skinner, if response Z in the presence of stimulus A results in a reinforcing stimulus, then response Z will come under the *control* of stimulus A. The operant conditioning of Thorndike and Skinner can therefore be considered a stimulus-response theory in that stimuli in the environment come to control the responses of the organism. In contrast to Pavlovian conditioning, the environment now serves only to *select* the appropriate behavior that must first be emitted by the animal. When an appropriate response is made, the environment will provide reinforcement such as food, warmth, or a mate, which will then increase the probability of this same response occurring the next time similar circumstances (stimuli) are encountered. The organism may then refine this behavior so that it is more effective or efficient in obtaining the environmental reward by repeated, cumulative rounds of behavioral variation and selection.

So in contrast to the seemingly instructive role of the environment in Pavlovian conditioning, operant conditioning is clearly a selectionist theory of learning, since the environment does not initially determine the adapted response, but rather selects it, by contingencies of reinforcement, from the many varied responses tried out by the organism. Thorndike, seeing the new discovery of neurons and their interconnections as additional evidence consistent with his connection-based view of learning, wrote of a very Darwinian-sounding "struggle for existence among neurone connections."[18]

Thorndike made only passing references to a selectionist view of learning. But Skinner put considerable effort into promoting operant conditioning as accomplishing over the lifetime of the individual animal what biological evolution accomplishes over the long evolution of a species.[19] Natural selection accounts for the existence of remarkably complex and adapted forms of life without providential recourse to the intentions or purposes of a designer. Thus Skinner saw the selection of behavior by its consequences through operant conditioning as an explanation for the development of remarkably complex and adapted forms of behavior over the lifetime of the individual animal without recourse to the purpose or intentions of the animal. Indeed, it is apparent that toward the latter part of his long and productive career, Skinner's principal objective was to do for psychology what Darwin had done for biology:

Compared with the experimental analysis of behavior, developmental psychology stands in the position of evolutionary theory before Darwin. By the early nineteenth century it was well known that species had undergone progressive changes toward more adaptive forms. They were developing or maturing, and improved adaptation to the environment suggested a kind of purpose. The question was not whether evolutionary changes occurred but why. Both Lamarck and Buffon appealed to the purpose supposedly shown by the individual in adapting to his environment—a purpose somehow transmitted to the species. It remained for Darwin to discover the selective action of the environment, as it remains for us to supplement developmentalism in behavioral science with an analysis of the selective action of the environment.[20]

Skinner discounts here the Lamarckian view of biological evolution, and at first appears to do the same for a Lamarckian view of learning. However, he curiously abandons Darwin and flirts with Lamarck in his discussion of the learning of human culture:

Cultural evolution is Lamarckian in the sense that acquired practices are transmitted. To use a well-worn example, the giraffe does not stretch its neck to reach food which is otherwise out of reach and then pass on a longer neck to its offspring; instead, those giraffes in whom mutation has produced longer necks are more likely to reach available food and transmit the mutation. But a change of culture which develops a practice permitting it to use otherwise inaccessible sources of food can transmit that practice not only to new members but to contemporaries or surviving members of an earlier generation.[21]

This Lamarckian interpretation of cultural learning appears fundamentally inconsistent with Skinner's belief that learning always results from certain spontaneously emitted behaviors being selected by contingencies of reinforcement. From the perspective of operant conditioning, cultural practices cannot be simply transmitted from one person to another, although it may certainly appear that such transmission occurs when we see children adopt the linguistic and cultural practices of their social environment provided by parents and peers. Just as the pattern and color of the tree bark appear to instruct the pattern and color of the back of the well-camouflaged tree toad, it also appears as if behaviors can be transmitted from one generation or individual to another. But in Skinner's theory, no such instructionist transmission of behavior ever takes place. Certainly, the natural selection of learning is different in important ways from the natural selection of biological evolution. As Skinner stated above, cultural practices can spread quickly throughout a community in a way that biological adaptations cannot. But this difference is not due to a Lamarckian transmission of behavior, but is rather a consequence of learning involving selective, psychological processes operating *within* organisms on a short time scale, and not an evolutionary selection process operating *among* organisms on a much longer time scale. (We will return to the problem of accounting for the adapted and adaptive nature of culture and cultural change in chapters 10 and 15.)

Stimulus-response theories of behavioral change appeared to hold great promise during the first half of this century as objective, scientific explanations of the adapted nature of behavior and the adaptive nature of learning. But as we now approach the century's end, these theories are much less popular, particularly as applied to human behavior. Part of the reason for their decline has to do with their disregard of cognitive processes in learning coupled with the continuing cognitive revolution in psychology that began in the 1970s. We will see in chapter 9 that much of the adaptive modification of behavior in humans (and in the more intelligent mammals such as apes) results not from the cumulative variation and selection by the *environment* of *overt* responses, as Skinner insisted, but rather from the cumulative variation and selection by the *animal* of *mental* representations that serve as substitutes or proxies for overt actions. In addition, by viewing all behavior as determined by the environment, and failing to take into account the purposeful, goal-directed aspect of behavior, the principles espoused by behaviorists such as Skinner have been found to be inadequate in explaining human behavior and unreliable for modifying it.[22] Skinner made an important contribution in emphasizing the cumulative variation and selection involved in learning new behaviors. But we will see in the next chapter that he was off the mark concerning both what is selected and what does the selecting as animals continually adapt their behavior to conditions imposed by an unpredictably changing and often uncooperative, even quite hostile, environment.

8

Adapted Behavior as the Control of Perception

What we have is a circuit, not an arc or broken segment of a circle. This circuit is more truly termed organic than reflex, because the motor response determines the stimulus, just as truly as sensory stimulus determines movement. Indeed, the movement is only for the sake of determining the stimulus, of fixing what kind of a stimulus it is, of interpreting it.
—John Dewey[1]

In chapters 3 and 7 we examined several of attempts by philosophers, biologists, ethologists, and psychologists to provide accounts for the fit of organisms' behavior to their environment. We first examined instinctive behavior and showed how attempts to explain its fit progressed from the providential theories of Aquinas and Paley to the instructionist theories of Lamarck and Erasmus Darwin, and then to the selectionist theories of Charles Darwin, Wallace, and Lorenz. We considered learned behavior and saw how the Pavlovian conditioning theory of Pavlov and Watson and the operant conditioning theory of Thorndike and Skinner attempted to explain how behavior can adapt to environmental conditions over the lifetime of an organism through the formation of new stimulus-response connections.

But we also noted one serious omission in all of these theories—they provide no adequate explanation for the goal-oriented, *purposeful* or *intentional* nature of adapted behavior. In addition, Pavlovian conditioning emphasizes a one-way transmission of instructions from stimulus to response, and both Pavlovian and operant conditioning see the role of the environment as a sort of behavioral conductor that orchestrates all of the adaptive changes in behavior. In this chapter we will consider these issues in greater detail and introduce a radically different approach to understanding the adaptive nature of behavior.

The Insights of James, Dewey, and Tolman

It was over 100 years ago in 1890 that the influential American philosopher and psychologist William James (1842–1910) strikingly characterized the purposeful nature of the behavior of living things in contrast with the movements of inanimate objects:

> Romeo wants Juliet as the [iron] filings want a magnet; and if no obstacles intervene he moves toward her by as straight a line as they do. But Romeo and Juliet, if a wall be built between them, do not remain idiotically pressing their faces against its opposite sides like the magnet and the filings with the [obstructing] card. Romeo soon finds a circuitous way, by scaling the wall or otherwise, of touching Juliet's lips directly. With the filings the path is fixed; whether it reaches the end depends on accidents. With the lover it is the end which is fixed, the path may be modified indefinitely.[2]

To make it clear that such purposeful behavior was not uniquely human, James provided an example from the amphibious world:

> Suppose a living frog . . . at the bottom of a jar of water. The want of breath will soon make him also long to rejoin the mother-atmosphere, and he will take the shortest path to his end by swimming straight upwards. But if a jar full of water be inverted over him, he will not, like the bubbles, perpetually press his nose against its unyielding roof, but will restlessly explore the neighborhood until by re-descending again he has discovered a path round its brim to the goal of his desires. Again the fixed end, the varying means![3]

Even though James's *Principles of Psychology*, from which these passages are taken, was considered to be the most important psychological work of its day, the stimulus-response, conditioning theories of Pavlov, Watson, Thorndike, and Skinner all totally discounted purpose as having a role in a scientific account of behavior. In fact, Skinner repeatedly used the analogy of biological evolution to argue against purpose in behavior. For example:

> Evolutionary theory moved the purpose which seemed to be displayed by the human genetic endowment from antecedent design to subsequent selection by contingencies of survival. Operant theory moved the purpose which seemed to be displayed by human action from antecedent intention or plan to subsequent selection by contingencies of reinforcement. A person disposed to act because he has been

reinforced for acting may feel the condition of his body at such time and call it "felt purpose," but what behaviorism rejects is the causal efficacy of that feeling.[4]

Darwin showed how natural selection leads to adapted complexity in the structure and instinctive behavior of organisms, without purpose either on the part of the organism or on the part of a supernatural provider. So Skinner is arguing here that the selection of behavior by the environment can similarly explain the adapted complexity of learned behaviors without recourse to purpose. This is accomplished by forming new stimulus-response connections strengthened by environmental reinforcement.

But such a stimulus-response view of learning is both seriously incomplete and misleading. In addition to the problems mentioned in the previous chapter, American philosopher and educator John Dewey (1859–1952) provided another. He observed that the stimulus-response interpretation of behavior was flawed since it recognizes that stimuli influence responses, but it neglects the equally important fact that *responses also influence stimuli*. Consequently, he criticized the concept of the stimulus-response reflex arc by noting that

what we have is a circuit, not an arc or broken segment of a circle. This circuit is more truly termed organic then reflex, because the motor response determines the stimulus, just as truly as sensory stimulus determines movement. Indeed, the movement is only for the sake of determining the stimulus, of fixing what kind of a stimulus it is, of interpreting it.[5]

To understand how behavior can adapt to its environment we have to consider just what Dewey meant by his statement that behavior determines stimulus just as much as stimulus determines behavior, an insight that was almost totally ignored until the middle of this century. As we have seen, Pavlovian and operant conditioning theories view all adapted behavior as responses to external stimuli, including those caused by the behavior of other organisms. Acid is placed in a dog's mouth and it salivates. A hungry rat is placed in a familiar cage where in the past food was made available whenever it pressed the bar within two seconds after a bell sounded, and so it immediately proceeds to push the bar after hearing yet again the familiar peal of the dinner bell. To use the more technical terms of psychological jargon, the stimulus is considered to be the independent variable and the response the dependent variable; that is, the response *depends* on the stimulus, and the stimulus is *independent* of the response.

Another way to conceptualize this one-way view of the relationship is to see the stimulus as the sole determining *cause* of the response, and the stimulus to be isolated from any effects of the response. But Dewey took exception to this view, stating that the *response also causes the stimulus*. But how can this be? How can the dog's salivating influence the presence of acid in its mouth? How can the rat's pressing of the bar in any way cause the bell to be rung, the sounding of which is controlled by the experimenter?

To appreciate Dewey's important insight, we have to do something that is rarely done in experimental psychology. We have to abandon the point of view of an objective outside observer and instead attempt to imagine how things appear from the organism's point of view. This shift in perspective brings with it a realization that a stimulus can have an effect on an organism only insofar as it is experienced or *perceived* by the organism. A dog who cannot perceive that acid has been (or is about to be) placed in its mouth will not salivate when the acid is so placed (or about to be). Once we imagine how the world appears to the organism, Dewey's point about response influencing stimulus begins to make sense. Actually, it is not necessary to consider the world as it may appear to dogs and rats. A cursory look around our own world will quickly and clearly reveal that behavior causes changes in perception as much as perception causes changes in behavior.

As you move your eyes across this page you cannot fail to notice that as you do so your perception of the page changes. We might reasonably conclude that we move our eyes for the purpose of bringing into view that which we want to read next, and that this in fact is the very reason for the behavior. So response does influence stimulus and so behavior does influence perception. But this is not to deny that stimulus also exerts an influence on response. A poorly written sentence or complex sentence may well lead to your returning to it in a second attempt to decipher its meaning. And a loud sound coming from behind your head will likely have you quickly turning around to see what has happened. So proper understanding of behavior has to take into account the reciprocal give-and-take relationship of stimulus and response, something neither Pavlovian conditioning nor operant conditioning proposed to do.

The purposeful nature of animal behavior was clearly demonstrated by the research conducted by psychologist Edward C. Tolman (1886–1959) and his students at the University of California at Berkeley from the 1920s to the 1950s. Among the best known of these studies was one conducted by Tolman's student D. A. Macfarlane in which rats learned to swim through a maze to obtain a food reward.[6] After they had learned to do this well, a raised floor was installed in the maze so that the rats now had to wade through the maze to get to the goal box. It was hypothesized that if the rats' learning consisted of acquiring specific swimming behaviors (that is, specific responses to specific stimuli), they would have to relearn the maze in the wading condition, as the movements and stimuli involved in wading are very different from those involved in swimming. It was found instead that after a very brief period of adjustment to the new situation (just one "run" through the maze), the rats performed as well in the new wading condition as they had in the old swimming condition. This was a clear demonstration that what the rats had learned while swimming the maze could not be described as the acquisition of stimulus-response connections but rather as more general knowledge about the location of the goal box, since it made little difference to the rats whether they swam or waded to their destination. Similarly, once a person knows how to get to a specific location by driving a car, he can also get there by bicycle (if he knows how to ride one) or by walking (if not too far), regardless of the fact that the stimuli and responses differ greatly from one mode of transportation to another.

Regardless of these findings and many others like them, Tolman was never able to eliminate the concept of stimulus-response connections from the very core of his theory of purposeful behavior. Indeed, his attempt to explain how behavior can vary and yet reach a consistent goal involves imagining long, complicated, invisible chains of such connections existing within the organism in the form of *intervening variables*, and conceiving of responses not as specific muscular contractions but rather as a *performance*. With respect to the latter Tolman wrote:

It is to be stressed . . . that for me the type of response I am interested in is always to be identified as a pattern of *organism-environment rearrangements* and not as a detailed set of muscular or glandular activities. These latter may vary from trial to trial and yet the total "performance" remains the same. Thus, for example, "going towards a light" is a *performance* in my sense of the term and is not properly a response (a set of muscular contractions).[7]

But substituting the word "performance" for "response" does nothing to explain how an organism is able to accomplish a repeatable "organism-environment rearrangement" by responding to stimuli; it simply states that it somehow happens. If "behavior may vary from trial to trial and yet the total 'performance' remains the same," how is it that the organism is able to vary its behavior to arrive at a desired goal?

Nonetheless Tolman made an important initial step toward solving this problem in his realization that sensory *feedback* was important; that is, the rat's behavior changed the stimuli it perceived and this feedback was essential in guiding the organism toward the final goal.[8] But Tolman never provided an explicit model for just how such a system would work, and so he never broke out of the behaviorist tradition of considering stimuli as causes of behavior. The first successful attempt to develop a working model of purposeful behavior would have to await the development of negative-feedback control systems in engineering and their application to the life sciences by a few bold pioneers with interests and expertise in both engineering and psychology.

An Introduction to Control Systems

Macfarlane's swim-first-wade-later experiment provided clear evidence that the rat's behavior can be purposeful. Careful observation of the naturally occurring behavior of animals and people is all that is necessary to lead us to the same conclusion. Environments do not normally remain cooperatively still as behavior takes place, particularly not when they consist of other competing organisms. No fixed, predetermined pattern of muscular contractions and tongue movements will guarantee the frog's success in hunting flies. Nor will some unvarying pattern of wing movements ensure the fly's success in avoiding the frog's lunging tongue. To facilitate an animal's survival and reproduction, behavior must somehow take account of an ever-changing and unpredictable environment. A particularly striking example is the nest-provisioning behavior of the solitary wasp:

Sometimes she drops the fly behind her, and then turning around, pulls it in [the nest] with her mandibles. In other cases, where a longer portion of the tunnel has been filled with earth, the fly is left lying on the ground while the wasp clears the way. The dirt that is kicked out sometimes covers it so that when the way is clear the careless proprietor must search it out and clean it off before she can store it away. In one instance, in which we had been opening a nest close by, the tunnel was entirely blocked by the loose earth which we had disturbed, and the wasp worked for ten minutes before she cleared a way to her nest.[9]

Countless examples of purposeful human behavior are easy to find. One with which many readers have daily personal experience is driving a car from home to work. Such behavior requires a remarkably complex, coordinated pattern of behavior of the fingers, hands, and arms operating on the steering wheel, shift lever, and turn indicator lever, while the legs and feet operate the accelerator, clutch, and brake pedals. Many patterns of behavior will get the driver and car safely to work (depending on the speed of the car and the particular route taken), but just slight changes in any one may have fatal consequences. And yet, no one set pattern of behavior will always be successful. Traffic may be heavy or light, fast moving or slow. The road surface may be dry and firm or wet and slippery. The engine may be responsive or balky. Road construction and traffic accidents may make deviations from the normally preferred route necessary. Because of these changeable conditions and other unpredictable disturbances, an exact replay of the driving behavior that was successful in getting you to work on Monday would certainly not get you to work on Tuesday. What is true for the person driving a car to work is also the case for the person walking from bedroom to bathroom, the bee finding and collecting nectar, the fox pursuing the hare and the hare avoiding the fox, the monarch butterfly migrating from Mexico to Canada, and the greylag gosling closely following the steps of its mother (or Konrad Lorenz). It is difficult to see how any complex behavior can remain adapted to some function if the organism does not continually make adjustments to it *while it is performing the behavior.*

But how is this possible? Can we imagine the functioning of an organism, made up of sense organs connected by a nervous system and brain to muscles, that is able to pursue a goal by continually modifying its behavior to adjust for environmental disturbances that would cause any fixed pattern of behavior to miss its mark? Yes, we can. The explanation comes to us not from the life sciences of biology, ethology, or psychology, but from electrical and mechanical engineers who in the 1930s began to make devices that could duplicate the purposeful behavior of humans.

These devices are known as *control systems,* and they work using the same type of stimulus-response circuit, or *loop,* that Dewey first mentioned in 1896. They also behave in the same variable-path but fixed-goal manner James described over a century ago as characteristic of animal and human behavior. Such devices are now commonly found in a multitude of electronic devices from the simple thermostat of a home heating system to the highly complex guidance system of antiaircraft missiles. To see how they work, we will consider one that is now commonly found on automobiles—the cruise control that automatically maintains a steady driving speed with no assistance from the driver.[10]

The cruise control system is engaged by first turning it on and then pushing the "set" button after the car has reached the desired speed. This speed then becomes the system's goal, or what control system engineers refer to as the reference level, and the system will then increase or decrease the amount of fuel it delivers to the motor as necessary to maintain this speed. If the car begins to climb a hill, the cruise control system will sense a reduction in speed (being equipped with a speedometer that measures the speed of rotation of the wheels) and will provide more fuel to the engine through a mechanical linkage to the throttle, causing it to increase its power output to maintain the speed despite the hill. As the car begins to descend the other side of the hill, the cruise control system will sense the increased speed, which will cause it to close the throttle, reducing the amount of fuel delivered to the engine so that again the desired speed is maintained. Because the system responds to too-high speeds by reducing the amount of fuel delivered to the motor and to too-low speeds by increasing the flow of fuel, it is referred to as a *negative feedback* system.

A clearer idea of the nature and functioning of a simple control system can be obtained by examining figure 8.1. The *sensor* converts some variable aspect of the environment (for example, light, sound, or speed) into a sensor signal (s), which varies from zero to some higher positive value. This sensor signal (s) is then compared with a *reference signal* (r) in the *comparator*, which subtracts s from r yielding an error signal e. This error signal is then amplified and converted by the *activator* into behavior (o for output). This behavior then acts on the *environment*, changing it in the intended direction, which again provides input (i) to the *sensor*, thereby closing the loop. However, it is not only the control system's output that influences the input to sensor, but also disturbances (d) emanating from the environment. So the feedback resulting from the control system's own behavior and the current disturbance from the environment combine to provide the input to the sensor.

A cruise control system acts very much like a human driver, with the goal of maintaining a given speed. To do this, the driver must attentively monitor the speedometer. If the speed drops, the driver must press down on the accelerator pedal. If the speed rises, the driver must reduce pressure on the accelerator. While the car is moving at the desired speed, no action is called for. It should come as no surprise that the cruise control system mimics the functioning of a human driver so well, as this is exactly what it was designed to do.

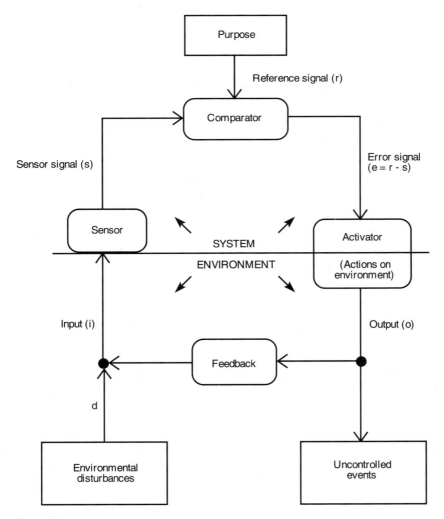

Figure 8.1
A basic control system (after McClelland, 1991).

The cruise control system has a number of intriguing aspects that are shared by all properly functioning control systems. First, it does not perceive the actual disturbances for which it must compensate. It has no way of determining whether the road is climbing or descending. It cannot tell if there is a stiff headwind or tailwind. It cannot know if a heavy trailer was attached to the car at the last stop, if a tire is losing air and offering steadily increasing rolling resistance, or if a spark plug has fouled causing a cylinder to fail and the engine to lose power. All it can sense, and therefore control, is the car's speed. Yet despite its complete ignorance of the multitude of interacting influences, it does a very good job of maintaining the desired speed.

Second, *a control system does not control what it does; it controls what it senses.* The word *control* is used here in its precise technical sense of *maintaining some variable at or near specified fixed or changing values* regardless of the disturbances that would otherwise influence it to vary. The cruise control system can only control what it senses to be the speed of the vehicle, and it does so by changing its output as required, that is, by delivering varying amounts of fuel to the engine. The only way it can maintain its sensing of the car's speed close to the reference level speed in the face of disturbances is to vary its output (change its behavior) as necessary. So we see that it controls its *input* (what it senses) and not its *output* (or behavior). Consequently, using a cruise control system to maintain a constant speed on a trip will allow you to predict accurately how long it will take to cover a certain distance. It not let you predict how much fuel will be used in getting there, because fuel consumption is not controlled, varying as it must to compensate for unpredictable disturbances. Since a control system controls what it senses, and since an organism's sensing of the environment is generally referred to as perception in the behavioral sciences, the application of control theory to the behavior of living organisms is known as *perceptual control theory* to distinguish it from the control theory applied by engineers and physicists to the inanimate world.

Finally, it is important to realize that whereas a control system's behavior is clearly influenced by the environment, it is not determined solely by the environment. Rather, its behavior is determined by what it senses (or perceives) *in comparison with its internal goal or reference level*. And it is here that we find a crucial difference between the nonliving control systems designed by engineers and the living ones fashioned by biological evolution. An engineered control system is usually designed so that its reference level can be manipulated by the operator, for example, by pushing the "accelerate" button of the cruise control system or by turning up the house thermostat. No such direct manipulation of the reference levels of living control systems is usually possible. We can certainly ask our taxi driver to drive more slowly or our child to be home before midnight, but there is no way to guarantee that the person will comply with our wishes.

Control systems were specifically designed to replace human operators in jobs calling for the control of important variables (for example, steam pressure in a boiler) and have been in wide use since the 1930s. They were almost completely ignored by the behavioral sciences, however, until the appearance of Norbert Wiener's ground-breaking book *Cybernetics* in 1948, which was quickly followed by important related works by W. Ross Ashby.[11] William T. Powers, an American control system engineer, was also struck by how the behavior of such systems resembled the purposeful behavior of living organisms, and it is primarily due to his work that a control system theory of human and animal behavior exists today.[12]

Perceptual Control Theory

Perceptual control theory does what no behaviorist stimulus-response theory has ever been able to do—it provides an explicit, working model that accounts for goal-oriented, purposeful behavior. Behavior must be purposeful if it is to enable an organism to survive and reproduce despite the unpredictable disturbances the organism continually encounters. Indeed, we could consider behavior to be adaptedly complex only insofar as it is able to achieve its purposes regardless of the continual challenges posed by environmental disturbances. The perspective provided by perceptual control theory, however, has not been received enthusiastically by behavioral scientists. The principal reason for its neglect appears to be that it turns the traditional analysis of behavior on its head. Instead of the still-dominant view of seeing stimuli (both past and present) controlling responses (or perceptions controlling behavior, or environment instructing organism), the theory offers the unorthodox view of behavior as controlling perception through the organism's *control of its environment*. Hence the title of Powers's seminal book, *Behavior: The Control of Perception*.[13]

Diagrams can be useful in providing some basic understanding of the functioning of control systems, but they have some serious limitations. For one thing, they may easily lead one to interpret the functioning of a control system as a series of sequential steps, with each step waiting for the completion of the previous one. Instead, in functioning control systems, both nonliving and living, all parts are active simultaneously so that both perception and behavior are happening *at the same time.*[14] Also, such diagrams may give the impression that control systems are limited to quite simple variables. In fact, complex control systems composed of many simple control systems can be designed to control quite complex variables, that is, variables that are computed composites of values of lower-order perceptual variables.

Powers proposes that a complex hierarchy of control systems underlies human and animal perception and behavior. Currently, 11 levels are envisaged. From lowest to highest they are intensity, sensation, configuration, transition, event, relationship, category, sequence, program, principle, and system concept. These levels will not be described in detail here,[15] but it should at least be noted that higher-level perceptions such as a configuration (for example, the visually perceived printed letter A) depend on a particular set of combinations of lower-order sensations, which in turn depend on a set of particular combinations of still lower-order intensities. But whereas higher-level perceptions depend on lower-order ones, the control of higher-level perceptions is achieved by manipulating lower-order reference signals. Thus as you write a note to a friend, you are controlling for the appearance of certain letters on the page. But to produce those letters, you must vary the reference levels for the positions and movements of your arm, hand, and fingers. In the same way, a higher-level reference level for reading results in varying the lower-level reference levels for the movements and positions of your eyes. That higher-level perceptions are made up of combinations of lower-order ones and that higher-order systems control their perceptions by varying the reference levels of subordinate systems is illustrated in figure 8.2. Notice, however, that this hierarchy is not a typical chain of *command*, since higher-order control systems do not tell lower-order ones what to *do*, but what to *perceive*. (We will return to the control system hierarchy at the end of chapter 12 in our discussion of education.)

CONNECTIONS TO HIGHER LEVELS

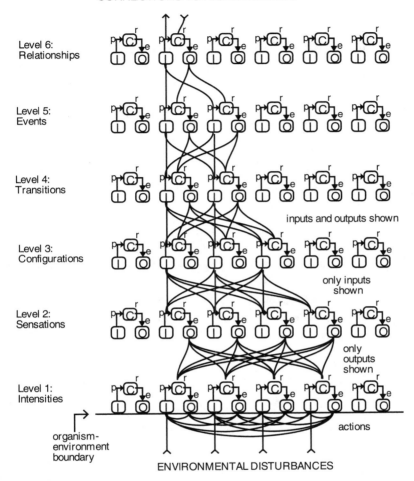

Level 6:
Relationships

Level 5:
Events

Level 4:
Transitions

inputs and outputs shown

Level 3:
Configurations

only inputs
shown

Level 2:
Sensations

only
outputs
shown

Level 1:
Intensities

organism-
environment
boundary

actions

ENVIRONMENTAL DISTURBANCES

Figure 8.2
A hierarchy of control systems (after McClelland, 1991).

Let us now turn away from the complexities of perceiving and controlling higher-order perceptions and consider the relatively simple variable of distance to see how perceptual control theory can be used to understand the behavior of the greylag gosling. As described in chapter 3, Lorenz discovered that the greylag gosling will imprint on the first large object it sees after hatching, which in natural settings is its mother. Thereafter it will maintain close contact with this object throughout its goslinghood. From the perspective of perceptual control theory, we would say that the gosling develops a control system that permits it to maintain a relatively fixed distance between it and its mother despite the disturbances caused by the mother goose's own walking, wading, and swimming, and regardless of obstacles such as bushes, trees, rocks, and other geese that may come between them. Such a living control system is a great deal more complex than the one used in an automobile cruise control system. We can nevertheless see how all of its properties can apply to this situation, the two most important being the presence of a reference level specifying the distance to be maintained and the negative-feedback loop connecting perception to action and action back to perception.

Powers and sociologists Clark McPhail and Charles Tucker developed a computer program that uses control systems to simulate just this type of collective behavior.[16] Figure 8.3 shows the results of one such simulation in which four individuals (each indicated by the letter G) maintain close proximity to another individual (M) who is moving toward a goal location indicated by a large circle. Obstacles for all individuals to avoid are indicated by small circles. By examining the paths of the individuals (shown by the meandering lines), it is seen that the four Gs are successful in both main-

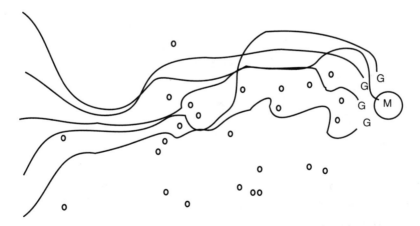

Figure 8.3
Simulation of four individuals (G) following another (M) (after McPhail, Powers, & Tucker, 1992).

taining close contact with M and avoiding the obstacles as M moves to its destination. Although these computer simulations were developed to model a particular type of collective human behavior, they also provide a striking simulation of the behavior of a group of goslings (Gs) maintaining close contact with their mother (M) as she herself moves to some destination, with all of them avoiding obstacles along the way.

A strong argument can therefore be made that it is not behavior in the form of fixed motor patterns as conceived by Lorenz that evolves and becomes instinctive. Rather, the basis of instinctive behavior is an interacting network of control systems that permits organisms to maintain certain perceptual goals (reference levels) despite the disturbances that they encounter. In much the same way that we instinctively know how to keep our body temperature at or near a constant 37°C through such automatic, unlearned behaviors as shivering and sweating, organisms also appear to inherit other control systems that underlie adaptive, species-specific behaviors. It is not that the spider is programmed with a fixed behavior pattern that will result in the construction of a web, but rather the spider is able to control its perception of its environment to match an inherited internal goal specification (reference level) by varying its behavior as necessary to construct its web. And in keeping with Lorenz's original insights concerning the central role of natural selection in the evolution of instinctive behavior, it is only through selecting those organisms with fit control systems and eliminating those with less fit ones that the adapted complexity of instinctive behavior can be explained. As Powers remarked:

. . . it's the capacity to perceive and control that evolves, not the specific acts by which control is effected. Behavioral acts achieve repeatable results only if they change appropriately with every disturbance, every change in initial conditions. There's no way to inherit behavioral outputs, because the outputs must remain adjustable to current circumstances, which never repeat exactly. All that can be inherited are control systems, and at the highest existing level perhaps some reference signals.[17]

But what about learned behavior? Can perceptual control theory account for adaptive changes in behavior resulting from experiences during the lifetime of an organism? To see how the theory views learning, we first must take a fresh look at both Pavlovian and operant conditioning. With respect to the former, we must reconceptualize the unconditional response to an unconditional stimulus not as the functioning of a stimulus-response connection but rather as that of a control system.[18] To take the example of Pavlov's dog into whose mouth acid (the unconditional stimulus) is introduced leading to salivation (the unconditional response), we can conceive of the stimulus as a disturbance to a reference level for oral acidity and the response of salivation as an action designed to restore oral acidity to its normal level. These and other unconditioned responses (for example, eye-blinking to a puff of air, the startle response to a sudden loud noise) can be seen as very basic, inherited control systems having obvious survival value that evolved through natural selection to protect the organism from harmful environmental disturbances.

Now, if a certain neutral stimulus, such as the sounding of a bell, regularly precedes the unconditional stimulus, it can be used by the organism as a signal that a disturbance is about to occur. So by producing the response after the neutral stimulus (now a conditional stimulus) but before the unconditional stimulus, the organism can prevent or at least lessen the disturbing effect of the unconditional stimulus. By salivating at the sound of the bell that precedes the introduction of acid into its mouth, the dog can buffer its mouth against the effect of the acid.

Considering now the operant conditioning of Thorndike and Skinner, we have seen how stimulus-response conceptualizations of learning cannot account for the purposeful nature of behavior as noted by William James in 1890—the ability to achieve fixed ends by varied means. Construing learning as the acquisition of fixed patterns of behavior cannot explain how organisms can be successful in achieving important goals, such as finding food, mates, and shelter, in the face of unpredictable disturbances.

Another problem with operant conditioning theory is that it provides no explanation for why certain events reinforce the organism's behavior and others do not.[19] Why is it that a hungry but well-watered rat will work a lever to obtain food but not water, while a thirsty but well-fed one will do the opposite? Perceptual control theory answers this question by seeing the reward as a controlled variable, that is, a variable that is controlled by the organism by varying its behavior. If a hungry rat pushes a lever to obtain food, it is only to bring his perceived rate of food intake close to its reference level, which has been chosen through natural selection during the evolution of the rat as a species.

Finally, perceptual control theory explains an intriguing pattern of behavior observed by Skinner that contradicts the basic notion that reinforcement increases the probability of the response preceding the reinforcing stimulus. Skinner found that he could obtain very high rates of operant conditioned behavior (such as a hungry pigeon pecking at a key to obtain food) by gradually decreasing the rate of reinforcement. Very high rates of behavior could be shaped by starting out with an easy reinforcement schedule that provided a speck of food for each key peck, and gradually moving toward more and more demanding schedules requiring more and more pecks (2, 5, 10, 50, 100) for each reward. Skinner was thereby "able to get the animals to peck thousands of times for each food pellet, over long enough periods to wear their beaks down to stubs. They would do this even though they were getting only a small fraction of the reinforcements initially obtained."[20]

But if, according to Thorndike's law of effect and Skinner's theory of operant conditioning, more reinforcement is supposed to cause more of the type of behavior that resulted in the reinforcement,[21] how could it also be that less reinforcement could also cause more of the behavior? This problem is effectively solved when we see reinforcement not as an environmental event that increases the probability of the specific behavior that preceded it, but rather as a means by which the organism can achieve a goal by controlling a perception. If the circumstances are arranged so that the hungry rat must perform more bar presses to be fed, and it has no other way to obtain food, the rat will adapt by increasing its rate of pressing to obtain its desired amount of food. And if the rate of reinforcement is increased to the point at which the rat can maintain its normal body weight, a control system model of behavior would predict that further increases in reinforcement should lead to decreases in the rate of behavior. Indeed, this is exactly what happens.[22]

It should now be obvious that a perceptual control theory interpretation of adapted behavior is radically different from a behaviorist view of operant conditioning. Whereas behaviorism sees the environment in control of the behavior of the organism, perceptual control theory sees the organism in control of its environment by means of varying its behavior. In other words, to behaviorists, behavior is controlled by the environment; to perceptual control theorists, behavior controls the environment. This is not to say that the environment has no influence on behavior. Rather, behavior can be adapted only if it is part of a larger control process that varies behavior to produce the perceptions specified by internal reference signals leading to the accomplishment of goals important for survival and reproduction.

Perceptual control theory also makes an important distinction between changes in performance and learning that is not made by behaviorist theories. According to the behaviorist theory of operant conditioning, the rat's increase in rate of bar pressing in response to a decrease in reinforcement is an example of learning—the animal has somehow learned that more presses are required to obtain a pellet of food and will consequently increase its rate of behavior in response to this new demand. But according to perceptual control theory, no learning has taken place, as this is just the normal functioning of a control system. Recall from our earlier description of an automobile cruise control system that control systems vary their output to control their input to compensate for environmental disturbances. No rewiring or other modification of the control system is necessary for this to happen. Consequently, the rat's increase in the rate of bar pressing is a change of performance that can be explained as the functioning of an existing control system, in the same way that the cruise control system will deliver more and more fuel to the car's engine as external disturbances (hill, headwind, the added weight of accumulating ice and snow) act to slow it down.

However, instead of decreasing the rate of reinforcement for the rat, let us imagine that the situation is changed so that food is delivered not when the rat presses the bar down but when it pushes the bar up. Now, the rat's existing control system will no longer prove effective in controlling the amount of food obtained. This is analogous to reversing certain electrical connections in the cruise control system so that more fuel (instead of less, as before) is delivered to the motor when the speed rises above the reference-level speed, changing a negative-feedback closed loop to a positive-feedback one. A striking difference emerges between the rat as a complex living control system and the cruise control as a much simpler artificial one. In the case of the latter, the car will accelerate until the throttle is wide open and its maximum obtainable speed is reached, a maximum speed that will not be controlled but will vary as a function of the external disturbances. In other words, the cruise control system will simply fail to do what it is supposed to do, and the car's speed will be literally out of control. But the rat will act differently. Although immediately after the change it will be unsuccessful in obtaining food, it will start to reorganize its pattern of behavior so that after a while it will be busily pushing up the same lever that it was busily pressing down just a short while ago.

But how does this reorganization take place? How is the rat to know that pressing the lever down is no longer going to do any good, and that it must push the bar up to be fed? Of course it cannot know this in advance. But the persistent, increased error caused by this change in the environment initiates blind, random changes in the control systems. In this case the reorganization involves a rather simple change in the direction of force applied to the lever. But more complicated situations can be imagined in which changes in the environment require more elaborate reorganizations involving the perception and control of new variables and the resetting of reference levels. As described by Powers:

> Reorganization is a process akin to rewiring or microprogramming a computer so that those operations it can perform are changed. Reorganization alters behavior, but does not produce *specific behaviors*. It changes the parameters of behavior, not the content. Reorganization of a perceptual function results in a perceptual signal altering its *meaning*, owing to a change in the way it is derived from lower-order signals. Reorganization of an output function results in a different choice of means, a new distribution of lower-order reference signals as a result of a given error signal. Reorganization is an operation *on* a system, not *by* a system.[23]

Reorganization, then, is also the functioning of a control system, but it is different in one crucially important respect from the functioning behavioral control systems we have considered up to now. The latter are organized so that response to error tends to remove the error. In a certain sense, the cruise control system knows to open the throttle if the speed of the car drops below the reference level, and the rat knows it must produce more bar presses if more are required for each reward. But the special type of control system designed to monitor and reorganize other working control systems cannot know what to do when it begins to show chronic uncorrected error where there was little error before. All it can do in this case is to modify the control system blindly in some way. If the change results in a reduction of error, any further modifications will be delayed. But if the change has no effect on the error or actually increases it, the next modification will come quickly.

In this respect, the reorganization system that rewires control systems to eliminate the error that the existing systems cannot reduce must act very much like *E. coli*. This common microorganism can either swim in a more or less straight line or tumble blindly. If it senses that it is getting closer to food it will continue along its merry way. But if it senses that it is not doing so, it will stop in its tracks, tumble a while, and then head off in a new, randomly chosen direction. If this new heading is perceived as better than the previous one, it will continue moving in this direction, and the next tumbling act will be put off for a while longer. If, however, the error has not been reduced, it will soon tumble again. Although this method of loco-motion may initially appear quite crude, it turns out to be a remarkably useful and virtually foolproof way for the bacterium to move around its environment.[24]

Accordingly, the reorganization of control systems is hypothesized to be an evolutionary process dependent on cumulative blind variation and selec-tion. If existing control systems are ineffective in controlling important per-ceptions, they will be randomly modified until the error is reduced. Learning to play the piano, ice skate, or speak a foreign language requires reorganization of control systems that is achieved by cumulatively modify-ing the systems that produce error and selectively retaining those modifica-tions that produce less error. And although the environment certainly plays an important role in influencing the state of the control systems that will be retained, it does not determine the organism's behavior. Instead, the organ-ism is actively involved in the selection process, with the chosen control sys-tem parameters and controlled variables depending on the higher-level goals that were selected as being important, and the effectiveness of these parameters and reference levels in the organism's current environment. That higher thought processes play a determining role in what appear at first to be simple Pavlovian and operant conditioning in humans was men-tioned in the discussion of Brewer's review in the previous chapter.[25]

Although first developed by Powers and his colleagues over 30 years ago[26] and given a detailed description by Powers in 1973, perceptual control theory is only now becoming more widely known, appreciated, and applied as a theory and research tool in the behavioral and social sciences. The insight that behavior is adapted to its environment only to the extent that it allows an organism to control crucial aspects of its environment has far-reaching implications for all aspects of the life sciences. In addition to a rapidly growing body of experimental psychological research,[27] it is now being applied to clinical psychology,[28] sociology,[29] law,[30] ethology,[31] business administration,[32] and philosophical and educational issues.[33] Such work is still in its infancy, but the power of perceptual control theory for understanding human behavior has been demonstrated in the construction of generative models of behavior that account for over 95% of the variance in a variety of tasks.[34]

But of most importance for our purposes, the theory provides a plausible explanation for how behavior can become *and remain* adaptedly complex. It is clearly not the case, as believed by Darwin and Lorenz, that organisms with useful fixed behaviors are selected during the course of evolution, resulting in innate, fixed patterns of behavior known as instincts. And it is also not the case that specific behaviors are selected by the environment by contingencies of reward during the life of the organism, as believed by Thorndike and Skinner. It is the selection of organisms with useful, adapted perceptual control systems over the course of evolution, coupled with the organism's cumulative variation and selection of its own perceptual control systems during its relatively brief life, that accounts for the adapted nature of behavior. There is no instruction by the environment, no stamping in of stimulus-response connections within the nervous system. Rather, we find a very Darwinian process of selection, not of behaviors, but of closed, negative-feedback loops encompassing perception, comparison with the reference level, and action, which allow patterns of behavior to remain functional, not only from one occasion to the next, but also within the continually changing environment of the behavior itself.

What may seem mysteriously ironic in all this is to realize that the purposeless process of natural selection has led to the evolution of purposeful organisms. But the irony fades when one considers the great survival and reproductive advantages of organisms that are able consistently to achieve goals essential to their survival and reproduction despite an unpredictable, uncaring, and often hostile environment.

9

The Development and Functioning of Thought

But the same term [thinking] is also used when thinking does become an achievement, that is, when it is productive. This happens when it changes our mental environment by solving problems which this environment offers. . . . The range of such achievements is tremendous. It extends from the solution of very simple problems in everyday life to veritable mental revolutions such as sometimes occur in the minds of great scientists and may then affect the lives of human beings forever afterward.
—Wolfgang Köhler[1]

One striking characteristic of our species is the degree to which we use thought processes to solve the many problems we encounter daily. Whether planning a vacation, balancing a checkbook, debugging a computer program, or making a scientific discovery that will ultimately affect the lives of millions of people, much more than just overt behavior is involved. The behaviorist theories of Pavlov, Thorndike, Watson, and Skinner dominated American psychology during much of the first two-thirds of the twentieth century. Some psychologists, however, particularly in Europe, continued their attempts to understand the development and functioning of the animal and human minds, including the development and functioning of thought itself. When behaviorism finally began to wane in the second half of this century, there began what has been called a cognitive revolution as the disciplines of psychology, linguistics, anthropology, philosophy, neuroscience, and computer science joined forces in an attempt to shed light on the mysterious and powerful inner workings of the brain.[2]

But if the process of thinking is adapted to solving the problems posed by the physical and social environments of a person or animal, this presents another example of a puzzle of fit that demands explanation. How is it that merely thinking about a problem can lead to its solution? If a solution is not evident when a problem is first encountered, what does thinking do to find one? And how is it that an adult is able to perceive things and solve problems that he could not as a child? These are the types of puzzles of fit that we will now consider.

The Problem of Problem Solving: Köhler's Chimps

At the same time that Thorndike and Watson were using a behavioral approach to animal (including human) learning that shunned any consideration of mental operations (see chapter 7), three German psychologists— Max Wertheimer, Franz Koffka, and Wolfgang Köhler—were taking a quite different approach to understanding animal learning. Of these three, Köhler (1887–1967) is probably the best known for his study of chimpanzees between 1913 and 1920 on Tenerife, one of the Canary Islands off the coast of Spain.[3]

Köhler devised a number of tasks to examine the problem-solving abilities of his chimps. One task involved suspending a banana high out of reach so that the only way to obtain it was to stack one or more boxes underneath on which the animal could climb and grab the fruit. A second task involved putting a stick inside the chimp's cage and a banana outside, so that the banana could be had by using the stick to pull it within reach. A variation of this required inserting one end of a rod into the end of another to make a tool of sufficient length. In the *Umweg* ("detour") task, a desired object was placed behind bars or a window, making it necessary for the animal first to move away from the object to circumnavigate the barrier between it and the object. Köhler used this last task not only with his chimpanzees but also with dogs, chickens, and children of various ages.

The chimpanzees demonstrated varying degrees of success on these problems. Almost all of them were able to solve the single box-stacking problem, and one was able to stack up to four boxes. They all eventually discovered how to use a stick to pull bananas within reach of the cage, but only two hit on the solution of joining two rods together to make a longer one. And whereas chimps (as well as children and dogs) were able to solve various *Umweg* tasks intelligently, the chickens were successful only when their frantic movements brought them by chance to a spot where they could see the detour around the obstacle.

The intelligent behavior demonstrated by the chimpanzees appeared very different from the gradual, trial-and-error solutions to the puzzle-box problems demonstrated by Thorndike's dogs and cats as described in chapter 7. Indeed, the apes would often pause and appear to think about the problem before suddenly coming up with a solution that was then immediately implemented as "a single continuous occurrence, a unity, as it were, in space as in time . . . as one continuous run, without a second's stop, right up to the objective."[4] It was this pause preceding a rapidly implemented solution to a problem that Köhler saw as an indication of truly "intelligent behavior," as opposed to the "mechanized behavior" demonstrated by the chickens in his *Umweg* task, and the cats and rats in Thorndike's and Watson's puzzle boxes and mazes.

Köhler believed that for his chimps to solve these and other problems intelligently, they had to be able to visualize the problem mentally in a new way. That is, a perceptual reorganization had to take place that brought to the chimp's attention certain relationships and possibilities that the chimp had not noticed before. Seeing a stick lying in the cage and a banana outside the cage will not suggest a solution unless a certain relationship is perceived between them, namely, that the stick can be used as a means to rake in the fruit. Indeed, Köhler found that such problems were often solved more quickly if the stick was placed where it and the banana could be seen at the same time. Once this new vision was mentally realized, the chimp then only had to act on it to solve the problem. This process of perceptual reorganization followed by a recognition that a particular reorganization provides a solution was referred to by Köhler as "insight" (*Einsicht* in German) and is also known as the aha! phenomenon.

But calling such behavior insightful does little to help us understand the thought processes involved, as indeed Köhler admitted in his later years:

Insight is insight into relations that emerge when certain parts of a situation are inspected. . . . In the solution of a problem . . . we suddenly become aware of new relations, but these new relations appear only after we have mentally changed, amplified, or restructured the given material.[5]

As an example of the role of insight in human problem solving, Köhler offers a problem similar to the one shown in figure 9.1. Here, the task is to determine the length of line *a* relative to some aspect of the circle. Some readers may see the answer to this problem almost immediately, but others

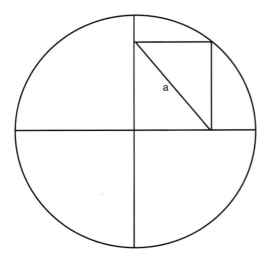

Figure 9.1
Problem of finding the length of line *a* (after Köhler, 1969).

may not. The latter readers would profit from taking time to try to solve the problem before reading further, while at the same time noting the thoughts that occur while wrestling with the problem.

The answer is that the length of line *a* is the same as the radius of the circle. This becomes obvious when the other diagonal of the rectangle is imagined. This other diagonal is found to extend from the center of the circle to its circumference, and therefore is equal to the circle's radius. Since the two diagonals of a rectangle are equal in length, line *a* must also be equal to the radius. Köhler continues his discussion of the role of insight making reference to this problem:

Thus, when we dealt with the diagonal within a certain rectangle constructed within a circle . . . everything was, of course, clear once we had drawn the second diagonal, which then proved to be identical to the radius of the circle. But why, after inspecting the situation as first given, did we ever think of drawing new lines, and particularly that special line, the second diagonal? . . . After it had happened, we understood, of course that this was the right procedure. But we could not realize this until the grouping had been done. What, then made us introduce this particular structuring or grouping at a time when we could not yet be aware of its consequences?[6]

Here he appears to be admitting that insight cannot explain problem solving, but is rather a consequence of first producing and then selecting a useful perceptual reorganization. But how can we (or our brain) know which reorganization will lead to the solution? Indeed, if we knew in advance how to solve it, we wouldn't have had the problem in the first place! This, of course, is just another instance of Meno's dilemma described in chapter 6.

One possible solution would be to postulate that the brain produces a series of varied perceptual reorganizations until one is produced that is recognized as leading to the solution and therefore selected. If you were unable to solve the problem presented in figure 9.1 immediately, you may recall experiencing just such a series of variations in your perception of it. Although such thoughts would no doubt be constrained by your previous knowledge of geometry and similar problems,[7] they would in an important sense be *blind* variations in that you could not know beforehand which particular perceptual organization would lead to the solution. But this is not the reasoning that Köhler uses. Instead, he concludes his discussion of problem solving by simply restating the problem, this time using the word "revolutions" to refer to perceptual reorganizations:

Why do such revolutions which occur in certain brains tend to be the right revolutions? . . . [W]hy do brain processes tend to produce perceptual organizations of remarkable clearness of structure? At least this part of nature, the human brain, seems to operate in a most selective fashion. It is the *direction* of its operations which is truly remarkable.[8]

This last sentence suggests that Köhler believed that some degree of foresight is involved since the brain seems to know in what direction the solution of a new problem lies. But of course this conclusion simply begs the question of the origin of this "directional" knowledge.

Köhler and his Gestalt psychology colleagues made many important contributions to our knowledge of animal and human perception and problem solving. They provided much in the way of evidence and arguments that the problem-solving behavior of animals and humans could not be easily accounted for by the concepts of stimulus, response, and reward of American behaviorism that was then so popular. The contrast between the two schools of thought was noticed by British mathematician and philosopher Bertrand Russell, who observed that "animals studied by Americans rush about frantically, with an incredible display of hustle and pep, and at last achieve the desired result by chance [whereas those] observed by Germans sit still and think, and at last evolve the solution out of their inner consciousness."[9] However, in their haste to discard behaviorist, "mechanistic" explanations of problem solving, the Germans may have also unfortunately discarded the solution to the puzzle of problem solving, a solution hinted at by Russell's use of the phrase "evolve the solution out of their inner consciousness."

Piaget's Genetic Epistemology

Another European who had an even greater impact on the study of thought processes was psychologist Jean Piaget (1896–1980). Prolific in research and writing from the age of 10 years until shortly before his death,[10] Piaget began his career as a biologist specializing in mollusks such as the snails inhabiting the lakes of his native Switzerland. However, a job in Paris administering intelligence tests to children sparked a life-long interest in the development of mental abilities and knowledge. Piaget called this study "genetic epistemology," the word *genetic* in this case referring not to the genes but rather to a conceptualization of the development of thought as a process of internally guided cognitive growth.

Piaget employed an ingenious mélange of questioning and simple experiments that led to a number of fascinating discoveries concerning children's thought and cognitive development. For example, he would spread two rows of eight coins each in front of a young child so that each coin in the bottom row was directly beneath the corresponding coin in the top row. When asked which row had more coins, the child would correctly answer that each row was the same. But if the distance between each coin in the bottom row was increased (without adding any coins) while the child watched so that the bottom row became longer, the child would now answer that the bottom row had more coins than the top row.

Piaget also demonstrated that children were able to solve certain concrete, hands-on tasks before they could solve the same problem at a more abstract, logical, and verbal level. So, for example, given a short red stick (1) and a longer green one (2), and then shown the same green one (2) and a still longer yellow one (3), a nine-year-old would have no difficulty in stating that the yellow stick (3) was longer than the red stick (1). However, when told and not shown that stick 1 is shorter than 2, and stick 2 is in turn shorter than 3 (that is, 1 < 2 and 2 < 3), and asked whether stick 1 or 3 was longer, the same child would have great difficulty arriving at the correct answer that 3 is longer than 1 (1 < 3).

These and many other observations and experiments by Piaget clearly demonstrated that the thought of the young child is different not only *in degree* from that of the adult, but also *in kind*, and he concluded that each child goes through an invariant series of cognitive stages, with each stage requiring a major overhaul of the preceding one.[11] For example, for a young infant, an object exists only if it can be presently seen, felt, heard, or smelled. At this age, removing a desired object from view usually results in the infant abandoning all efforts to find and obtain it. But the child soon develops "object permanence," so that she is now able to seek and find objects that were hidden while she was watching. From a Piagetian perspective, the developing child is like a little scientist who is constantly developing and testing new theories about the world, rejecting old theories when a new one proves better suited to making sense of the world and meeting her needs.

But if, as Piaget demonstrated, the child's thought becomes better and better adapted to the world, we must ask ourselves how this increase in fit is possible. Piaget was very clear in his rejection of both providential and instructionist views. His rejection of a genetic form of providentialism can be seen in his criticisms of the view that some innate, preformed knowledge must exist that cannot be explained by a process of cognitive growth. And his rejection of instruction is made clear in his criticisms of empiricist psychological theories that attempt to account for knowledge growth as a process of taking in and internalizing sensory experience. Instead, he considered mental growth to be a *constructive* process that cannot be accounted for by innately provided knowledge or instructive sensory experience, either working separately or in combination.

It might reasonably be expected that Piaget's realization of the inadequacy of providential (innatist) and instructionist (empiricist) accounts of mental development, coupled with his early training as a biologist, would have led him quite naturally to a selectionist view of the development of thought. He rejected the theory of Darwinian evolution,[12] however, reasoning that:

Either chance and selection can explain everything or else behavior is the motor of evolution. The choice is between an alarming waste in the shape of multitudinous and fruitless trials preceding any success no matter how modest, and a dynamics with an internal logic deriving from those general characteristics of organization and self-regulation peculiar to all living beings.[13]

But what exactly does Piaget mean by his appeal to behavior as the "motor of evolution" guided by "an internal logic" and "general characteristics of organization and self-regulation peculiar to all living beings"; and where did this logic, organization, and self-organization come from? Unfortunately, he provided no clear, compelling answers to these questions, leaving himself vulnerable, as we will see, to the innatist arguments of Noam Chomsky.

Having rejected Darwin's selectionist theory of evolution, it should not be surprising that Piaget saw little role for variation and selection in his theory of cognitive development. But this put him in a curious predicament in essentially rejecting all three major explanations for the growth in adapted complexity of human thought processes: providence (both divine and genetic), instruction, and selection. He described an alternative theory of development using terms such as "assimilation," "accommodation," "equilibration," "reflective abstraction," and "autoregulation"—terms many psychologists and students of psychology have struggled to understand.[14]

Despite anti-Darwinian sentiments, the major themes of his theory can be understood from a selectionist perspective. According to Piaget, the two major ways in which children (as well as adults) interact with their world are through assimilation and accommodation. Assimilation refers to an incorporation of some sensory experience into a preexisting thought structure or schema (we will ignore for now the origin of this preexisting schema). For example, a child having seen sparrows and blackbirds and able to recognize them as members of the category "bird" would likely assimilate the sighting of a starling into this same category. The child might also attempt to assimilate the first observed butterfly into the bird schema since it shares certain similarities with other members of this category. However, calling a butterfly a bird would very likely result in a correction by an adult or older child, "That's not a bird, it's a butterfly!" This would then require accommodation of the child's thought so that butterflies and birds would be treated as different types of objects, each with its own label and distinguishing characteristics.[15]

But an adult cannot directly instruct the child with respect to the conventional use of words such as *bird* and *butterfly*. This means that a father cannot simply transmit the meanings of new words to his child. Indeed, the child only knows that some sort of error has been made and that, according to his father, the current object in view is not a bird but a butterfly. But the father's remark does not tell the child *why* it is a butterfly and not a bird. Is it because it is yellow and the other flying organisms he has seen are brown and black (but then what of canaries?)? Is it because it stops to sip nectar from flowers, and the other flying creatures do not (but then what of hummingbirds?)? Or is it because birds have only been seen in the afternoon and it is now morning (but then what of the birds that get the worms?)? Clearly, the child must make some sort of guess as to how to modify his bird schema and create a new butterfly one. This guess may well be initially wrong, but by continuing to use and test additional guesses, the child will eventually come to the same notions of bird and butterfly that are shared by most adults of the speech community. (More will be said about the child's acquisition of vocabulary in chapter 11.)

The Providential Innatism of Chomsky and Fodor

The person who probably did more than any other to change the course of psychology in the second half of the twentieth century was not even a psychologist. In 1959 Chomsky published a devastating review of Skinner's book *Verbal Behavior* that is often cited as marking the beginning of the cognitive revolution in psychology. The review legitimized the study of thought and cognitive processes after half a century of behaviorist domination in North America that had banned all such attempts as unscientific and subjective.

In this and other writings, Chomsky convincingly maintained that language behavior could not be accounted for by reinforcement of certain responses under particular stimulus conditions. Rather, any attempt to understand how we can acquire, use, and understand language must take into consideration that most of language consists of novel utterances that the speaker has never heard or uttered before. Thus the only way a person with a finite brain can produce and understand a potentially infinite number of novel sentences is by using an internal mental grammar of abstract, generative rules and principles.

Chomsky was fascinated by how children are able to master their native language so quickly and with so little difficulty despite both their informal, unsystematic exposure to language and its staggering complexity. To account for this remarkable feat, he came to the conclusion that much conceptual and linguistic knowledge is innate. As he states:

> It seems that the child approaches the task of acquiring a language with a rich conceptual framework already in place and also with a rich system of assumptions about sound structure and the structure of more complex utterances. These constitute the parts of our knowledge that come "from the original hand of nature," in Hume's phrase. They constitute part of the human biological endowment, to be awakened by experience and to be sharpened and enriched in the course of the child's interactions with the human and material world.[16]

Chomsky revolutionized the field of linguistics in pursuit of his ambitious goal to find a system of innate, universal grammar that would make the task of aquiring language as easy as possible while still leaving enough leeway for a child to acquire any of the 5000 or so of the world's languages fate may have chosen as his mother tongue.

But Chomsky's powerful intellect and formidable powers of rhetoric were not to be limited to attacks on the instructionist theories of the behaviorists. In a series of lively debates with Piaget, he and philosopher Jerry Fodor attacked Piaget's constructivist theory of knowledge as well, and it is here that Piaget's reluctance to embrace a selectionist theory of cognitive development caused him considerable difficulty in countering these assaults. The innatist position and the basic argument against the constructivism of Piaget is perhaps best summed up by Fodor:

> There literally isn't such a thing as the notion of learning a conceptual system richer than the one that one already has; we simply have no idea of what it would be like to get from a conceptually impoverished to a conceptually richer system by anything like a process of learning.[17]

How could Piaget have best countered these arguments? If we substitute the words "less adaptedly complex" and "more adaptedly complex" or "less fit" and "more fit" for Fodor's "conceptually impoverished" and "conceptually richer," respectively, we can see how an evolutionary process involving cumulative cognitive variation and selection could in principle provide a selectionist answer to Chomsky and Fodor, thus allowing Piaget to reject their innatist views of mental development for a constructionist one. Piaget's failure to do so and his vain attempt to provide a Lamarckian explanation of how an organism's experience can be transmitted to its genome[18] contributed to Chomsky's looking very much the winner in the published version of this debate.[19]

The issues concerning language, thought, and cognitive development are exceedingly complex and therefore risk being seriously oversimplified and misunderstood in such a cursory presentation. I tolerate this risk only because I believe that the core of this dispute involving three of this century's most influential thinkers about thought can be understood as an avoidable consequence of the rejection of the creative potential of Darwinian evolution, and failure to appreciate how a within-organism selectionist mechanism could in principle account for the construction of knowledge and thought without requiring either the preformed, innate knowledge of Chomsky and Fodor or the convoluted constructive mechanisms of Piaget.

Chomsky and Fodor offered a genetically providential explanation for the remarkable fit between the language and concepts of the child and that of his environment. As with all providential theories, however, theirs encounters serious difficulties when one inquires as to the origin of the provided knowledge. If they accepted that biological evolution could in principle lead to the emergence of this complex and adapted knowledge, they would be undermining their basic position that simpler systems cannot on their own give rise to more adaptedly complex ones. The two may be right in their assertion that all human conceptual knowledge is innate. But their basic argument that such knowledge *must* be innate because new, more complex, adapted systems cannot emerge from preexisting, less complex adapted systems (as clearly happens in the process of adaptive organic evolution) is obviously flawed. The evidence discussed in chapter 4 that the mammalian brain undergoes a type of adaptive evolution during the lifetime of the animal by way of variation and selection of synapses provides an important additional reason to doubt Chomsky and Fodor's innatist views.

Piaget also rejected an evolutionary account of thought and knowledge as well as the providential innatism of Chomsky and Fodor. This left him in a particularly difficult position to account for the child's construction of knowledge and advanced thought processes, and led him to flirt with Lamarckian explanations for cognitive development. (We will return to Chomsky and Fodor and these issues in our discussion of human language in chapter 11.)

Early Selectionist Theories of Thought

Regardless of the rejection of Darwinian explanations of cognitive development and thought by Piaget, Chomsky, and Fodor during the second half of this century, selectionist views of knowledge, thought, creativity, and invention have a long and impressive history. Since proponents of such views include some of the most insightful and influential psychologists, scientists, and philosophers of their day, it will be well worth our time to become acquainted with some of these views.

Alexander Bain (1818–1903), a Scottish philosopher and psychologist, emphasized the great number of trials required for scientific discoveries and the importance of the potential discoverer's commitment and fascination with the subject of his problem:

The invention of Daguerre [of the first photographic process] illustrates—by a modern instance—the probable method whereby some of the most ancient inventions were arrived at. The inventions of the scarlet dye, of glass, of soap, or gunpowder, could have come only by accident; but the accident, in most of them, would probably fall into the hands of men engaged in numerous trials upon the materials involved. Intense application—"days of watching, nights of waking,"—went with ancient discoveries as well as with modern. In the historical instances, we know as much. The mental absorption of Archimedes is a proverb.[20]

The number of trials necessary to arrive at a new construction, is commonly so great, that, without something of an affection, or fascination, for the subject, one grows weary of the task. The patient thought of the naturalist desirous of rising to new classifications, grows out of his liking for the subject, which makes it to him a sweet morsel rolled under the tongue, and gives an enjoyment even to fruitless endeavours.[21]

English economist and logician W. Stanley Jevons (1835–1882), in his rejection of Sir Francis Bacon's instructionist view of science (to be discussed in chapter 10), also offered a selectionist account of the human mind's ability to provide new insights into its environment.

I hold that in all cases of inductive inference we must invent hypotheses until we fall upon some hypothesis which yields deductive results in accordance with experience.

It would be an error to suppose that the great discoverer seizes at once upon the truth or has any unerring method of divining it. In all probability the errors of the great mind exceed in number those of the less vigorous one. Fertility of imagination and abundance of guesses at truth are among the first requisites of discovery; but the erroneous guesses must be many times as numerous as those which prove well founded. The weakest analogies, the most whimsical notions, the most apparently absurd theories, may pass through the teeming brain, and no record remain of more than the hundredth part. There is nothing really absurd except that which proves contrary to logic and experience. The truest theories involve suppositions which are inconceivable, and no limit can really be placed to the freedom of hypothesis.[22]

Regardless of Jevons's selectionist insight, it surely cannot be the case that "the truest theories involve suppositions which are inconceivable," but rather that suppositions that would be inconceivable if based firmly on already achieved knowledge are conceivable for selectionist, guess-based thought mechanisms of Jevons's "teeming brain."

Chauncey Wright (1830–1875), an American mathematician and philosopher of the pragmatic school, had an important impact on American philosophy, primarily through his interactions with William James and Charles Sanders Peirce. The social event of Wright's life was his trip to England in 1872 to visit Charles Darwin, and Darwin was sufficiently impressed with Wright's defense of Darwinism that he reprinted and distributed at his own expense one of Wright's essays entitled "The Genesis of Species." Of most relevance to the current chapter is that Wright also applied Darwin's principles of variation and selection to an attempt to understand the functioning of the human mind:

In further illustration of the range of the explanation afforded by the principle of Natural Selection. . . we may instance an application of it to the more special psychological problem of the development of the individual mind by its own experiences. . . . Here, then, is a close analogy, at least, to those fundamental facts of the organic world on which the law of Natural Selection is based; the facts, namely, of the "rapid increase of organisms," limited only by "the conditions of existence," and by competition in that "struggle for existence" which results in the "survival of the fittest." As the tendency to an unlimited increase in existing organisms is held in check only by those conditions of their existence which are chiefly comprised in the like tendencies of other organisms to unlimited increase, and is thus maintained (so long as external conditions remain unchanged) in an unvarying balance of life; and as this balance adjusts itself to slowly changing external conditions, so, in the history of the individual mind, beliefs which sprang spontaneously from simple and single experiences, and from a naturally unlimited tendency to generalization, are held mutually in check, and in their harmony represent the properly balanced experiences and knowledges of the mind, and by adaptive changes are kept in accordance with changing external conditions, or with the varying total results in the memory of special experiences.[23]

This extract reveals Wright to be no master of lucid English prose, but the use of the very Darwinian ideas of variation, competition, and selection are evident here. Instead of a Darwinian competition among organisms, Wright describes a mental competition among beliefs, with both other current beliefs and the environment acting to eliminate those less fit and leaving the better-adapted ones.

Already described in chapter 8 as having an important insight into the purposeful nature of the behavior of humans and other animals, William James was very much influenced by Darwinian ideas in his formulation of psychological theories concerning the development and function of thought.[24]

... new conceptions, emotions, and active tendencies which evolve are originally *produced* in the shape of random images, fancies, accidental outbirths of sponta- neous variation in the functional activity of the excessively unstable human brain, which the outer environment simply confirms or refutes, preserves or destroys— selects, in short, just as it selects morphological and social variations due to molecu- lar accidents of an analogous sort. ...

The conception of the [newly discovered scientific] law is a spontaneous variation in the strictest sense of the term. It flashes out of one brain, and no other, because the instability of that brain is such as to tip and upset itself in just that particular direc- tion. But the important thing to notice is that the good flashes and the bad flashes, the triumphant hypotheses and the absurd conceits, are on an exact equality in re- spect of their origin.[25]

Frenchman Paul Souriau in 1881 mentioned the central role played by chance in invention:

A problem is posed for which we must invent a solution. We know the conditions to be met by the sought idea; but we do not know what series of ideas will lead us there. In other words, we know how the series of our thoughts must end, but not how it should begin. In this case it is evident that there is no way to begin except at random. Our mind takes up the first path that it finds open before it, perceives that it is a false route, retraces its steps and takes another direction. Perhaps it will arrive immediately at the sought idea, perhaps it will arrive very belatedly; it is entirely impossible to know in advance. In these conditions we are reduced to dependence on chance.

By a kind of artificial selection, we can in addition substantially perfect our thought and make it more and more logical. Of all the ideas which present themselves to our mind, we note only those which have some value and can be utilized in reasoning. For every single idea of a judicious and reasonable nature which offers itself to us, what hosts of frivolous, bizarre, and absurd ideas cross our mind. Those persons who, upon considering the marvelous results at which knowledge has arrived, can- not imagine that the human mind could achieve this by a simple fumbling, do not bear in mind the great number of scholars working at the same time on the same problem, and how much time even the smallest discovery costs them. Even genius has need of patience. It is after hours and years of meditation that the sought-after idea presents itself to the inventor. He does not succeed without going astray many times; and if he thinks himself to have succeeded without effort, it is only because the joy of having succeeded has made him forget all the fatigues, all of the false leads, all of the agonies, with which he has paid for his success. . . .

If his memory is strong enough to retain all of the amassed details, he evokes them in turn with such rapidity that they seem to appear simultaneously; he groups them by chance in all the possible ways; his ideas, thus shaken up and agitated in his mind, form numerous unstable aggregates which destroy themselves, and finish up by stopping on the most simple and solid combination.[26]

American psychologist James Mark Baldwin (1861–1934) is of particular interest since he spent considerable time in France and is believed to have had an influence on the psychological theorizing of Piaget. Unlike Piaget, however, he believed that Darwinian views of cumulative variation and selection could be applied fruitfully not only to psychology but also to ethics, logic, philosophy, religion, judgment, and logic. In a volume published in 1909 to mark the fiftieth anniversary of the publication of Darwin's *Origin*, Baldwin treated all these subjects and expressed the hope that his book would also stimulate Darwinian thinking in other fields of the humanities such as anthropology, philology, political science, and literary criticism. Baldwin thought the selectionist theory of adaptive evolution had potential application to all these disciplines, but as a psychologist he was particularly interested in its application to human thought and intelligence:

And how far the method of law called by Darwin "natural selection" goes, what its range really is, we are now beginning to see in its varied applications in the sciences of life and mind. It seems to be—unless future investigations set positive limits to its application—a universal principle; for the intelligence itself, in its procedure of tentative experimentation, or "trial and error," appears to operate in accordance with it.[27]

Austrian physicist and philosopher Ernst Mach, who lived from 1838 to 1916, and whose name remains synonymous with the speed of sound, was intrigued with the function of thought, particularly as it related to the advancement of science. When he assumed the new professorship in "The History and Theory of Inductive Sciences" created for him at the University of Vienna in 1895, he offered a clearly selectionist account of scientific and artistic creativity:

The disclosure of new provinces of facts before unknown can only be brought about by accidental circumstances. . . .

After the repeated survey of a field has afforded opportunity for the interposition of advantageous accidents, has rendered all the traits that suit with the word or the dominant thought more vivid, and has gradually relegated to the background all things that are inappropriate, making their future appearance impossible; then, from the teeming, swelling host of fancies which a free and highflown imagination calls forth, suddenly that particular form arises to the light which harmonizes perfectly with the ruling idea, mood, or design. Then it is that which has resulted slowly as the result of a gradual selection, appears as if it were the outcome of a deliberate act of creation. Thus are to be explained the statements of Newton, Mozart, Richard Wagner, and others, when they say that thoughts, melodies, and harmonies had poured in upon them, and that they had simply retained the right ones.[28]

Particularly significant in Mach's explanation of creativity is the notion of a cumulative selection process that gradually leads to fit "thoughts, melodies, or harmonies," but that nevertheless may appear both to the thinker and observer as an inexplicably sudden and insightful creative act.

Renowned French mathematician Henri Poincaré (1854–1912) wrote an essay on mathematical creativity derived from his own creative experiences in which he emphasized the blind recombination of elements and the selection of those products in the unconscious mind according to criteria of harmony, beauty, and usefulness:

One evening, contrary to my custom, I drank black coffee and could not sleep. Ideas rose in crowds; I felt them collide until pairs interlocked, so to speak, making a stable combination. . . .

What happens then? Among the great numbers of combinations blindly formed by the subliminal self, almost all are without interest and without utility; but just for that reason they are also without effect upon the esthetic sensibility. Consciousness will never know them; only certain ones are harmonious, and, consequently, at once useful and beautiful. . . .

Perhaps we ought to seek explanations in that preliminary period of conscious work which always precedes all fruitful unconscious labor. Permit me a rough comparison. Figure the future elements of our combinations as something like the hooked atoms of Epicurus. During the complete repose of the mind, these atoms are motionless, they are, so to speak, hooked to the wall; so this complete rest may be indefinitely prolonged without the atoms meeting, and consequently without any combination between them.

On the other hand, during a period of apparent rest and unconscious work, certain of them are detached from the wall and put in motion. They flash in every direction through the space . . . where they are enclosed, as would, for example, a swarm of gnats. . . . Then their mutual impacts may produce new combinations. . . .

In the subliminal self . . . reigns what I should call liberty, if we might give this name to the simple absence of discipline and to the disorder born of chance. Only this disorder itself permits unexpected combination.[29]

It is noteworthy that although these eight individuals worked in quite divergent fields of inquiry, from psychology for Bain to mathematics for Poincaré, they all were in remarkable agreement with respect to their selectionist perspective on human thought. Central to them all is the idea that useful thoughts (beliefs, ideas) can be found only if the thinker produces a large number of varied and blind guesses. Thus, Bain is convinced that "the inventions of the scarlet dye, of glass, of soap, or gunpowder, could have come only by accident." Jevons argued that "fertility of imagination and abundance of guesses at truth are among the first requisites of discovery." Wright wrote of the "unlimited, expansive power of repetition" of memories, experiences, and beliefs. James explained how "new conceptions, emotions, and active tendencies which evolve are originally *produced* in the shape of random images, fancies, accidental outbirths of spontaneous variation in the functional activity of the excessively unstable human brain." For Souriau "there is no way to begin except at random." Mach wrote about "the teeming, swelling host of fancies which a free and highflown imagination calls forth." And Poincaré described how "among the great numbers of combinations blindly formed by the subliminal self, almost all are without interest and without utility."

But of course simply churning out a great variety of thoughts (commonly referred to as brainstorming today) cannot in itself lead to useful ones. For this, as in biological evolution, the problem at hand must somehow be involved in selecting and retaining the useful thoughts while eliminating the useless ones. Thus Wright wrote of the "'struggle for existence' which results in the 'survival of the fittest'" ideas. James remarked that thoughts are something which "the outer environment simply confirms or refutes, preserves or destroys—selects, in short, just as it selects morphological and social variations due to molecular accidents of an analogous sort." Souriau appeared to make a very clear analogy to natural selection in stating that "by a kind of artificial selection, we can in addition substantially perfect our thought and make it more and more logical" and that "of all the ideas which present themselves to our mind, we note only those which have some value and can be utilized in reasoning." Mach noted that "from the teeming, swelling host of fancies which a free and highflown imagination calls forth, suddenly that particular form arises to the light which harmonizes perfectly with the ruling idea, mood, or design." And Poincaré found that of the multitude of ideas which come to mind as one struggles with a problem, "only certain ones are harmonious, and, consequently, at once useful and beautiful."

But whereas these individuals all shared what we might call the selectionist insight concerning thought, they did not provide particularly strong logical or philosophical arguments, or cogent historical or psychological evidence to support their theories. In addition, the rise of behaviorism in the United States at the beginning of the twentieth century made it very unpopular to base any psychological theory on subjective and unobservable mental processes. Scientists such as Köhler, Piaget, and Chomsky who swam against the behaviorist tide and rejected trial-and-error theories of overt behavior also rejected trial-and-error theories of thought. It may well be for these reasons that selectionist views of thinking practically disappeared during the first half of the twentieth century and did not make a comeback until the second half.

Thought as Substitute Trial and Error: Campbell's Hierarchy of Knowledge Processes

It is largely due to the work of American psychologist Donald T. Campbell (born in 1916 and now professor emeritus at Lehigh University in Bethlehem, Pennsylvania) that a general selection theory of thought is known (and contested) today. Campbell has three major accomplishments to his credit in this regard. First, he documented the discovery and use of selectionist theories of thought by philosophers, psychologists, mathematicians, and other scientists since the time of Darwin.[30] Second, over a period of more than 35 years he provided strong arguments that the Darwinian process of blind variation and selective retention underlies all achievements of fit, including the fit of our perceptions to the world they represent, the fit of our thoughts and mental processes to the real-world problems we confront and successfully solve, and the fit of our scientific theories and predictions to the universe they describe. Finally, he provided a hierarchy of knowledge processes in an attempt to show how the evolution and ontogeny of all forms of knowledge can be accounted for within a general selectionist framework.

Campbell lays out four basic premises which provide the foundation for his 11-level hierarchy of knowledge processes.[31]

1. A blind-variation-and-selective-retention process is fundamental to all inductive achievements, to all genuine increases in knowledge, to all increases in fit of system to environment.

Here we see the centerpiece of Campbell's epistemology in the bold claim that blind variation and selection is the only natural (that is, nonmiraculous) explanation for any increase in adapted complexity, for any increase in fit of one system with respect to another.

2. In such a process there are three essentials: (a) Mechanisms for introducing variation; (b) Consistent selection processes; and (c) Mechanisms for preserving and/or propagating the selected variations. Note that in general the preservation and generation mechanisms are inherently at odds, and each must be compromised.

While the first point may appear to be a dogmatic and unsupportable claim, the second point is quite different in that it posits mechanisms for variation, selection, and retention. Thus, Campbell makes the testable prediction that for any system that shows an increase in adapted complexity over time, careful examination will reveal mechanistic processes of variation, selection, and retention underlying this adaptation. In this regard it is of interest to note that Darwin believed that these three processes were responsible for biological evolution. He could argue convincingly only for mechanisms of selection ("survival of the fittest"), however, since he had no knowledge of the mechanisms of either genetic variation or retention. If Campbell's conjectures are correct, we would expect that research into natural and artificial adaptive systems would eventually yield concrete evidence of the underlying mechanisms of variation, selection, and retention. This indeed turned out to be the case in the fields of immunology and neurology, as we already saw in chapters 4 and 5. Chapters 14 and 15 address how the same appears true in computer science and molecular design.

3. The many processes which shortcut a more full blind-variation-and-selective-retention process are themselves inductive achievements, containing wisdom about the environment achieved originally by blind variation and selective retention.
4. In addition, such shortcut processes contain in their own operation a blind-variation-and-selective-retention process at some level, substituting for overt locomotor exploration or the life-and-death winnowing of organic evolution.

This concept of shortcut or vicarious[32] processes that substitute for longer, more tedious, and more costly blind variation and selective retention is central to Campbell's hierarchy and so merits further elaboration. Let us take a blind man as an example. If placed in an unfamiliar house, the blind man can exit only by moving to and searching the walls with his hands until he finds the door leading outside. In the process he may well trip on furniture and other objects, or tumble down a flight of stairs. This is obviously very costly in terms of energy and time, and a dangerous way to solve his problem of exiting the house.

But now imagine that the man is given a long cane that he can use to probe the space around him more efficiently and safely. Note that although he is not in any sense less blind than he was without the cane, he is in much better circumstances since by tapping his cane he can substitute for the more costly and dangerous manual groping he would have to do without the cane. The cane has a rather limited range of effectiveness, however, and it is still necessary for the man to move about the room as he taps.

Let us now suppose that the man is equipped with an infrared sensing device that, by emitting and timing the arrival of reflected infrared waves, can compute the distance of an object and present this information by producing an audible tone that covaries in pitch with this distance so that close objects sound higher than more distant ones. Now he is able to obtain information from the entire room without having to move around it at all. Slight dips in frequency would indicate the corners of the room, with a dramatic dip indicating an open window, doorway, or stairway. Notice that such a system begins to resemble vision in its operation. A hostage who is transported to a strange house while blindfolded and then manages to remove his blindfold is in much the same situation as the blind man with the infrared device. Initially, the hostage has no idea where a window or door might be, and can only look around "blindly" until he finds something that might provide a means of escape.

But can it make any sense to refer to vision as blind? We certainly don't just look anywhere and everywhere in using vision to solve problems. We normally don't search for doors on the ceiling or floor. Unlike the drunkard, we don't look for our lost keys under a streetlight if we know we dropped them in a dark part of the street. When faced with a problem, we don't normally just try *anything*. To do so would be very inefficient and make it quite unlikely that we would find the solution to our problem. That is, we use *constraints* (or *restraints*) to solve problems and make new discoveries.

Undoubtedly such constraints can be very helpful in finding solutions to problems and discovering new knowledge, and they may be necessary if we want to be able to find a solution in a reasonable length of time. An experienced mechanic will not check the muffler of a car if the motor will not start, and will not examine the fuel injection system if the car runs well but makes excessive noise. Experience permits the physician to eliminate many possible causes of disease based on the patient's symptoms. Knowledge based on previous similar experiences can constrain the search space of possible solutions that make useless variations *less* likely and therefore potentially useful ones *more* likely. But we must keep in mind three important things concerning constraints. First, insofar as they are useful and therefore fit the problem at hand, *this fit of the constraints themselves must be explained*. Thus if we wish to avoid providential or instructionist explanations for the fit, the existence of constraints must be explained as the result of previous blind variation and selection.

Second, no matter how useful they may be, constraints alone cannot account for solutions to new problems. As argued by Campbell:

Intelligent variations require an explanation for how these variations or hypotheses came to be wise-in-advance. That most hypotheses *are* wise, I have no doubt. As such, they reflect already achieved knowledge or, at very least, wise restrictions on the search space. Such wisdom does not, however, explain further advances in knowledge. That hypotheses, even if not wise, are far from random, I agree. But wise or stupid, *restraints* on the search space do not explain novel solutions.[33]

Third, there is no guarantee that constraints that proved useful in the past will continue to do so in the future. Indeed, in many respects progress in science can be seen as the sweeping away of old constraints that are no longer considered valid. The early conception of the universe revolving about the earth at its center constrained astronomy in ways that were originally useful, but this constraint was completely rejected by Copernicus and later astronomers. The everyday ideas that light is unaffected by gravity and mass is unaffected by heat and velocity were constraints important in Newton's day, but were discarded by Einstein and later physicists. And the constraints provided by religious doctrine that viewed the design of all creatures as unchallengeable proof of the handiwork of a supreme creator were dramatically overturned by Darwin.

Now that we have examined the basic ideas motivating Campbell's thoroughly selectionist view of knowledge, we are in a better position to examine his proposed hierarchy of knowledge processes (table 9.1) and see how and where thought fits in. The first level is genetic adaptation. This, of course, refers to the processes of cumulative genetic variation and selection that underlie biological evolution. We already saw in chapters 2 and 3 how biological evolution can lead to remarkable fit in both the structures and behavior of organisms. Genetic adaptation has a rather serious limitation, however; it normally takes a considerable amount of time and thus cannot provide adaptive changes during the lifetime of any individual organism.

But as we have already considered in our discussion of learning and perceptual control theory in chapters 7 and 8, changing environmental conditions and unpredictable disturbances continually present new problems and make it advantageous, if not necessary, for an organism to be able to learn and adapt during its own lifetime, that is, ontogenetically. Accordingly, the second level of Campbell's hierarchy is the simplest conceptualization of such an ability, that is, *nonmnemonic*[34] *problem solving*. We had an example of nonmnemonic problem solving in our brief encounter with *E. coli* in the previous chapter. Recall that the bacterium can only either swim in a relatively straight line or tumble about randomly until it sets off in a new random direction. If it senses that it is swimming in the right direction, toward food or away from danger, it will continue to move in that direction. If, however, it senses that it is not getting closer to food or farther from danger, it will tumble for a while and then set off in a new direction. So whereas *E. coli* is able to solve the problem of finding food

Table 9.1
Donald T. Campbell's (1974a) hierarchy of selectionist knowledge processes

11. Science
10. Cultural cumulation
9. Language (overlapping 6 and 7)
8. Socially vicarious exploration; Observational learning and imitation
7. Mnemonically supported thought
6. Visually supported thought
5. Habit
4. Instinct
3. Vicarious locomotor devices
2. Nonmnemonic problem solving
1. Genetic adaptation

and avoiding danger, and can do this in complex, changing environments, it cannot profit from experience. As it begins to tumble, it has no memory of the direction in which it was last traveling and therefore is just as likely to set off in this same direction as any other. Of course, it is much more likely to choose a new bearing, since there are many more new directions than the single old one.

Campbell's third level is *vicarious locomotor devices*. As we learned with regard to the blind man, exploring one's environment with a remote sensory system has important advantages over locomotion and direct physical contact. Particularly striking examples of this are the sophisticated echolocation systems of bats, porpoises, and cave birds. Without such a sensory system, a bat could only make its way out of a dark cave by repeatedly flying into the walls of the cave until it eventually discovered the entrance. Its echolocation system makes such costly (in terms of time and energy) and dangerous fumblings unnecessary, as it provides a type of substitute locomotor device so that the bat can find its way into the open faster, with less energy, and more safety than by flying into the cave walls. Campbell admits that it is difficult for most people to conceptualize vision itself as a vicarious locomotor device that operates using blind variation and selection. He maintains that this is, however, the case, as suggested by the blind man with the infrared distance detector.

We now move to Campbell's fourth and fifth levels, *habit* and *instinct*. As described in chapter 3, instinct is generally understood as resulting from the evolutionary selection of organisms with useful behaviors, and habit is typically considered to result from behavioral consequences of the experiences of an individual organism. As shown in chapter 7, however, serious problems arise if instinct and habit are considered as being linked to specific behaviors, since behavior must remain constantly adaptive (and therefore variable) for it to be useful in achieving the goals of an organism living in an unpredictable and disturbance-rich environment. We can nonetheless keep the essential character of Campbell's thought if we reconsider instinct to be the result of the selection of organisms with useful (fit) perceptual control systems, and substitute for habit the reorganization of these control systems that an organism accomplishes to solve problems.

As we move into the next level of Campbell's hierarchy we finally meet the principal concern of this chapter—thought. The sixth level, *visually supported thought*, is:

the dominant form of insightful problem solving in animals, e.g., as described by Köhler [and] requires the support of a visually present environment. With the environment represented vicariously through visual search, there is a substitute trial and error of potential locomotions in thought. The "successful" locomotions at this substitute level, with its substitute selective criteria, are then put into overt locomotion, where they appear "intelligent," "purposeful," "insightful," even if still subject to further editing in the more direct contact with the environment.[35]

To provide an example of visually supported human thought, imagine attempting to rearrange the furniture in your living room to accommodate an upright piano. In looking over the room as currently furnished, you could readily imagine other possible arrangements. You might think, "The sofa could be moved from the back wall to under the window, freeing up wall space for the piano, and the two chairs currently under the window could be moved to the empty corner." Of course, this plan may not prove to be the most acceptable if you then realize that the piano would block access to the built-in bookcase, but other arrangements could easily be imagined as you observe the room's current configuration and contents. Once a decision is made, however, you could implement it directly, without having to try out physically each of the arrangements you considered and then rejected in your thinking.

As mentioned by Campbell above and as we considered earlier, this appears to be the type of thought process used by Köhler's chimpanzees to solve various problems. However, the chimps did not seem capable of Campbell's seventh level of knowledge processes, that is, *mnemonically supported thought*:

At this level the environment being searched is vicariously represented in memory or "knowledge," rather than visually, the blindly emitted vicarious thought trials being selected by a vicarious criterion substituting for an external state of affairs. The net result is the "intelligent," "creative," and "foresightful" product of thought, our admiration of which makes us extremely reluctant to subsume it under the blind-variation-and-selective-retention model.[36]

To extend our example, imagine that you are now contemplating the purchase of a piano in the dealer's showroom and imagining how it could be incorporated into your living room. Since you are not in your living room and therefore cannot see it, you must rely on your memory of it to determine where the piano could be put. This memory represents the knowledge you must have to solve the problem of piano placement. In this case, knowledge may well exist as some type of mental image, although this is most likely only one of many forms that it can take (language being another to be discussed later in this chapter). As such, it substitutes for the visual perception of the room, which in turn substitutes for locomotion-based examination of the room and its contents. The comments of Bain, Baldwin, James, Jevons, Mach, and Poincaré given earlier would all appear to apply to this type of mnemonically supported thought. Another way to conceptualize both visually and mnemonically supported thought is to consider them as forms of *simulation*. (The current use of computers for simulation will be examined in chapter 13.)

Campbell's eighth level comprises *socially vicarious exploration* and *observational learning and imitation*. We now consider the social aspects of knowledge processes and how other organisms can greatly facilitate an individual's acquisition of knowledge. In socially vicarious exploration

the trial-and-error exploration of one member of a group substitutes for, renders unnecessary, trial-and-error exploration on the part of other members. The use of trial and error by scouts on the part of migrating social insects and human bands illustrates this general knowledge process. . . . [Observational learning and imitation] are procedures whereby one animal can profit from observing the consequences to another of that other's acts, even or especially when these acts are fatal to the model. The aversion which apes show to dismembered ape bodies, and their avoidance of the associated locations, illustrates such a process.[37]

But even in the case of learning by imitation, Campbell is careful to discount an instruction-based explanation in stating that

There is no "direct" infusion or transference of knowledge or habit, just as there is no "direct" acquisition of knowledge by observation or induction. As pointed out by Baldwin, what the child acquires is a criterion image, which he learns to match by a trial and error of matchings. He hears a tune, for example, and then learns to make that sound by a trial and error of vocalizations, which he checks against the memory of the sound pattern. Recent studies of the learning of bird song confirm and elaborate the same model.[38]

A particularly striking characteristic of human knowledge processes is the role that language plays in thought and the acquisition of knowledge. Although Campbell puts language at the ninth level, he notes that it overlaps with the two previous levels, since it is generally believed to be the most important tool for both mnemonically supported thought and socially dependent knowledge processes.

He notes that through our use of language "the outcome of explorations can be relayed from scout to follower with neither the illustrative locomotion nor the environment explored being present, not even visually-vicariously present."[39] He further observes:

From the social-functional point of view, it is quite appropriate to speak of the "language" of the bees, even though the wagging dance by which the scout bee conveys the direction, distance, and richness of his find is an innate response tendency automatically elicited without conscious intent to communicate. This bee language has the social function of economy of cognition in a way quite analogous to human language. The vicarious representabilities of geographical direction (relative to the sun and plane of polarization of sunlight), of distance, and of richness by features of the dance such as direction on a vertical wall, length of to-and-fro movements, rapidity of movements, etc., are all invented and contingent equivalences neither entailed nor perfect, but tremendously reductive of flight lengths on the part of the observing or listening worker bees.[40]

Of course, the richness and expressive power of human language far surpass the one-track communication of bees, but the major function remains the same. Language allows us to gain knowledge through someone else's experiences and thoughts, insofar as they can be expressed in language. Yet even our knowledge of language, as in the meanings of words, must proceed by trial and error. As we considered in our discussion of Piaget, concepts such as bird and butterfly, and hence the meanings of their associated words, cannot be transmitted from one person to another, but must be created and tested by each individual. We can therefore never be certain that what we understand by a certain word is the same meaning understood by someone else. "Just as certain knowledge is never achieved in science, so certain equivalence of word meanings is never achieved in the iterative trial and error meanings in language learning."[41] (We will return to the puzzle of language learning and use in chapter 11.)

But despite the tentative, unjustified nature of the meanings shared in language, language has clearly played a major role in the cultural and technological evolution of our species. Through the relatively recent expression of written language, we can learn from the experiences and thoughts of those who lived centuries ago, and share this knowledge and our own discoveries with generations yet to come. The *cultural cumulation* greatly facilitated by language is the tenth level of knowledge processes in which "there are a variety of variation and selective retention processes leading to advances or changes in technology and culture."[42] And as the most impressive example of the accumulation of human knowledge, science is Campbell's eleventh level.

The grand breadth of this hierarchy of knowledge processes clearly goes beyond the scope of a chapter dealing primarily with thought. It is presented here in some detail since Campbell's radically selectionist epistemology is perhaps unique in its outright rejection of all providential and instructionist accounts of knowledge acquisition at all levels, from biological evolution and human perception and thought through the progress of science. Wherever Campbell sees evidence of fit, he is quick to point out how the mechanisms of variation and selection provide the only explanation for such fit that does not require the assistance of a supernatural provider or the workings of unfathomable mechanisms of passive sensory instruction.

Campbell has offered philosophical, logical, historical, and anecdotal reasons for considering thinking as the blind variation and selection of thought trials for over 35 years. But he has not undertaken psychological research to provide evidence for his claims and has admitted that it is very difficult to test them through scientific studies of thinking and problem solving. This may be one reason why Campbell's hierarchy has not had very much impact on mainstream psychological theory. This may well change, however, as the selectionist discoveries in immunology and neuroscience, together with the many practical uses to which selection theory is now being put (and to be discussed in part IV), become more widely known and appreciated.

10

Cultural Knowledge as the Evolution of Tradition, Technology, and Science

From the amœba to Einstein, the growth of knowledge is always the same: we try to solve our problems, and to obtain, by a process of elimination, something approaching adequacy in our tentative solutions.
—Karl Popper[1]

The analogy that relates the evolution of organisms to the evolution of scientific ideas can easily be pushed too far. But with respect to the issues of this closing section it is very nearly perfect. . . . Successive stages in that developmental process are marked by an increase in articulation and specialization. And the entire process may have occurred, as we now suppose biological evolution did, without benefit of a set goal, a permanent fixed scientific truth, of which each stage in the development of scientific knowledge is a better exemplar.
—Thomas Kuhn[2]

We have now considered in some detail two major ways in which biological structures, processes, and behaviors become adaptedly complex. We saw how through biological evolution those organisms that by blind chance possess variations in biological structure or perceptual control systems that provide survival and reproductive advantages will be more successful in passing on these structures and systems to their offspring. The cumulative selection of such advantageous modifications from one generation to the next (by the elimination of less successful organisms) can have dramatic effects on the evolution of a species and is also responsible for the origin of new species. This is phylogenetic selection among organisms.

We also concluded, however, that natural selection among organisms suffers from two notable drawbacks—it takes a considerable amount of time,[3] and it cannot be used by any *individual* organism to cope with an unpredictably changing environment or to learn from experience. An animal of a species that for millions of years depended on a particular food will have great difficulty surviving and reproducing if the food becomes unavailable, unless it can learn to find and use a substitute. Similarly, a mammal with a fixed repertoire of antibodies will be unlikely to survive if it encounters a new virulent strain of virus. But through ontogenetic variation and selection occurring within an animal's nervous and immune systems, it is possible for the animal to survive and reproduce even if its environment is quite different from that of its ancestors. The adaptation resulting from such lifespan evolution can occur quite rapidly and permit an organism to cope with a complex, changing environment, an important part of which is made of its fellow organisms with which it must compete for food, shelter, and mates. We noted in previous chapters, however, that such adaptation appears to be based on the same basic mechanisms of cumulative blind variation and selection that underlie the much slower and less responsive process of phylogenetic evolution.

But there is a third way, which we considered only briefly in chapter 9, in which adapted complexity can arise. Although strictly speaking it is a form of learning, it differs in some important respects from the type of learning we have already considered. Many organisms, particularly mammals and birds, are able to make good use of the experiences of their fellow creatures. Even the honeybee is able to profit from the experiences of hivemates who have discovered a new source of food. This learning from the learning of others is nowhere more developed than in humans where it is generally referred to as *culture*. Its role in both the remarkable proliferation of our species to all corners of the globe and in our continued survival demands that we consider its adaptedness and explain its origin and evolution.

Tradition

Of Rice and Religion
Rice is humankind's single most important food, providing nearly a third of all the calories we consume.[4] In tropical climates where three crops can be grown each year, as much as six tons of rice can be produced annually from one hectare (2.47 acres) of land. And this can be accomplished from the same land year after year for centuries without the addition of chemical fertilizers and pesticides.

To achieve and maintain these impressive yields, special, labor-intensive cultivation practices are required. First, seedlings are obtained by sowing rice into specially prepared seedbeds. After several weeks the seedlings are transplanted by hand to a flooded field or terrace where they are left to grow in 5 to 20 cm (2 to 8 inches) of water for several months. Two to three weeks before harvest, the paddy is drained. The rice stalks are cut and tied into bundles and allowed to dry before threshing separates the grain from the rest of the plant.

The most striking aspect of rice cultivation is the use of flooded fields or terraces, and this is the key to obtaining high, recurring yields. Flooding helps to keep essential nutrients in the soil that would otherwise be lost through leaching and exposure to the atmosphere. Under flooded cultivation other nutrients containing iron, aluminum, manganese, and calcium release phosphorus for use by the plants, instead of it remaining chemically bound to these minerals as it is when dry. More nitrogen is also available, since the oxidation of ammonia into nitrates is retarded, with the result that nitrates are produced at about the same rate as the plants can assimilate the nitrogen they contain (much of this nitrogen would otherwise be lost into the atmosphere). In addition, blue-green algae and a small water fern called azolla establish residence in the rice paddies. These plants have the rare ability to take in nitrogen directly from the atmosphere, which is then made available to the rice plant as the algae and azolla die and decompose. Finally, the ability of the plants to thrive in flooded fields despite the lack of oxygen at their waterlogged roots is due to their remarkably efficient system of air passages connecting their leaves above water to the roots below.

The Indonesian island of Bali is in many ways ideally suited to the cultivation of rice. Lying almost directly on the equator, temperatures vary little throughout the year from its 26° C (79° F) average. Rain can be expected to fall about 200 days each year, and humidity remains at a fairly constant 70% to 80%. The majority of the population inhabits the south side of the island on the slopes of mountains whose lakes and rivers provide a reliable and controllable supply of water for irrigation. It should therefore come as no surprise that rice and water are key elements in Balinese traditional life. Every morning the women lay out a few grains of cooked rice on squares made from banana leaves as offerings of thanks to Wisnu, the water god, and Dewi Sri, the goddess of rice and fertility.

Because wetland rice cultivation requires periodic flooding and draining, an extensive network of hillside terraces and irrigation systems exists on the mountainsides. But more than this is required to ensure efficient rice production. Although the climate is such that rice can be grown at any time of the year, the timing and movement of water down the terrace slopes must be carefully planned. Ideally, planting should be arranged so that the water can be moved down the slope, providing for periodic flooding and draining that corresponds to the growing cycle of each terrace. If decisions concerning planting and harvesting times were left up to each individual farmer, there would be the risk of taxing the water supply beyond its limits during one part of the year while seriously underusing the available water during other periods.

To coordinate this use of water, the Balinese have formed the *subak* or "irrigation society." This is a secular organization with taxing rights and a written constitution that provides one vote for each member; however, the ultimate authority of the subak is vested in Dewi Sri.[5] So the operation of the irrigation system makes best use of the available water, but the context in which the appropriate dams and sluice gates are opened and closed and the terraces are prepared for planting and harvesting is a religious one:

The schedules of the various stages and water-openings are all set by the cycle of religious occasions that farmers are obliged to acknowledge with prayers and offerings. Included among these occasions are the days on which the ground should be broken (and not before), when the field should be flooded, when seed should be planted, and when the young seedlings should be transplanted. A large ceremony is held in the main subak temple when the rice is about to flower, and a full scale three-day festival takes place when the seed is set. There are lesser ceremonies, (pleading for good growth and for protection from pests, for instance), made at shrines in every field, and other large ceremonies are held as the crop ripens, when it is harvested, and when it is placed in the granary. . . . [There are between 9 and 16] specific religious occasions on which every subak member must make obeisance to the appropriate gods.[6]

As a result:

In Bali, where maintaining high levels of wet-rice production in a relatively small area is made more complex by the rugged nature of the terrain, farmers have made a religion of their activities. Rice in Bali is not grown according to any production timetable that the modern agronomist might work out, but according to the stipulations of the temples and the rice goddess—and with very good effect. No Balinese rice farmer ever needs to consider the technical details of how rice should be grown to produce maximum crops from his land—precisely when to plough, when to flood, when to plant, when to drain and so forth. All he has to do is follow the calendar of *Dewi Sri*, the goddess of rice and fertility, and the crops are virtually guaranteed. Rice cultivation is the ultimate expression of the Balinese readiness to follow the edicts of some greater authority: the cult of the rice goddess not only demonstrates the integration of the secular and the spiritual worlds of Bali at the most fundamental level, it also provides an eloquent example of the functional significance of religion in human ecology.[7]

The fit between the cultural practices of the Balinese and the growing requirements of rice is well evident. But what is the explanation of this fit? If one were to ask a typical Balinese rice farmer, he would no doubt say that rice cultivation practices were provided by the rice goddess, since it is only by observing her calendar that continued good crops are ensured. Indeed, the fact that rice cultivation is so successful would appear to be quite convincing evidence that Dewi Sri is both knowledgeable and kind, and deserves respect, thanks, and praise, all of which she does in fact receive.

But quite a different picture emerges if we try to understand the farmers' knowledge of rice cultivation from a naturalistic perspective, that is, as somehow originating and developed without divine providence. This may first seem problematic, since it is quite unlikely that they would know anything about the extraordinary oxygen transport system that permits rice plants to thrive in flooded terraces, or the intricate biochemistry of wet-rice cultivation that ensures the plants an adequate supply of essential nutrients without depleting the resources of the land. So it might first appear that their farming knowledge must have been provided to them from some wiser source. But if we recall the remarkably fit products that biological evolution has been able to produce through the processes of cumulative blind variation and natural selection, we can appreciate that the rice-farming practices could have evolved similarly without the Balinese having explicit knowledge of the underlying reasons for their effectiveness. In much the same way birds have no formal understanding of physics, yet are able to use Bernoulli's principle quite successfully to fly. If over a long period of time biological evolution has produced palm trees, lobsters, panda bears, and humans, it should not seem unlikely that cultural evolution could result in an efficient system for the Balinese to produce their staple food.

Indeed, the view that culture originates and changes over time in a manner analogous to the cumulative trial and error of biological evolution has become increasingly popular among social scientists. As anthropologist John Reader explains:

The farmers who founded and refined the wet-rice system and maintained its high levels of production for centuries knew nothing of nitrogen cycles and oxygen transportation in plants. They worked purely by trial and error. In the process, however, they acquired a sound appreciation of just what made the system work, and of how to keep it working.[8]

It cannot be doubted that the Balinese farmers are successful in cultivating rice, but is it actually the case that they "acquired a sound appreciation of just what made the system work"? For the traditional farmers it is observance of Dewi Sri's calendar together with participation in the many religious activities that are responsible for their success. Yet it can be easily shown that such religious observances are in no way essential to obtaining continued good rice harvests, since good harvests are obtained elsewhere in the world where the biological requirements of the rice plant are met and Dewi Sri and her calendar are totally unknown. So clearly a fit exists between Balinese farmers' agricultural practices and the requirements of the rice plant, although individual farmers may not know (and need not know) the underlying scientific reasons for it.

Regardless of a lack of technological or scientific understanding of rice cultivation, the society in which the rice farmer lives is structured in such a way to ensure continuation of the farming practices found over the centuries to be effective. By making it appear that these practices have divine origin and guidance, it is less likely that an individual farmer would challenge the system. So although daily rice offerings to the gods and frequent temple ceremonies in themselves have no direct causal link to the success of the crop, these traditional activities are well adapted in a larger sense since they ensure that traditional agricultural methods which have proved effective over the centuries will continue.

Does his lack of scientific understanding mean that the traditional Balinese farmer is in any way irrational or illogical in his adoption of the centuries-old methods of rice cultivation? Hardly, since for him rice cultivation and religious practices form one integrated system. It would be well nigh impossible for him to determine which particular aspects of his way of life are essential for obtaining continued good harvests and which are not. Indeed, such experiments (for example, refusing to participate in religious activities to see if this reduces rice yield) would possibly result in the radical farmer being ostracized from his community and make it impossible for him to obtain the water supply on which his crop depends. Instead, there are important advantages for individuals to adopt the agricultural and other traditional practices of the majority of their community.[9]

So since many aspects of traditional rice cultivation are not individually testable, we should not be surprised to find that some of them are not functional or are even maladapted to the requirements of rice production. The difficulty that an individual would encounter in attempting to analyze which aspects are actually well adapted and which are not, save for the fact that they may play important social functions, also argues for the rationality of accepting and observing the total cultural package.

Hidden Adapted Aspects of Tradition

On the other hand, many traditional customs, rituals, and taboos that may at first appear to lack any adapted qualities or may even seem maladapted, may on closer examination turn out to have interesting adapted characteristics. A classic example is the sacred cattle of India. With a population second only to that of China, India has difficulty feeding her people. Yet in the midst of this great need for food, millions of cattle are allowed to roam the countryside, trample gardens, and snarl traffic. According to traditional Hindu belief, the body of a cow is home to millions of deities, and the next reincarnation of the cow's soul will take human form; killing the cow sends its soul back to square one, and 86 more reincarnations are required to achieve cow status again. Thus Indians put up with the inconvenience posed by the cattle and would never consider butchering and eating one. This would appear to be an example of how traditional cultural beliefs and practices can be maladapted with respect to basic biological needs. "Orthodox Hindu opinion regards the killing of cattle with abhorrence, even though the refusal to kill the vast number of useless cattle which exist in India today is detrimental to the nation."[10]

If we take a more careful look, however, we can see that good reasons exist for the apparent folly of India's sacred cattle. In a classic paper published in 1966, American anthropologist Marvin Harris proposed that these cattle are on the whole beneficial to Indian society. They provide significant amounts of milk and meat (the latter eaten by Moslems, Christians, and lower-caste Hindus) as well as hides for India's leather industry, the world's largest. More than 300 million tons of manure are collected each year for use as cooking fuel in rural households, 90% of which have no other source of fuel. In addition, the cattle are regularly pressed into service as oxen, allowing millions of farmers simultaneously and quickly to plow their parcels of land to take full advantage of the sudden monsoon rains for their crops.

So we see that aspects of traditional culture that may at first appear to be at odds with the requirements of human life and reproduction may actually turn out to be in some respects quite functional, even if individuals are unaware of these adapted aspects and offer other reasons for their practices, such as concerns about reincarnation. This does not mean that the cattle situation in India could not be improved, only that if a striking although puzzling cultural practice has persisted over a long period of time, we may well expect to find some functional reasons for its existence if we look long and hard enough. But even if we do look long and hard, we may still fail to uncover functional explanations for many practices and beliefs found throughout the world in both so-called traditional and advanced societies. Can a theory of cultural evolution based on cumulative variation and selection of useful beliefs and practices account for this?

To attempt to make sense of the puzzle of apparently maladapted traditions, another problem will have to be considered. Many aspects of traditional culture may be beneficial to the society as a whole, but may be detrimental to the individuals who practice them. For example, in most societies people are expected to respect the property of others and to do their fair share of work. It is easy to appreciate how this is beneficial for the society as a whole, since if everyone simply stole what they needed no one would produce the required goods and services. Yet to the individual, it is advantageous, at least in a biological, evolutionary sense, to be selfish in doing the least amount of work possible, and to devote as much of one's energies as possible to staying healthy, mating, reproducing, and ensuring the survival and reproduction of one's own children. The cost involved to the individual in cooperating with society's conventions is most dramatically apparent in organized warfare, where men and women are expected to lay down their lives for the good of their community. Individual costs are involved in all instances of altruistic behavior, that is, when one individual pays the price of reduced reproductive success that benefits another's reproductive success. Thus we have the dilemma of explaining how individuals who are in reproductive (genetic) competition nonetheless quite often cooperate.

One way of understanding how such altruistic and cooperative practices could have arisen is to consider the genetic relatedness of the potential cooperators. If closely related individuals share a higher proportion of their genes than do less closely related ones, we can understand why a mother would risk her life for her children, since she is also likely protecting the genes that predispose her to do so. W. D. Hamilton made the first convincing arguments for *kin selection* as an explanation for altruism in two papers published in 1964. These papers provided an explanation for how cooperative behavior could evolve genetically (that is, by biological evolution) among genetically close relatives. And there is considerable evidence that for humans, cooperation is much more likely to occur, and dangerous violence is much less likely to occur, among close relatives.[11]

But we also find cooperative behavior among quite unrelated members of a species, and even involving different species. How can this be explained? One explanation is that it can be advantageous for an interacting group of organisms to cooperate instead of compete if the risk of noncooperation is high (for example, death in combat or lack of food) and if the individuals involved are likely to meet (and recognize each other) again. Robert Trivers coined the term reciprocal altruism to describe this type of cooperation,[12] and political scientist Robert Axelrod demonstrated using game theory and computer simulations how cooperation among unrelated individuals can evolve and remain stable.[13]

For such cooperation to evolve and persist, however, it must be possible to restrict the receipt of altruistic deeds to those individuals who will likely reciprocate. Otherwise, cheating in the form of accepting the benefits of others' altruistic deeds while providing nothing or little in return could become rampant and eventually undermine the cooperative nature of the community (those attempting to cooperate become "suckers" if recipients do not reciprocate). It is here that otherwise apparently useless or even maladapted traditions may be employed to indicate membership in the same reciprocally altruistic community. In this respect, *"easily perceivable* homogeneities in dialect, dress, rituals, and scarification would be particularly useful. Thus the Luo of Kenya knock out two front teeth of their men, while the adjacent Kipsigis enlarge a hole pierced in their ears to a two-inch diameter."[14] Although in itself a particular manner of dress or speech may seem to have no value as an adaptation, it may well take on importance as an easily perceivable sign of group membership and thereby facilitate in-group cooperation. In many American cities today, the manner in which a youth wears his baseball cap can mean life or death as he passes through areas in which rival gangs, distinguished in part by how they sport their caps, compete for control.

From this perspective, a number of otherwise puzzling aspects of traditional cultures become somewhat less puzzling. For example, some striking similarities occurred among the first city-state civilizations to emerge in Africa, China, the Middle East, India, and the Americas, including:

1. A division of labor and political centralization headed by a supreme, tyrannical ruler.
2. A belief in a supernatural cosmology that provided authority for the ruler.
3. A set of moralizing preachments that "preached the value of *duty to the political organization* and its customs . . . the duty of self-sacrificial military heroism in defense of the state . . . within-group honesty" and "preached against self-interested deviations from duty (covetousness, jealousy, etc.)."[15]
4. A belief in an afterlife of "rewarding and punishing heavens, hells, and reincarnation."[16]

In marked contrast to such uniformity, these civilizations had strikingly divergent beliefs concerning the *specifics* of the supernatural forces that appear to have played roles in the daily lives of their inhabitants. Different gods were praised and placated by various rituals and taboos. Such diversity argues for the independent origin of these beliefs and practices, although the function of each seems quite the same—to inhibit selfish behaviors that could undermine the cooperative structure of the community, and to encourage cooperative acts. To this end, all of these cosmologies (as do almost all existing religions) appear to have included a belief in an afterlife involving rewarding heavens and punishing hells. Wasteful royal funerals, in which "fully useful horses, soldiers, wives, weapons, jewels, and money were interred"[17] are a dramatic affirmation of such a belief. These funerals are a striking example of a practice that is downright wasteful in squandering hard-earned community resources, but that may be adapted at the social level as evidence of the   reality of an afterlife. Thus they might motivate individuals to forego selfishness in this life to obtain the rewards of their socially adaptive cooperation in the next.

So it does appear possible to account for at least some of the apparently adapted and maladapted aspects of traditional cultural beliefs and practices using a model of cultural evolution that works at the level of the group. It also depends on cumulative variation, selection, and propagation not of genes, but rather of shared beliefs and practices. Because such aspects of tradition have proved themselves by the very fact that they have been around for a long time and that most people of a community observe them, it makes sense for individuals also to adopt them. But they are not confirmable as adaptive by the individual; indeed, even expressing doubt as to the truth of cherished traditional beliefs is often reason enough for banishment from the society or worse. In addition, traditions that distinguish members of one community from rival ones distinguish cooperative individuals from selfish ones. Therefore, we can expect some of tradition to appear useless and even maladapted to the biological demands of survival and reproduction, such as exorbitantly wasteful royal funerals.[18]

Technological Change

The second aspect of culture that we will consider is technology. Although today we tend to think of technology as involving powerful machines and sophisticated electronic equipment, in its more general sense it refers to any tools or methods that allow an easier or more efficient production of useful goods and services. A horse-drawn plow is a more effective and productive way of preparing soil for crops than is a shovel or hoe; nets for fishing have advantages over spears; and for covering long distances, air travel is superior in speed, safety, and comfort (if perhaps not in cuisine) to land and sea travel. Other species may use tools to a limited extent,[19] but none compares with us in terms of the variety and sophistication of our technological achievements. Each generation inherits crucial knowledge in the form of tools and procedures for providing food, clothing, shelter, health care, and protection from enemies. The acquisition of technology by individuals through either formal or informal education is essential to a society's continued survival and prosperity.

Not all would accept the claim that all, or at least many, cultural traditions are well adapted in any obvious material sense, although it was argued above that numerous traditional practices and beliefs are well adapted in at least some respects. The adapted nature of technology and its progress is harder to doubt. To increase production or provide better services at less cost and effort, technology must achieve an unlikely fit between its tools and techniques on one hand and the materials with which it is to work, the constraints of the physical world, and the limitations and preferences of human operators and consumers on the other. Nonetheless, human history is replete with examples of innovations that successfully (and increasingly) met the formidable odds imposed by these demands. Stone tools, the wheel, the horse collar, the plow, iron, paper, the spinning wheel, the clock, the cotton gin, the printing press, steam and internal combustion engines, barbed wire, the electric motor, the vacuum tube, the transistor, and the computer microchip are just a few examples of items developed in China, Europe, and the Americas. Because every successful innovation provides a demonstrably better way of producing some good or providing some service, each represents yet another puzzle of fit.

The similarity between the biological evolution of organisms and the technological evolution of tools, machines, and instruments has not escaped the attention of those attempting to describe, explain, and predict technological progress. As early as 1863 English writer and critic Samuel Butler explored the theory that machines develop in a way that resembles the evolution of living organisms. Indeed, Butler was so impressed by the rapid evolution of machines during the Industrial Revolution that he predicted that they would eventually constitute a new class of living things that would surpass the sophistication of humans and relegate us to second-class status.[20]

Although Butler's account of the evolution of machines was more satire than science, there have been serious attempts to understand the development of technology as an evolutionary process. Three recent books provide good examples.

The Evolution of Artifacts

In *The Evolution of Technology*,[21] George Basalla continues Butler's consideration of made things (artifacts) as analogous to living organisms. First, he argues for the continuity of technological innovation: "Any new thing that appears in the made world is based on some object already in existence."[22] New inventions do not emerge magically from the minds of great inventors, but rather are modifications of previously existing artifacts. In 1793 Eli Whitney's cotton gin that removed seeds from short-stapled cotton was based on the Indian *charka*, which had been in use for thousands of years to remove seeds from long-stapled cotton. Joseph Henry's electric motor of 1831 copied many of the mechanisms employed in the steam engine. The development of the first transistor at Bell Laboratories in 1947 by Bardeen, Brattain, and Shockley owed much to the work of German physicist Ferdinand Braun who, in the 1870s, found that certain crystals conduct electricity in only one direction.

Second, Basalla points out the diversity of artifacts that are available to any society, recognizing that certain cultures display more diversity than others. The source of this diversity can be found in psychological, socioeconomic, and cultural factors including the fertility of human imagination, our proclivity to play and fantasize, and the imperfect copying that invariably results when a person attempts to make an artifact based on an already existing one. An example of this diversity is the more than 1000 smokestack designs that were patented in the United States during the nineteenth century in the unsuccessful attempt to prevent the escape of embers and sparks from wood-burning locomotives.[23]

In the same way that organisms produce more offspring than can survive, there are often more variations of a given tool or machine than can survive and be taken up by the next generation of users. Thus selection becomes the third essential element of Basalla's view of technological evolution.

From the vast pool of human-designed variant artifacts, a few are selected to become part of the material life of society. In nature it is the ability of the species to survive that counts—the fact that the organism, and especially its kind, can thrive and reproduce in the world in which it finds itself. The artifact may also be said to survive and pass on its form to subsequent generations of made things. This process requires the intervention of human intermediaries who select the artifact for replication in workshop or factory.[24]

Technology as Knowledge

Joel Mokyr's *The Lever of Riches*[25] is an engaging historical account of technological advances in Europe and China. Mokyr sees technological innovation as the major motor of economic growth, a way of providing occasional "free lunches" and many "cheap lunches" in that the cost of a successful innovation is paid for many times over by the increase in productivity it makes possible. Like Basalla, Mokyr suggests an evolutionary, selectionist account of technological development, but instead of giving a primary role to the artifacts themselves, he emphasizes the growth of human knowledge as the basic unit and gives analogues in technological evolution for the genotype and phenotype of living species.

> The approach I adopt here is that techniques—in the narrow sense of the word, namely, the knowledge of how to produce a good or service in a specific way—are analogues of species, and that changes in them have an evolutionary character. The idea or conceptualization of how to produce a commodity may be thought of as the genotype, whereas the actual technique utilized by the firm in producing the commodity may be thought of as the phenotype of the member of a species. The phenotype of every organism is determined in part by its genotype, but environment plays a role as well. Similarly, the idea constrains the forms a technique can take, but adaptability and adjustment to circumstances help determine its exact shape. Invention, the emergence of a new technique, is thus equivalent to speciation, the emergence of a new species.[26]

Mokyr's approach is therefore more psychological and epistemological than Basalla's in insisting that technology "is not something that somehow 'exists' outside of people's brains."[27] His perspective is also more concerned with economics in the priority he gives to "lowest quality-adjusted cost" as the major criterion for selecting among the competing technologies present in a society at any one time.

Technology as Vicarious Variation and Selection

Certainly one of the most marvelous of all technological achievements is the creation of machines that fly. In *What Engineers Know and How They Know It*, Walter Vincenti presents five detailed case histories of important innovations that paved the way for the success of the modern airplane.[28] In his last chapter, he uses these cases to garner evidence for a "variation-selection model for the growth of engineering knowledge." Drawing heavily from Campbell's work, he breaks away from the limits imposed by the rather strict biological analogy employed by Basalla and Mokyr and makes a compelling case for a more general selectionist account of technological development.

Vincenti emphasizes the advantages of vicarious over direct trials in arriving at successful designs for technological innovations. Direct trials refer to the actual building of a new device and trying it out for its intended purpose to determine its degree of success or failure. Such direct trials of overt variations characterize the attempts of the French between 1901 and 1908 to develop flying machines.[29] During this time the French built a wide variety of aircraft and tested one after the other, usually with disastrous results. Not basing their trials on any systematic attempt to discover the basic principles of flight, there was little in the failure of one design to inform the design of the next, and the continuous building (and subsequent crashing) of working prototypes consumed considerable time and resources.

In contrast, vicarious trials for designing new technologies permit both expansion of the variations to be considered as well as a much more efficient (in terms of time and energy) selection process. Vicarious trials can be either experimental or analytical. Experimental ones consist of "the substitution of partial experiments or complete simulation tests for proof test or everyday use."[30] An example is the use of wind tunnel testing in aeronautical engineering. The Wright brothers' extensive use of such tests gave them an important advantage in the race to build the first airplane, since during the time it would take their French competitors to build and launch yet another complete prototype, the Americans built, tested, eliminated, and designed many different scale models and therefore made more rapid progress in their search for a successful design.

Analytical vicarious trials are even further removed from direct trials in that they use more abstract tools to test the worthiness of a particular design variation. They can be considered a kind of "test run on paper,"[31] although since computers are increasingly used for such tests today, they are more often test runs on computer silicon. "As each hypothetical arrangement of parts is sketched either literally or figuratively on the calculation pad or computer screen, the candidate structure must be checked by analysis. The analysis consists of series of questions about the behavior of the parts *under the imagined conditions of use after construction.*"[32] It is through such tests that scientific knowledge can be exploited in technological development.[33] The Wright brothers profited from an analytical approach due to their knowledge of the principles of fixed-wing flight developed in the early 1800s by Sir George Cayley. This knowledge permitted the brothers to analyze the problem of flight into three subproblems consisting of lift, thrust, and control. By attacking each of these problems separately using vicarious experiments, they made rapid progress in designing the first machine capable of sustained manned flight.[34]

In emphasizing the importance of vicarious variation and selection, Vincenti in effect has described a technology of technology, this meta-technology providing better and better tools for developing new technology and making possible the ever-quickening pace of development. As these tools develop, they permit more efficient and reliable vicarious testing of variations on existing ideas. This also permits a broader scope of variations that can be tested efficiently. In effect the range of variations can actually be increased without detriment and possibly with great advantage, since the new vicarious means of variation and selection are so rapid and efficient in weeding out the unfit variations. As Vincenti has remarked, "engineers have freedom to be increasingly blind in their trial variations as their means of vicarious selection become more reliable. One sees engineers today, for example, using computer models to explore a much wider field of possibilities than they were able to select from just a decade ago."[35]

The Development of Science

The final aspect of culture to be considered here is scientific knowledge. Technological knowledge can be put to *practical use* to solve problems concerning the production of goods and services. Scientific knowledge provides *explanations* for observed phenomena in terms of underlying mechanisms, thus providing a basis for predictions and perhaps ultimately controlling the phenomena under study. Technology may help to advance science, especially in providing new tools for exploration such as orbiting telescopes, electron microscopes, and powerful computers. And science can aid technology, as particle physics made possible nuclear weapons and nuclear power plants. But the two can be seen as distinct. The Australian bushman's ability to produce boomerangs is technological knowledge that does not depend on scientific knowledge of physics and aerodynamics. And the greatest scientific discovery of the nineteenth century owed little to technology, other than that involved in building and navigating the *Beagle*, the ship that allowed Darwin to experience firsthand the world's remarkable diversity of life, in particular the subtle differences in species found scattered among the Galápagos Islands.[36]

Science provides what many consider to be the most striking and undeniable instances of fit of one system—human knowledge—to another—the universe in which we live. It is for this reason that countless philosophers and scientists have attempted to understand how this remarkably complex and fit aspect of our culture is achieved.

Bacon's Systematic Method of Induction

Sir Francis Bacon (1561–1626) lived in England during the dawn of the modern Western scientific and technological era. He was greatly impressed by the astronomical discoveries of Copernicus and Galileo, which provided a radically new view of the universe, and the technological achievements represented by such inventions as gunpowder and the printing press. But Bacon had little patience for the speculative philosophers of his time who, he believed, had made little if any progress in understanding nature since the time of the Greeks. He felt strongly that it was only through a carefully implemented scientific methodology that we would be able to understand the world around us and regain the mastery over nature that was lost at the time of the fall of Adam and Eve.

But Bacon believed that more than just mastery over nature had been lost. The human mind was also corrupted by false images or idols, which caused it to perceive the world in unreliable and subjective ways. These idols included the tendency to seek out and see only confirmations of already held beliefs, and the proclivity to notice the more striking aspects of the world while failing to see the subtler ones. Bacon held that these impediments to objective and accurate observation had to be swept away, leaving the mind as a blank slate so that the world could accurately transmit and impress its true nature on the human mind.

To this end he offered a systematic method of induction based on careful observation, comparison, and experimentation. For instance, to understand the true nature of heat, one has to find many instances of heat and determine by observation what they have in common. But positive instances alone do not suffice. One must also find negative instances, in particular, instances that are very similar to the positive ones but that lack the quality of interest. It is only by listing and comparing instances in which the phenomenon of interest is present with those in which it is absent that the true nature of the phenomenon can be ascertained.[37] By mechanically following this method, Bacon believed that one would eventually arrive at valid inductions concerning the makeup and behavior of the natural world that could then be applied deductively to predict and control natural phenomena. Such a view of scientific knowledge is referred to today as empirical positivism, since it is based on the belief that careful empirical observation can provide us with certain (or positive), justified knowledge of the universe's content and laws.

The Problem of Induction

Bacon's writings on science and its method had a great impact in England and on the European continent. Both Newton and Darwin acknowledged their debt to Bacon, and shortly after his death various scientific societies, such as the Royal Society in England and similar institutions on the continent, were established to undertake the type of systematic, scientific research that Bacon had advocated. Indeed, the remarkable success of Newton in discovering the laws of nature that govern the movements of both terrestrial and celestial bodies hinted that it was only a matter of time before all of nature's secrets would eventually be uncovered using the empirical, inductive method based on unbiased observation and objective comparison.

But this was not to happen in the manner envisaged by Bacon and the new breed of empirical scientist whom he influenced. A century after his death, David Hume (see chapter 6) pulled the epistemological rug out from under all attempts to arrive at a foolproof method of induction by which general laws and theories could be discovered and justified by unprejudiced observation of the natural world. In fact, it would appear that Bacon was at least partially aware of the impending problem, as indicated by the importance he placed on negative instances, as he shows in this account of the power of prayer:

And therefore it was a good answer that was made by one who when they showed him hanging in a temple a picture of those who had paid their vows as having escaped shipwreck, and would have him say whether he did not now acknowledge the power of the gods,—"Aye," asked he again, "but where are they painted that were drowned after their vows?" And such is the way of all superstition, whether in astrology, dreams, omens, divine judgements, or the like.[38]

Bacon's concern with negative instances anticipates Popper's similar emphasis on the essential role of refutation in science (to be discussed next). But it apparently did not occur to Bacon to wonder, as it did later to Hume, that if a negative instance (for example, experiencing silent lightning) can lead one to reject a belief (that all lightning produces thunder) previously supported by countless observations, how could one be sure that for any general belief supported by observation that a negative instance does not exist somewhere? Inasmuch as it is not possible to prove that negative instances do not exist somewhere, it is not possible to be absolutely certain of the truth of any general belief, scientific or otherwise, no matter how well the belief has been supported by past empirical observations.

Despite this serious logical and methodological difficulty, however, science does appear to make progress in ways in which other belief systems, such as religion, astrology, and palmistry, do not. If science is indeed able to attain progressively better fit to the world it describes, we are left with the puzzle of how this fit is achieved, since it is evident that empirically based induction of the type advocated by Bacon and found wanting by Hume is simply not up to the task of providing infallibly true knowledge of the world.

Popper and Falsification
Science continued to make important breakthroughs at an accelerating pace from the seventeenth through the twentieth centuries. It could be argued, however, that no comparable breakthroughs occurred in understanding how science was able to continue to make progress, that is, achieve better and better fit with the objects and phenomena it described, until Sir Karl Popper (1902–1994) confronted the problem. Popper grew up in Vienna during a time of great intellectual and scientific activity in Europe in general and in the Austrian capital in particular. As a student of the philosophy of science, he was fascinated by the ability of science to achieve better and better fits to the world it described, and consequently he set his mind to determining what it was that set science apart in this respect from nonscientific domains. What was it that allowed Newton and Einstein to propose theories that were convincingly better than those provided by their predecessors, whereas the political and economic theory of communism offered by Marx and the psychoanalytic theory developed by Freud were not demonstrably better than competing theories? Popper proposed a simple yet bold solution to this problem of the "demarcation" of science from nonscience and in so doing offered a solution to the vexing problem of induction raised by Hume.

According to Popper, what sets science apart from nonscientific beliefs and theories is that scientific theories are *falsifiable*. For instance, one of Newton's laws of physics is that force is equal to mass times acceleration. This theory can in principle be falsified by experimentation, since it makes specific testable predictions. If the theory is correct, it should take the same force to impart a certain acceleration to an object as it takes to impart twice this acceleration to an object which has half the mass. If this is found not to be the case, and no methodological errors have been made, the theory must be rejected and a better one formulated and tested in its place. This contrasts with Freud's psychoanalytic theory, which is formulated in such a way as to make it immune from falsification. If psychoanalytic theory says that all males are jealous of their father and covet their mother but a certain male denies having these feelings, a confirmed Freudian would argue that the male was repressing his true feelings of paternal jealousy and maternal desire. Similarly, fields such as astrology and palmistry are not scientific, since if a prediction does not prove accurate, reasons can always be found after the fact why things did not turn out as foretold.

Popper's discovery of the importance of falsification also had a side benefit in that it solved the problem of induction. As already noted, a scientific theory that proposes a general, universal law of nature can never be rationally justified, since by virtue of its universality it must go far beyond the limited observations of mortal scientists. So no matter how many times it is observed that event A is followed by event B (for example, heating water to 100° C causes it to boil), it cannot be proved logically that all A are followed by B. But whereas apparently confirming cases cannot justify a scientific theory, disconfirming cases do allow us to refute it. Finding a clear instance of A that is not followed by B (for example, finding that in Death Valley water at 100° C does not boil) means that our theory must be revised or abandoned. Then a new theory must be proposed that accounts for all that the old theory accounted for as the well as new findings that it could not handle. According to Popper, the fit of science is not due to observation and induction of true, justified (or justifiable) theories, that is, the accurate, instructive transmission of knowledge from the environment to the scientist. Rather, science progresses through the creation of conjectures (guesses) and the subsequent weeding out of inadequate hypotheses, leaving those that are better than the ones eliminated only because they have not yet been eliminated themselves.[39]

It did not escape Popper's attention that his view of the process of scientific achievement had much in common with Darwin's selectionist theory of biological evolution:

The growth of our knowledge is the result of a process closely resembling what Darwin called "natural selection"; that is, *the natural selection of hypotheses*: our knowledge consists, at every moment, of those hypotheses which have shown their (comparative) fitness by surviving so far in their struggle for existence; a competitive struggle which eliminates those hypotheses which are unfit.[40]

Popper's view of scientific progress as a cumulative selection process throws interesting new light on science and its achievements. First, in the same way that biological evolution depends on the existence of blind variation in the structure and behavior of organisms, science depends on similar blind variation in hypotheses that are proposed. This does not mean that the hypotheses are not constrained by the knowledge already achieved. No respectable scientist is going to propose that the core of the earth is made up of strawberry jam or the moon's surface is Swiss cheese. And no whale is likely to give birth to a horse. In both biological evolution and science, such constraints reflect the past accumulation of knowledge by prior blind variation and selection and are essential in narrowing down the types of future variations that appear. But the constraints alone cannot account for the emergence of new and better fits of organism to the environment, and scientific theory to the universe.

Second, an evolutionary perspective accounts for the tentative nature of scientific theories. Each now-extinct species, which together make up a much larger number than those species still extant, had before its extinction been successful in surviving for quite some time, in some cases hundreds of millions of years. But in none of these cases was this long period of survival (which is clear evidence of fit to the environment) able to guarantee the future success of the species. Similarly, the long-term, popular acceptance of a scientific theory in no way guarantees that it will not become extinct as better theories evolve to compete with it and eventually replace it. The phlogiston theory of fire, the caloric theory of heat, the ether theory of outer space, Newton's theory of mechanics, and Lamarck's theory of evolution have all been eliminated in the struggle for survival described by Popper, despite the fact that each was the best and dominant theory of its day. So in addition to explaining how science can achieve progressively better fit to the universe, the selectionist view as proposed by Popper explains why we can never justify any particular theory as absolutely and infallibly true. In the same way that previously successful species become extinct, scientific theories are eliminated and replaced by better ones.

Finally, a selectionist view emphasizes the creative role of the scientist. Popper contended that the scientist's mind is not an epistemological bucket that is filled with knowledge from the environment through the eyes, ears, and other sensory organs. This is not unlike a Lamarckian view of evolution in which the environment somehow instructs the reproducing organism to create new adapted forms. Instead, the scientist actively constructs knowledge in the form of unjustified theoretical conjectures, which are then tested and compared to competing conjectures. In this way, the scientist's experience of the world does not provide the theories to be tested. Rather, observation is used to weed out the unsatisfactory ones already constructed. As it is not possible to predict the course of biological evolution, it is similarly impossible to predict the future course of science. And because technological and social changes are strongly influenced by scientific developments, it is similarly impossible to predict the future course of history.[41]

As falsification is the key ingredient to Popper's philosophy of science, it should come as no surprise that Popper valued serious attempts to falsify all proposed scientific hypotheses because it is only in this way that the better theories can be selected and the poorer ones eliminated. For this reason, Popper believed it is important that scientific theories be given an "objective" existence in the form of spoken or (even better) written words and other symbols that can be disseminated widely to other scientists for scrutiny, for example, in the form of conference presentations and publications.[42] The critical worldwide attention any important new scientific theory now receives makes it increasingly difficult for inadequate theories to survive for very long, as shown by the animated flurry of research and refutation that followed the announcement of cold nuclear fusion by two University of Utah physicists in 1989. Indeed, it will be argued later that the increasingly strong global selection pressure put on theories is an important factor in science's rapid progress, and that considerably less global selection pressure has been put on tradition and technology.

It should not be surprising that a philosophy of science as radically different as Popper's has attracted considerable criticism.[43] But whereas some philosophers continue their search for a completely reliable scientific method based on the foolproof induction of general scientific laws from observation and experiment, modern mainstream philosophy of science has joined Popper at least insofar as rejecting an empiricist, transmission perspective based on justified induction, and taking instead a much more cautious, probing, fallible, tentative, and often evolutionary view.[44]

Tradition, Technology, and Science as Forms of Adaptive Evolution

Genes, Memes, Replicators, and Interactors

We have now seen that current attempts to account for the adapted complexity of culture—whether it be in the form of traditional beliefs and behaviors, technology, or science—have increasingly turned away from providential and instructionist theories, toward evolutionary, selectionist ones. But what are the actual mechanisms by which culture evolves? Unfortunately, what we know of the details of cultural evolution does not begin to compare with what we know of the details of biological evolution.

Darwin was in much the same position when he proposed the theory of natural selection. He could see the competition among organisms and the merciless hammer of nature eliminating the less fit organisms unable to survive and reproduce in sufficient numbers. However, he had no adequate theory of inheritance, and knew nothing of genes, their molecular structure, and how their mutation and recombination provide the blind, probing variation on which selection operates. We know now that genes play the role of replicators, since they make copies of themselves that are handed down from parent to offspring, and that interact with the offspring's environment in determining the form and behavior of the individual. But it is the fit of the organism's form and behavior, with respect to its interaction with its inanimate and animate environment, that determines whether any particular organism will be successful in surviving to maturity and reproducing. In this sense, organisms or groups of organisms can be considered to be the *interactors* on the stage of evolution, and it is at the level of interactors that selection takes place.

The distinction between replicator (genotype) and interactor (phenotype) is important as it helps to clarify certain long-standing problems and controversies concerning the units of selection in biological evolution.[45] Similarly, the distinction may be of use in understanding cultural evolution as well. But then we must ask, what are the replicators in cultural evolution? For cultural evolution we require an entity that is analogous to the gene in biological evolution. This replicating entity is now referred to as the *meme*.[46]

The meme is a unit of cultural replication. As originally introduced by Dawkins:

Examples of memes are tunes, ideas, catch-phrases, clothes fashions, ways of making pots or building arches. Just as genes propagate themselves in the gene pool by leaping from body to body via sperms or eggs, so memes propagate themselves in the meme pool by leaping from brain to brain via a process which, in the broad sense, can be called imitation. If a scientist hears, or reads about, a good idea, he passes it on to his colleagues and students. He mentions it in his articles and his lectures. If the idea catches on, it can be said to propagate itself, spreading from brain to brain.[47]

Admittedly, the meme is not quite the tidy conceptual entity that the gene is. Whereas genes consist of sequences of four nucleotide bases comfortably nestled in spiraling molecules of DNA, memes can take on many different forms, from articles of clothing and spoken words to written pages and computer programs. These various types of memes can all replicate, and at a rate that leaves biological evolution far behind. A new word or phrase used by a character in a popular movie can result in millions of copies worldwide within a few weeks. The idea behind a new technology can also spread quickly in a short amount of time, which is why patents were invented to protect the rights of inventors. And a new and better scientific theory published in a prestigious scientific journal will quickly replicate in the minds of scientists and students throughout the world.

But memes, like genes, do not determine on their own whether and how quickly they will replicate. For this they require interactors who, in cultural evolution, are human agents who in interaction with their environments determine which memes are to be selected and, consequently, replicated. Any achieved cultural fit must be due to the ability of the interactor-environment interface to eliminate less-fit memes. If Einstein's theory of relativity replaced Newton's theory of mechanics because it is an improvement over Newton's, something in the interaction of scientists with their environment must have led to the elimination of the Newton meme and the propagation of the Einstein meme. Indeed, if culture is to achieve progressively better fit with its environment, this environment must somehow participate in selecting the interactors resulting in the subsequent differential replication of the interactors' memes.

Tradition, Technology, and Science: Similarities and Differences

Participation of the environment in the selection of memes would appear to be necessary for all forms of cultural fit, but there are important differences in how this participation takes place. Adapted traditional beliefs and behaviors, as seen in rice planting in Bali, take a very long time to develop and change. That is because the selection process does not depend on individuals as interactors, but rather on a much larger social group. Indeed, as already mentioned, individuals often do not even have the option of intentionally changing traditional beliefs and practices, since to do so might result in banishment from the group or perhaps even death. So the selection process for tradition operates very slowly, at least when compared with other forms of cultural evolution.

The situation is quite different for technological development. In experimenting with new techniques for producing goods and services, individual artisans and inventors are able to test the worth of various memes and select the ones that are best suited to the task at hand. So technological change based on individuals as interactors can occur much more quickly than traditional change based on societies as interactors. To the extent that technology evolves means of vicarious variation and selection (as in the Wright brothers' use of wind tunnels and scale models to test airplane designs), adaptive technological change can occur at even faster rates.

The same is true for science. But scientific development differs from technological development in another important way. In technological innovation, the individuals who come up with promising innovations are most often the same people who test them. It was the Wright brothers and they alone who tested all the variations in wings and propellers they imagined might be successful. But in science today, the individuals involved in testing new theories are not usually restricted to the originators of theories. Important new scientific theories are typically published in widely read scientific journals, and such publication requires stringent peer review by fellow scientists. The more abstract nature of science and the easily communicated mathematical and natural language in which its memes are expressed makes it possible for scientific memes to proliferate quickly to others worldwide. This puts enormous selection pressure on the memes since large numbers of scientists will be eager to falsify any new theory that receives widespread attention, and many of them may well propose theories of their own. Of course, exciting new technological innovations are also eventually tried and tested by individuals not involved in their creation, and much scientific testing requires new technology as well. But we can nonetheless expect a new scientific theory such as $E = MC^3$ published in a prestigious physics journal to undergo more rapid and more widespread attempts at falsification than a new design for automobile tires.

Although some important differences exist in the evolution of tradition, technology, and science with respect to the selection process, in another respect they are very much alike in that the new memes that are tested arise by the blind variation of previously existing memes. When humans first attempted to understand the invisible causes of visible events such as lightning or disease, they conjured up myriad gods, spirits, and other mysterious entities and forces, many of which remain with us to this day. When a physicist or biologist attempts to understand the same phenomena, entities and forces no less mysterious appear in the form of electromagnetic fields, unimaginably small viruses, and still tinier subatomic particles. At this point there is no fundamental difference between superstition and science. The difference lies in the selection procedure and in the fact that scientific memes must be in principle falsifiable in such a way that the phenomena to which they refer participate in the selection. We cannot falsify the belief that evil spirits cause a particular illness, since the spirits are by definition undetectable and make their presence known only through the disease. We can, however, reject the hypothesis that a particular virus or bacterium causes a certain disease. But although science can and does make significant progress in the cumulative introduction and winnowing of new memes, it can never lead us to the positive, infallible knowledge scientists continually seek.

The Myth of Cultural Transmission

Although attempts to understand adaptive cultural change as a selection process analogous in important ways to adaptive biological evolution have become increasingly common, it should not be overlooked that this approach has many critics. A discussion of the criticisms will be saved for chapter 15, but one of the most cited differences between biological and cultural evolution will be addressed here—the claim that biological evolution is Darwinian and cultural evolution is Lamarckian.

By this it is meant that biological evolution is a process of blind variation of genes and subsequent selection of organisms containing these genes by the environment, whereas cultural evolution proceeds by the transmission of acquired characteristics. On the surface, this may indeed appear to be the case. We currently know of no way by which any adaptive changes to an organism's form or behavior acquired during its lifetime can be encoded in its genes and consequently passed on to its offspring. As Weismann claimed (and molecular biology has not yet been able to refute), changes in genes can result in changes in the structure and behavior of organisms, but acquired changes in an organism's structure and behavior resulting from use, disuse, or learning cannot be encoded in the genes. In other words, information is not transmitted from environment to genome. The role of the environment is solely one of selection, not instruction or transmission.

In contrast, it is claimed that cultural evolution is Lamarckian in the sense that the cultural knowledge that one person acquires during a lifetime can be transmitted to another. If I discover a better way of growing potatoes, I can show you how it is done and then you can use the new method as well. My newly acquired potato-growing practices can be inherited culturally by my children, as well as by other individuals with whom I come in contact or who learn about my new, adapted memes through books, word of mouth, or other indirect means, and thus this process may initially seem Lamarckian in nature. In fact, many who have made the most valuable contributions in developing evolutionary models of cultural change speak of cultural transmission in this sense,[48] and Dawkins was quoted earlier describing how "memes propagate themselves . . . by leaping from brain to brain."

But to understand how cultural change might be Lamarckian, we must consider carefully just what is meant by transmission and what is transmitted. It should first be pointed out that a selectionist theory of adapted complexity does not rule out all forms of transmission. Indeed, one can certainly consider the replication of genes that is required for reproduction to be genetic transmission. This can be seen most clearly in asexual organisms whose offspring are almost always genetically identical to the parent. Thus, the parent's genetic code is transmitted to the offspring.[49] But what is essential to keep in mind is that transmission is not responsible for the fit of a genome to its environment. This fit comes about through Darwinian selection, not Lamarckian transmission or instruction. If we are interested in explaining puzzles of fit, we are primarily interested in how the fit comes about, and only secondarily interested in how it is replicated and propagated once it is achieved. So genetic transmission plays an indispensable role in biological evolution, but its role is a secondary one of preserving and propagating the fit *that has already been achieved through selection.*

Having made the distinction between the achievement of fit and the propagation or replication of fit, we can better assess the role that Lamarckian instructive transmission might play in cultural evolution. With respect to achieving the initial fit of cultural knowledge to environment, modern postpositivist philosophy such as Popper's, as well as much current psychological theory, convincingly maintains that this fit cannot be the result of instruction from the environment to the individual. If there is a fit of knowledge to the environment, the role of the environment is one of selecting from among the various and sundry knowledge memes that already were created by the individual. So like biological evolution, an evolutionary view of adaptive cultural development depends on selection for the achievement of fit, and not transmission or instruction from world to mind by way of the senses of the type that Bacon and other empirical positivists believed possible.

But what about the propagation of knowledge to other individuals? Do not transmission and instruction play a role here comparable to the transmission of genetic information from parent to offspring? A little reflection suggests that the answer is no. The only way that John could possibly obtain knowledge from Mary is through his senses. Mary cannot transmit her genes to John, nor can she give him her brain or parts of it, or replicate the patterns of its structure in John's brain. Like the rest of the physical world, Mary is simply part of John's environment, knowable to him only through his senses. This does not mean that John cannot learn from Mary and her experiences. But as we saw in chapter 7, learning appears to be no more possible through instruction than is the Lamarckian inheritance of acquired characteristics. Accordingly, we will see in the next two chapters how current theories of language and education also reject the notion of the instructionist transmission of knowledge from one individual to another (as from teacher to student), and instead view linguistic communication and education as processes dependent on cumulative blind variation and selection.

11

The Evolution, Acquisition, and Use of Language

It [language] certainly is not a true instinct, for every language has to be learned. It differs, however, widely from all ordinary arts, for man has an instinctive tendency to speak, as we see in the babble of our young children; while no child has an instinctive tendency to brew, bake, or write. Moreover, no philologist now supposes that any language has been deliberately invented; it has been slowly and unconsciously developed by many steps.
—Charles Darwin[1]

Of all the behaviors in which humans engage, probably none is so complex and yet commonplace as speaking and listening to language. Indeed, it is rather unusual to observe a gathering of two or more conscious humans who are not involved in the continuous use of language.

It is not difficult to think of many ways in which language is an important tool for the human species. Perhaps most important, language helps us to accomplish things that no individual could achieve alone. Organized hunting, warfare, communal agriculture, and the construction of dams, canals, roadways, buildings, and transport vehicles all depend on language to coordinate the activities of the individuals involved. Language allows people to share their experiences, successes, and failures with others, making it possible for knowledge to be shared among the members of a community. Language in its more permanent written form, first on stone, then paper, and now increasingly on computer disks, makes it possible for us to understand something about those who lived in other places and times. And language, both spoken and written, appears essential for the development of science and technology in that it allows individuals to make their ideas and theories public and thus subject to both the skepticism and further development of their peers.

We may normally first think of language as a tool for communication among individuals, but it also appears to be used silently as a medium of thought. Although its specific role in facilitating and shaping our thought processes remains the subject of lively debate among scholars, the subjective experience of being human suggests that language plays an important role in human cognition and consciousness.[2]

Human language is also remarkable for how it is put together, that is, its grammar. The structural complexity of all 5000 to 6000 extant languages is such that despite centuries of analysis, linguists have yet to come up with a complete and accurate grammatical description of any one of them. Linguists have been successful in discovering important regularities in many languages, and have also shown how different languages that may appear quite different on the surface share underlying similarities. Nonetheless, certain structural aspects of all languages have so far eluded formal understanding. Since we are so intimately familiar with our own native tongue we seldom consider this complexity, yet it becomes evident to anyone who has attempted to learn a foreign language as an adult.

This complexity would be just a curiosity, rather than a puzzle of fit demanding explanation, if it did not contribute to the usefulness and expressive power of language. We all experience occasional difficulty in putting our thoughts into words, but all languages, from the cautious, conservative French of the academicians to the ever-changing inner-city English of African-Americans, provide for a wide range of expressive possibilities, and we are seldom unable to describe or comment on what we consider to be an important aspect of our physical or social world. We have nouns to refer to both physical objects such as cars and horses as well as to abstract concepts such as justice and love. We have verbs such as *walk* to refer to actions, others such as *believe* to refer to the states of organisms and objects, and still others such as *sit* which can refer to either states or actions. We have adjectives such as *red* and adverbs such as *quickly* to modify nouns and adjectives. We have grammatical devices such as word endings and word orders to signal the relationships among nouns and verbs. We have tense and mood systems to specify further the time and manner in which events did take, will take, or would take place. We can use language deceptively to describe events that did not or will not take place, and we can even use it imaginatively and hypothetically to describe things that might exist, will exist, or that we would like to exist.

We also have to consider the human physiological characteristics that make language possible, and how language fits so well the physical and social environment in which we use it. The human vocal tract is unique among all animals in allowing the production of a rich variety of sounds that are used for speaking. This is accomplished by producing a continuous stream of sounds (most originating in the vocal cords) that is modified by actions of the tongue, lips, and teeth. Large portions of the brain are involved in producing and receiving speech, and newborn infants appear to be "prewired" for the categorical perception of speech sounds. The use of sound as a medium of language is well matched to the ocean of air around us. Speech sounds travel quickly through air and, unlike light, can penetrate walls and turn corners. Speaking, rather than gesturing manually, keeps the hands free for other tasks and is effective in both the light of day and dark of night.[3]

Human language thus appears to be very well designed for communicating with others and controlling our physical and social environments. Thus it provides another striking example of a puzzle of fit. We must consider at least three aspects of language. First we have its origin and evolution as a characteristic of the our species. Second is the puzzle of how it is possible that all normal children, who take so many years to develop fully their cognitive, physical, and social skills, are able to learn the language of their community with amazing speed and apparent ease. Finally, we must look into just how language allows us to communicate our thoughts, questions, requests, and desires to others.

The Origin and Evolution of Language

Attempting to reconstruct the evolution of human language is fraught with difficulties. Whereas the physical remains of our ancestors may endure many millions of years, unfortunately we have no records of early speech since "language leaves no bones."[4] However, this has not prevented many scholars from proposing accounts of the origin of language and how it evolved into the thousands of tongues that are spoken today. "For instance, it has been argued that language arose from mimicry of animal calls, imitations of physical sounds, or grunts of exertion—the infamous 'bow-wow,' 'ding-dong,' and 'heave-ho' theories."[5] Such unfounded speculations became so rampant in nineteenth-century Europe that in 1866 the Société de

Linguistique de Paris banned the topic altogether from its meetings and publications.

But although evolutionary evidence of a purely linguistic nature does not exist, recent research on hominid fossils and studies of the modern human brain and vocal tract are beginning to shed some light on the evolution of language that has begun to move us beyond the realm of pure speculation.

The Biological Prerequisites of Human Speech

Since speaking requires the production of a continuously and finely varied stream of sound, we can gain some knowledge concerning the evolution of language by studying the evolution of the vocal tract and comparing it with corresponding systems in other mammals. The human system is different from that of all other terrestrial mammals in one striking way. Darwin himself noted "the strange fact that every particle of food and drink which we swallow has to pass over the orifice of the trachea, with some risk of falling into the lungs."[6] So unlike mammals that maintain separate pathways for breathing and feeding, thus enabling them to breathe and drink at the same time, adult humans are at a much higher risk for having food enter their respiratory systems; indeed, many thousands die each year from choking.[7] In addition, our relatively short palate and lower jaw are less efficient for chewing than those of nonhuman primates and our early human-like ancestors, and provide less space for teeth.[8]

But if the design of the human throat and mouth is far from optimal for eating and breathing, it is superbly suited for producing speech sounds. All mammals produce oral sounds by passing air from the lungs through the vocal cords, which are housed in the larynx (or "Adam's apple"). The risk of choking to which we are exposed results from our larynx being located quite low in the throat. This low position permits us to use the large cavity above the larynx formed by the throat and mouth (supralaryngal tract) as a sound filter. By varying the position of the tongue and lips, we can vary the frequencies that are filtered and thus produce different vowel sounds such as the [i] of *seat*, the [u] of *stupid*, and the [a] of *mama*.[9] We thus see an interesting trade-off in the evolution of the throat and mouth, with safety and efficiency in eating and breathing sacrificed to a significant extent for the sake of speaking. This suggests that the evolution of language must have provided advantages for survival and reproduction that more than offset these other disadvantages. We will save for a bit later discussion of what these advantages might have been.

The importance of the evolution of the human vocal tract to fit the functions of speech are suggested by studies of Neanderthal man, who lived in Europe and the Middle East about 100,000 years ago. To find out what types of speech sounds Neanderthals might have been able to produce, Philip Lieberman, of Brown University in Rhode Island, and his associates reconstructed the vocal tract of these hominids based on fossil evidence and applied computer-modeling techniques to it. They concluded that Neanderthals had a relatively high larynx and relatively flat tongue, and could therefore have produced only a limited number of nasal vowel sounds. Most significant, they would have been unable to produce the sounds [i], [u], and [a], the three quantal vowel sounds that are most easily distinguished by human listeners. Thus Lieberman concludes that even "if Neanderthal hominids had had the full perceptual ability of modern human beings, their speech communications, at minimum, would have had an error rate of 30% higher than ours."[10]

Also existing in Europe at the same time were the Cro-Magnons, a separate hominid species who appear to have been slightly taller but with lighter bones and less powerful muscles than the Neanderthals. One could easily imagine that the Neanderthals' superior strength would have been an advantage for hunting and for any competitive encounters with their Cro-Magnon "cousins." But the Cro-Magnons appeared to have one important advantage in their favor—a modern vocal tract capable of producing all the sounds of human speech. It is therefore tempting to speculate that both the disappearance of Neanderthals about 35,000 years ago and the survival and continued evolution of Cro-Magnons into what we are today was due at least in part to the superior linguistic ability of our ancestors.[11]

The study of the evolution of the our vocal tract also provides hints concerning the evolution of our brain. Obviously, the throat and mouth would not have evolved the way they did to facilitate language production and comprehension while compromising eating and respiration if the brain had not been capable of producing and comprehending language.

Lieberman has proposed three major stages in the evolution of the neural bases for language. First was *lateralization* of the brain, meaning that each half became specialized for different functions. For most of us, the left hemisphere provides most of the neural circuitry required for language production. For about 90% it also controls the dominant (right) hand used for tasks involving fine motor control, suggesting that lateralization may have originally evolved in response to selection pressure for skilled hand movements. As neurologist Doreen Kimura noted,

[making and using tools] requires the asymmetric use of the two arms, and in modern man, this asymmetry is systematic. One hand, usually the left, acts as the stable balancing hand; the other, the right, acts as the moving hand in such acts as chopping, for example. When only one hand is needed, it is generally the right that is used. It seems not too farfetched to suppose that cerebral asymmetry of function developed in conjunction with the asymmetric activity of the two limbs during tool use, the left hemisphere, for reasons uncertain, becoming the hemisphere specialized for precise limb positioning. When a gestural system [for language] was employed, therefore it would presumably also be controlled primarily from the left hemisphere. If speech were indeed a later development, it would be reasonable to suppose that it would also come under the direction of the hemisphere already well developed for precise motor control.[12]

That brain lateralization had prelinguistic origins is supported by recent findings of handedness and lateralization among nonhuman primates.[13]

The second component of language evolution involved the evolution of brain structures responsible for the voluntary, intentional control of speech. Although we usually take the voluntary and intentional nature of language for granted, it is of interest to contrast human use of language with the communication systems of other animals. Of particular interest are the chimpanzee observations of Jane Goodall:

Chimpanzee vocalizations are closely tied to emotion. The production of a sound in the *absence* of the appropriate emotional state seems to be an almost impossible task for a chimpanzee. . . . A chimpanzee can learn to suppress calls in situations when the production of sounds might, by drawing attention to the signaler, place him in an unpleasant or dangerous position, but even this is not easy. On one occasion when Figan was an adolescent, he waited in camp until the senior male had left and we were able to give him some bananas (he had none before). His excited food calls quickly brought the big males racing back and Figan lost his fruit. A few days later he waited behind again, and once more received his bananas. He made no loud sounds, but the calls could be heard deep in his throat, almost causing him to gag.[14]

Figan's difficulty in concealing news of the bananas from his associates contrasts sharply with the ease with which humans can use language to deceive and manipulate others, to talk of the past, and to plan for the future. Lieberman attributes our control over language to certain changes in the brain, including the evolution of what is referred to as Broca's area, as well as the enlargement of the prefrontal cortex (the part of the brain just behind the forehead) and a rewiring of concentrations of neurons referred to as the basal ganglia.

The third component in the evolution of human language involved the ability to put sounds and words in specific orders and to perceive these orders as meaningful. In all languages, the order in which words and parts of words are produced and perceived is crucial to the meaning. The sentence *Mary saw John* conveys a different meaning from *John saw Mary*. Of all the communication systems used by the earth's animals, it appears that only human language derives its expressive power from the recombination of a finite (though large) number of words and word parts into an infinite number of different sequential orders. The ways in which words may be ordered and how these different orders relate to meaning is syntax.

Lieberman suggests that the evolution of motor control for speech itself provided the basis for the development of syntax. This is because articulating even a simple word such as *cat* requires the precisely timed sequential coordination of movements of the tongue, lips, and jaws. The order of sounds within words also makes a difference as to their meaning, with *cat* different from *tack*. Lieberman thus concludes that "speech motor control is the preadaptive basis, that is, the starting point [for syntax]. Once syntax became a factor in human communication, the selective advantages that it confers . . . would have set the stage for natural selection that specifically enhanced these abilities—independently of speech motor control."[15]

The importance of language and the advantages it provides us in communicating, coordinating our activities, and thinking suggests that, in addition to being a product of our evolution, it also played a large part in shaping our evolution, particularly that of our brain. Lieberman has consequently

. . . propose[d] that natural selection to enhance faster and more reliable communication is responsible for the second stage of the evolution of these mechanisms—the evolution of the modern human brain. Communication places the heaviest functional load on "circuitry" for both electronic devices and brains. The transistors and solid-state devices that made digital computers a useful tool were first developed for communication systems. Indeed one can argue that the demands of communication preempt the highest levels of technology and organization of a culture, whether couriers on horses or lasers and fiber-optic bundles are the means employed. In short, evolution for efficient, rapid communication resulted in a brain that has extremely efficient information-processing devices that enhance our ability to use syntax. These brain mechanisms also may be the key to human cognitive ability. As many scholars have noted, human language is creative; its rule-governed syntax and morphology allow us to express "new" sentences that describe novel situations or convey novel thoughts. The key to enhanced cognitive ability likewise seems to be our ability to apply prior knowledge and "rules" or principles to new problems.[16]

Given the progress that has been made in understanding the evolution of language, together with modern biology's acceptance of natural selection as the explanation for the appearance of design in the structure and behavior of organisms, it might come as somewhat of a surprise to learn that some scholars reject natural selection as an explanation for the appearance, structure, and use of language. What is particularly noteworthy about these critics of natural selection is that some of them are widely recognized as leading thinkers and researchers in their respective fields, among them Noam Chomsky and Stephen Jay Gould.

Chomsky not only rejects natural selection as an explanation for the evolution of human language, but also rejects Darwinian explanations for certain well-understood biological phenomena. He has stated that "evolutionary theory appears to have very little to say about speciation, or about any kind of innovation. It can explain how you get a different distribution of qualities that are already present, but it does not say much about how new qualities can emerge."[17] This is a curious statement, given that Darwin proposed natural selection to account for speciation specifically and that the essence of his theory is still accepted today by mainstream biologists as the sole explanation for adaptive evolution and speciation. As for evolution "not say[ing] much about how new qualities can emerge," those unconvinced by the biological evidence will see in part IV how artificial evolution is being applied increasingly to create useful, innovative products from drugs to computer programs. An attempt to understand better Chomsky's rationale for rejecting a selectionist account of language will be offered later in this chapter, where we take a look at how children learn the language of their community.

Gould's reluctance to accept a selectionist account of human language stems from his more general concern that adaptationist explanations of biological traits are often misapplied. In an important paper written with Lewontin,[18] they proposed that certain biological traits may not be solely due to adaptive natural selection, but rather may have their origins as side effects or by-products of other evolutionary changes, which are then "seized" at a certain time by a new function. We referred to this phenomenon of exaptation (originally called *preadaptation* by Darwin) in addressing the evolution of the human brain in chapter 5. The exaptationist perspective proposes that the ability to communicate by language had little or no role in the changes that made the brain capable of language. "The brain, in becoming large for whatever adaptive reasons, acquired a plethora of cooptable features. Why shouldn't the capacity for language be among them? Why not seize this possibility as something discrete at some later point in evolution, grafting upon it a range of conceptual capacities that achieve different expression in other species (and in our ancestry)?"[19]

Perhaps we can better understand the exaptationist view of language by leaving biology for a short time and considering an example of technological change.[20] The very first cameras had chemically treated plates of glass that captured the image projected by the lens. In attempts to make photography less costly and more convenient, celluloid sheets were introduced, followed by rolls of celluloid film. But whereas long rolls of flexible film were intended solely to facilitate still photography, they made it possible also to take many pictures in quick succession, thus leading to the development of the motion picture camera. We therefore cannot say that roll film evolved to make the motion picture camera possible, since the idea of a motion picture camera probably did not even exist when the first rolls of film were produced. Instead, to use Darwin's term, the roll film of the still camera was preadapted, although quite accidentally and unintentionally, for use in the motion picture camera. To use Gould's more neutral and more accurate terminology, this feature of the still camera was exapted for use in motion picture cameras. So, in effect, Chomsky and Gould assert that the human brain is analogous to roll film in that it evolved for reasons originally unrelated to language concerns; but once it reached a certain level of size and complexity, language was possible.

Exaptation is an important conceptual tool in understanding the evolution of biological structures and behaviors, but it alone cannot account for the continued evolution of adapted complexity. Although roll film made the first motion picture cameras possible, modern movie cameras and the film they use are more complex and better adapted to the production of high-quality movies than the very first movie cameras. And whereas some of these additional technological developments may have also been exapted from other fields, for example, developments in electronics and chemistry making possible more accurate light meters and more sensitive film, the way in which all the component parts of a modern movie camera work together can be understood only as resulting from selection processes operating at the level of the entire camera, not just its component parts.

Similarly, certain preexisting structures and functions of the human brain and vocal tract may have been taken over (or exapted) for use in language. However, this cannot by itself account for the ways in which the brain, the vocal tract, and language fit together to create a total system that is quite remarkably adapted to serve the functions for which language is used. The fact remains that the process of cumulative blind variation and selection is the only process currently understood that can account for the nonprovidential appearance of the adaptive complexity that is seen in the design of language, and the design of the human brain and vocal tract for language. As Pinker and Bloom point out in their important discussion of the role of natural selection in the evolution of language, "language shows signs of complex design for the communication of propositional structures, and the only explanation for the origin of organs with complex design is the process of natural selection."[21]

The Evolution of Language

But what about the sounds, structures, and rules that make up language? How did they originate and evolve over time, leading to the languages spoken throughout the world today? As already noted, we unfortunately have no records of how language was used by our prehistoric ancestors. Nonetheless, our current knowledge of evolution provides at least a general scenario of how it evolved. As Pinker and Bloom observed, for language to have evolved by natural selection:

There must have been genetic variation among individuals in their grammatical competence. There must have been a series of steps leading from no language at all to language as we now find it, each step small enough to have been produced by a random mutation or recombination, and each intermediate grammar useful to its possessor. Every detail of grammatical competence that we wish to ascribe to selection must have conferred a reproductive advantage on its speakers, and this advantage must be large enough to have become fixed in the ancestral population. And there must be enough evolutionary time and genomic space separating our species from nonlinguistic primate ancestors.[22]

But since there are so many conceivable ways in which language could have conferred a "reproductive advantage on its speakers" and so few conclusive data on this subject, we can only speculate on which ones actually were important. We already mentioned the use of language to coordinate human activity, and it is not difficult to imagine how the ability to plan and coordinate hunting, agricultural, and warfare activities would have conferred survival advantages to individuals and groups with language skills. Also, as mentioned earlier, language makes it possible for individuals to share knowledge, thereby avoiding the mistakes and errors that others have already made.

We cannot go back in time to see how language was used by early humans, but we can learn from the behavior of communities of hunter-gatherers who still live today in much the same way as all our ancestors did as recently as 12,000 years ago when the total human population of the earth was only about 10 million.[23] Among such groups are the !Kung of the Kalahari Desert in Namibia and Botswana who use language to discuss

everything from the location of food sources to the behavior of predators to the movements of migratory game. Not only stories, but great stores of knowledge are exchanged around the fire among the !Kung and the dramatizations—perhaps best of all—bear knowledge critical to survival. A way of life that is difficult enough would, without such knowledge, become simply impossible.[24]

Although sharing knowledge of the location of food would certainly seem to be a function of language providing important survival advantages, survival in itself cannot ensure that an individual's genes and the language abilities that go with them will be inherited. Inheritance requires reproduction, and human reproduction, as does all sexual reproduction, requires a partner of the opposite sex. It should not be surprising, therefore, to find that language plays an important role in sexual selection. "Just as female birds seem to have favored elaborate songs by males (not to mention long and shiny feathers) when choosing a mate, so prehuman females might have promoted a fancier form of language"[25] by preferring men with more impressive language skills.

Of course, we do not know, and will probably never know, the actual events and selection pressures that gradually transformed the hoots, grunts, and cries of our ancestors into our current remarkable vocal communication system. But two things are known: first, language is a highly complex and adaptive tool without which human life as we know it would be impossible; and second, at some time in the past it did not exist. And regardless of arguments that natural selection is not quite up to the task of explaining the emergence and refinement of language, the fact remains that Darwinian evolution is the only currently available, nonmiraculous explanation for the appearance of our most remarkable and useful ability, an ability that some believe may be responsible for a complete experience of human consciousness.[26]

The Child's Acquisition of Language

Fortunately, the child does not have to be concerned about the evolution of language, the brain, and the vocal organs that make language possible. These already exist in the community and in the biological structures found above her shoulders. The child's task is therefore "simply" to apply her biological endowment, provided by natural selection, to learn the language she hears spoken by her parents and community, whether it be Estonian, Eskimo, or English. We should not expect this to be too much of an ordeal, since after countless centuries of language use, natural selection should have provided a good fit between the child's abilities and the requirements of the task.

Language as Learned

The acquisition of language may not initially appear to be very different from the other things that children learn. It may seem unremarkable that an American child who hears countless hours of language spoken to and around her will eventually begin to produce the same sounds, words, and grammatical structures. Parents also provide considerable encouragement for their children to speak, as is evident in the smiles and hugs that typically follow the first utterance of "mama" or "dada." Children learn to do many things—put on their clothes, drink from cups, open and close doors, and even operate the television set and VCR—apparently from observing and imitating the actions they see others performing and from being reinforced by the satisfaction of the consequences of their actions. Why should language acquisition be any different?

This is essentially what was proposed by B. F. Skinner, introduced in our discussion of learning in chapter 7. It will be recalled that Skinner's theory of learning attempted to explain the acquisition of any new behavior as a process of operant conditioning by which new behaviors (such as a rat pushing a bar) would be learned to the extent that they were reinforced in some way by the environment (such as receiving food). He extended his behaviorist view of learning to human language.

> In all verbal behavior under stimulus control there are three important events to be taken into account: a stimulus, a response, and a reinforcement. These are contingent upon each other, as we have seen, in the following way: the stimulus, acting prior to the emission of the response, sets the occasion upon which the response is likely to be reinforced. Under this contingency, through a process of operant discrimination, the stimulus becomes the occasion upon which the response is likely to be emitted.[27]

But he recognized at the outset that the social use of language is in one important respect quite unlike nonverbal behaviors operating on an inanimate environment.

> When a man walks toward an object, he usually finds himself closer to it; if he reaches for it, physical contact is likely to follow; and if he grasps and lifts it, or pushes or pulls it, the object frequently changes position in appropriate directions. All this follows from simple geometrical and mechanical principles. . . . However, when we use language to act upon the world, as when we ask another for some water, the glass of water reaches the speaker only as the result of a complex series of events including the behavior of the listener. . . . Indeed, it is characteristic of such [verbal] behavior that it is impotent against the physical world.[28]

Consequently, Skinner determined that although the links among stimulus, response, and reinforcement may be less obvious and more indirect for verbal behavior, they nonetheless exist and can be used to explain language learning and use. For example, he noted that in a given language community certain verbal behaviors such as "Wait!" and "Sh-h!" are typically followed by certain consequences, such as someone waiting or being quiet. Such a result depends, of course, on the cooperation and behavior of the other person. But if the consequence is achieved, that particular verbal response will be strengthened and be more likely to occur in a similar instance in the future. Thus, like all other behaviors, language learning is completely dependent on contingencies of reinforcement.

In essence, Skinner's analysis of verbal behavior is an attempt to show how language is shaped by the environment in the same way that a rat's lever pushing or pigeon's key pecking can be controlled by providing and withholding food. By giving reinforcement for the sounds, words, and sentences the child produces that approximate the adult form of the language, and by withholding such reinforcement when an utterance is in some way deviant, the child's verbal behavior is gradually shaped over time to approximate the language of the community. Skinner argued that such contingencies of reinforcement are not only responsible for the child's learning language, but are the determining factors for all behavior, including adults' language behavior.

In insisting that reinforcement is the key to understanding language behavior, Skinner had to stretch the concept of reinforcement to cover situations that are quite unlike those found in studies of animal learning. For instance, he claimed that many verbal behaviors are "automatically self-reinforcing," as when "the child is reinforced automatically when he duplicates the sounds of airplanes, streetcars, automobiles, vacuum cleaners, birds, dogs, cats, and so on."[29] And "the young child alone in the nursery may automatically reinforce his own exploratory verbal behavior when he produces sounds which he has heard in the speech of others."[30] He stretched the concept of reinforcement to situations where the person producing language is not even present when the reinforcement takes place, as when a public speaker or writer is reinforced by "the fact that effects of verbal behavior may be multiplied by exposing many ears to the same sound waves or many eyes to the same page."[31]

Chomsky, in his influential and widely cited review[32] of Skinner's book, underscored these and many other problems with an operant conditioning analysis of human behavior, including language. Concerning the role of reinforcement, after having first cited the above and other examples, Chomsky concluded:

From this sample, it can be seen that the notion of reinforcement has totally lost whatever objective meaning it may ever have had. Running through these examples, we see that a person can be reinforced though he emits no response at all, and that the reinforcing *stimulus* need not impinge on the *reinforced person* or need not even exist (it is sufficient that it be imagined or hoped for). When we read that a person plays what music he likes, says what he likes, thinks what he likes, reads what books he likes, etc. BECAUSE he finds it reinforcing to do so, or that we write books or inform others of facts BECAUSE we are reinforced by what we hope will be the ultimate behavior of reader or listener, we can only conclude that the term *reinforcement* has a purely ritual function. The phrase "X is reinforced by Y" . . . is being used as a cover term for "*X* wants *Y*," "*X* likes *Y*," "*X* wishes that *Y* were the case," etc. Invoking the term reinforcement has no explanatory force, and any idea that this paraphrase introduces any new clarity or objectivity into the description of wishing, liking, etc., is a serious delusion.[33]

It is interesting to note that Chomsky's analysis is not inconsistent with perceptual control theory discussed in chapter 8, as he implied that people behave to satisfy their internal "wishes," "likes," and "wants," and not because certain behaviors were reinforced in the past.

Language as Innately Provided

In his subsequent writings on language acquisition, Chomsky has attacked Skinner's and other learning theories by focusing on the syntactic structure of language and the fact that all normal children show impressive knowledge of this structure despite considerable variation in their exposure to language. Chomsky's insights and his development of what is called generative grammar revolutionized our understanding of language. The particular generative grammars being developed to explain aspects of various languages are probably beyond the grasp of anyone who has not formally studied modern linguistics. The general notion of a generative grammar is fortunately more accessible.

Let us first consider what syntax is and why it is necessary for language. Due to the nature of the human vocal tract, we do not normally produce (or perceive) more than one speech sound at a time. This makes oral language a *serial* medium, meaning that sounds are strung together one after another like beads on a string. As already noted, the order in which sounds are uttered to form words (compare *pot* with *top*) and words are uttered to form sentences (compare *The dog ate the pig* with *The pig ate the dog*) are related to the meaning of a sentence. All languages differ in the degree to which word order is crucial for understanding, and English is particularly choosy, with most orders being meaningless, or nearly so. As linguist Derek Bickerton observed:

Try to rearrange any ordinary sentence consisting of ten words. There are, in principle, exactly 3,628,800 ways in which you could do this, but for the first sentence of this paragraph only one of them gives a correct and meaningful result. That means 3,628,799 of them are ungrammatical. How did we learn this? Certainly, no parent or teacher ever told us. The only way in which we can know it is by possessing, as it were, some recipe for how to construct sentences, a recipe so complex and exhaustive that it automatically rules out all 3,628,799 wrong ways of putting together a ten word sentence and allows only the right one. But since such a recipe must apply to all sentences, not just the example given, that recipe will, for every language, rule out more ungrammatical sentences than there are atoms in the cosmos—and there are at least five thousand different languages![34]

Although many languages are not as strict as English concerning word order, all of them require at least certain orders of basic sounds (called phonemes) to make up words, even if they are more flexible in the permissible orderings of words to form sentences.[35]

Syntax refers to the principles that govern the permissible orderings of words in a language and how these orderings are related to meaning. According to Chomsky, we produce syntactic—properly ordered—sentences not by memorizing a list of words and sentences, and not even by learning the general structural patterns that make up sentences and then using these as frames to create new sentences. Instead, we know quite abstract rules of the language that we use to generate sentences, most of which are novel in that they have neither been heard nor spoken before by the speaker. Let us look at a few simple examples of the generative rules of syntax proposed by Chomsky for English.

(1)   S → NP + VP

(2)   NP → Det + N

(3)   VP → V + NP

Rule (1) states that a sentence (S) may be composed of a noun phrase (NP) plus a verb phrase (VP). Noun phrase is defined in (2) as a determiner (Det) such as *a*, *the*, *this*, or *that* followed by a noun. And (3) defines a possible verb phrase (VP) as a verb (V) followed by another noun phrase. One learns these rules, together with some others including recursive rules that permit the use of a sentence as a noun phrase as in *I saw John throw the ball*. Once one also learns which words fit into which classes, one can produce an unlimited number of sentences, such as *The earthquake destroyed the city*, or *That dog has big ears*. In addition, transformational rules transform a sentence into related sentences, so that *The earthquake destroyed the city* can be transformed into the question *Did the earthquake destroy the city?* or into the passive voice sentence *The city was destroyed by the earthquake*.

The magnitude of the child's achievement in learning language can be appreciated by considering the complexity of these rules. Take, for example, a simple question such as *Did you see my toy?* To produce such a question the child must know that such a yes-no question is formed from the corresponding declarative (nonquestion) sentence by moving the first auxiliary verb before the subject or adding the auxiliary verb *do* to this position if no auxiliary already is present. This is a rather complex rule, and yet it is tacitly known by every English-speaking child who can ask a yes-no question! It is furthermore clear that such rules are not explicitly taught by parents to their children, since few parents could even state them, and it is highly unlikely that a typical three- or four-year-old could comprehend them even if the parents did.

To make the acquisition of syntax even more remarkable, Chomsky says that the examples of language that children hear are inadequate for them to figure out the underlying generative rules on their own. Children hear only a relatively small subset of sentences, and due to lapses of attention or memory, false starts, slips of the tongue, interruptions, and other disturbances, many of the sentences they do hear are not well-formed. Considering also noise from nearby machines, television sets, airplanes, and other sources together with the frequent ear infections that many young children experience, and occasions when a child hears clearly articulated, grammatical sentences become even less frequent. Chomsky has referred to these characteristics as the "poverty of the stimulus," implying that language as heard by a child is not sufficiently clear, accurate, and structured for the child to be able to deduce and learn its underlying generative rules.

It is also clear that children go well beyond what they hear in coming up with grammatical rules. For example, many if not all children living in an English-speaking environment will use words such as *breaked*, *drawed*, *holded*, and *cutted* despite never having heard adults use such words. But although they make certain types of errors, there are many other possible errors which they never seem to make. For example, no child has been heard to say "Kitchen the in is Daddy?" to turn "Daddy is in the kitchen" into a question.

Chomsky, together with many other linguists and child language researchers influenced by his theories, have thus concluded that children do not and could not acquire language by operant conditioning as proposed by Skinner, or indeed by any method of learning as learning is generally understood in psychology today. The fact that all normal children, regardless of intelligence level, are quite proficient speakers of the language to which they are exposed by the age of four years suggests to Chomsky that much of that knowledge is part of the human biological endowment. And although different languages certainly vary in terms of their sounds, words, and grammatical constructions, these differences can be understood as variations on a theme, the theme being characteristics that are shared by all natural languages and that linguists call "universal grammar." For Chomsky, knowledge of universal grammar is as much a part of Baby Sue's biological inheritance as is the nose on her face.

In effect, then, Chomsky believes that since language cannot be acquired by environmental instruction (due primarily to the poverty of the stimulus), to a very large extent it must be *innate*. To be sure, he and like-minded linguists recognize that children must be exposed to a language for it to be acquired, but rather minimal and haphazard exposure is all that is required to trigger its acquisition. In this respect spoken language is very different from other abilities such as mathematics and reading skills, which normally require special experiences for their development, such as formal schooling and prolonged practice. Even then many children and adults have difficulty acquiring these academic skills.

Chomsky is not alone in his innatist view of human cognitive abilites related to language. While he focused on syntactic knowledge, psycholinguist and philosopher Jerry Fodor has maintained for many years that all human conceptual knowledge is innate. Concepts such as TRIANGLE, DOG, and FREEDOM, and therefore the meaning we attach to these words, must be innate since it is impossible for someone to learn a concept that was not in some form already known before. It will be noted that this argument bears an uncanny resemblance to the one made by Plato in the dialogue with Meno discussed in chapter 6. For Fodor, what appears to be learning is actually "fixation of belief,"[36] using experience to select among a host of innate ideas. This may initially seem consistent with a selectionist (and therefore evolutionary and constructive) theory of learning and cognitive development. But Fodor places some severe limitations on what he believes such selection can achieve, as seen in his statement cited in chapter 9:

There literally isn't such a thing as the notion of learning a conceptual system richer than the one that one already has; we simply have no idea of what it would be like to get from a conceptually impoverished to a conceptually richer system by anything like a process of learning.[37]

But if evolution itself can be considered to be a form of learning in which organisms over phylogenetic time acquire knowledge about their environment,[38] it turns out that, despite Fodor, we *do* have a theory of learning, namely, natural selection, that explains how complex, adapted systems such as organisms and components of organisms can emerge from simpler or more "impoverished" ones.

Indeed, the argument that knowledge in the form of new and richer concepts cannot be constructed, but must rather already exist and be innately provided, creates a serious problem for Fodor and others who use this can't-get-there-from-here logic. Obviously, at some time (to be safe, let's say one second after the Big Bang), concepts such as RED, MOTHER, CONSERVATION OF ENERGY, and INTERROGATIVE SENTENCE did not exist, yet today they do exist in the minds of humans. How did they originate? Certainly, biological evolution must have had something to do with it since these concepts are clearly complex and functional characteristics of our species. But if Fodor is right that there is no way "to get from a conceptually impoverished to a conceptually richer system," he must also believe biological evolution to be impossible. Once it is recognized that the evolutionary process of cumulative blind variation and selection has in fact resulted in the emergence of more complex systems from simpler ones, and that an evolutionary process involving the cumulative variation and selection of ideas, thoughts, and concepts could also be an essential and universal part of human cognitive development, Fodor's argument for the impossibility of learning appears seriously flawed.

Mark Bickhard, a cognitive scientist at Lehigh University, made just such a critique, using in the following quotation the word *representation* to refer to Fodor's conceptualization of knowledge:

If representations cannot emerge, however, then they cannot come into being at all. A narrow focus on this point yields Fodor's innatism: neither learning nor development, as currently understood, can construct emergent representation; therefore the basic representational atoms must be already present genetically. Unfortunately, this conclusion does not follow. If representation cannot emerge, then it cannot emerge in evolution any more than it can in development. The problem is logical in nature, and is not specific to the individual. Conversely, if some way were posited in which evolution *could* yield emergent representation, then there is no a priori reason why that emergence would not be just as available in the development of the individual. Fodor's innatism, then, simply misses the basic issue. If representation cannot emerge, then it is impossible for it to exist, and evolution is in no better position in this respect than is individual development; on the other hand, if representation *can* emerge, then there is something wrong with the models of learning and development that cannot account for that emergence. When those models are corrected, that emergence should be as available to the individual as to evolution. In either case, Fodor's strong innatism does not follow.[39]

But isn't it true that things can happen during evolution that cannot happen during the lifetime of an individual human? Human hearts and arms evolved over a very long period of time through among-organism selection. Yet no human can "learn" to grow another heart or arm during a lifetime because such biological structures are determined by genes that do not change during one's lifetime.[40] In contrast, the cognitive abilities underlying language and conceptual knowledge are dependent on the structure of the brain, and the brain is remarkably adaptive ontogenetically, whereas the genome is not.[41] As explained in chapter 5, the brain retains the ability to make adaptive changes through a variation and selection of synapses. Thus it is at least conceivable that new linguistic and conceptual knowledge could emerge as a result of such within-organism selection.

None of this, of course, proves that Chomsky and Fodor are wrong in their assertions that our knowledge of language and concepts is innately determined. But it does argue against their reasoning, and the reasoning of other cognitive and linguistic innatists, that this knowledge must be innate. If the human genome could have acquired such knowledge by way of the among-organism selection of human evolution, then it must be considered at least a possibility that the brain could acquire similar knowledge by way of the within-organism selection of synapses. The implications of within-organism selection for such innatist views of human cognition will be considered again at the end of chapter 15.

Language Acquisition as Selection

Chomsky's and Fodor's views of language acquisition are undeniably very popular among linguists and cognitive scientists today. Nonetheless, noteworthy opposition to their perspective exists, much of it coming from psychologists who take a less linguistic and more functional perspective on language, its acquisition by children, and its use. A number of these individuals have adopted, either explicitly or implicitly, a selectionist view of language learning.

Let us leave syntax for a while and consider what is involved in learning the meanings of individual words. Since children appear to learn new words and their meanings so quickly, it might first appear that it is simply a matter of forming an association between each new word they hear and some object (for example, *cat*), quality (*black*), relation (*on*), action or state (*eating*) that the child can perceive and that is perhaps even pointed out by a helpful adult. But further reflection indicates that learning vocabulary cannot be quite that simple.

The difficulty inherent in determining what a word means was pointed out by W. V. Quine, arguably America's most influential living philosopher.[42] He uses the example of a linguist visiting a strange country whose language he does not know. During his visit, the linguist hears the word *gavagai* used in the presence of a small, furry mammal with long ears and initially assumes that *gavagai* is equivalent to the English *rabbit*. But on further reflection he realizes that *gavagai* could actually refer to the concept ANIMAL or MAMMAL or HEAD or FUR or RABBIT-LIKE SHAPE or HOP, or perhaps even something quite unrelated to the rabbit such as the time of day. *Gavagai* could even be the proper name of a person who in some way resembles a rabbit, or an expletive to curse the appearance of yet another garden pest. Quine argues that no matter how much evidence our linguist collects, he simply has no way of ever being certain that two words from different languages have the same meaning.

The child faces essentially the same conundrum.[43] Even if a helpful mother points to an animal and says "cat," how is the child to know that the word *cat* refers to the animal itself (actually, a species of animal) and not to its color, its fur, its relationship to the carpet, the cat-plus-the-carpet-it-is-sitting-on, the sound it is making, or its current behavior of scratching itself? When one realizes the infinite possibilities concerning the meaning of any word, it becomes clear that it is not possible for an adult to provide information that would reliably transmit the meaning of a word to a child (or to any other individual, for that matter). The child can only suppose that the word *cat* refers to some concept already in her mind, since she surely has no direct access to those concepts and meanings in the adult's mind, but she can never be absolutely sure which it is. That children do make guesses and jump to unjustified, tentative conclusions is clear when a child refers to a small black dog as a "cat" or, perhaps more humorously, when she refers to the visiting parish priest as "Daddy."

It is informative to compare the child in this situation with that of the scientist testing theories, such as that water must be heated to 100° C for it to boil. For the scientist, no amount of evidence can be taken as conclusive proof of the theory. One can boil water using heat from burning gas, electrical resistance, or solar energy and find that the source of heat makes no difference in the boiling point. One can boil water in vessels made of steel, iron, aluminum, stone, or plastic, or do the experiment at different times of day and during different phases of the moon and obtain what appears to be additional evidence for the theory. These findings may appear to lend support to the theory, but they cannot prove it. Indeed, a water-boiling experiment conducted using an accurate thermometer at an altitude of 2000 meters above sea level will show that the theory is in fact false, since the boiling point of water depends on air pressure, which is reduced at higher altitudes.

The child's situation with respect to words and their meanings appears analogous. A young child growing up with little or no contact with nonsibling children may over a period of many years be presented with *absolutely overwhelming evidence* that *mommy* refers to the one particular woman who is almost always close by and who feeds, clothes, bathes, and cuddles him. Imagine little Johnny's surprise when during his first visit to kindergarten at the age of five he hears another child using the word *mommy* to refer to a woman whom he has never seen before! Johnny then has no option other than to reject his initially "well-supported" hypothesis about the meaning of *mommy*. Eventually he will replace it with the theory that *mommy* refers to not one particular person, or even a class of persons, but rather to a special kind of *relationship* between one human being and another. And of course, even this meaning is subject to revision as young Johnny hears or reads about a cow, dog, or cat who is the "mommy" of a calf, puppy, or kitten.

But although learning word meanings must necessarily proceed through a process of theory construction, rejection, and revision, it is also clear that the meaning theories that children entertain are either often accurate or quite close to the adult meaning, since it doesn't seem to take many cycles of trial-and-error elimination to arrive at the accepted meaning. If this were not the case, children would be hard pressed to learn new vocabulary as quickly and easily as they do.

Donald T. Campbell, whose variation-and-selection perspective on learning and thought was described in chapter 9, has theorized that children are aided in their guesses as to the meanings of new words by an innate expectation that words refer to the more easily perceivable, stable aspects (or entities) of their environment, a characteristic he refers to as "entitativity."[44] Thus, since a cup is perceived as a single entity that can be separated from and used independent of the rest of its environment, a child will expect that there is a word that refers to a cup, and not one that refers to the combination of both cup and saucer or to just the handle and bottom of the cup. Similarly, the child will expect that *cat* is more likely to refer to the animal she sees moving across the rug rather than to a combination of the cat and rug, or to just the cat's head and tail. As Steven Pinker concluded (referring to Quine's *gavagai* example):

... humans [are] innately constrained to make only certain kinds of guesses—probably correct kinds—about how the world and its occupants work. Let's say the word-learning baby has a brain that carves the world into discrete, bounded, cohesive objects and into the actions they undergo, and that the baby forms mental categories that lump together objects that are of the same kind. Let's also say that babies are designed to expect a language to contain words for kinds of objects and words for kinds of actions—nouns and verbs, more or less. Then the undetached rabbit parts, rabbit-trod ground, intermittent rabbiting, and other accurate descriptions of the scene will, fortunately, not occur to them as possible meanings of *gavagai*.[45]

Such strategies, and likely many others,[46] are very useful in constraining or biasing the child's theories of word meaning, but the fact remains that neither the child nor the adult can ever be absolutely confident that his meaning for a word is identical to that of any other person. Even consulting a dictionary provides no absolute assurance, since a dictionary can only define words through the meanings of other words whose meanings are also unverifiable. But the more interaction a person has with other speakers of the language, the more confident (though never certain) he can be that meanings are shared, since such interaction provides for increased opportunities for the rejection, revision, and resulting fine-tuning of meanings.[47] And as the child's vocabulary increases, already learned words can be used effectively to narrow down the meanings of new words.

It turns out that acquiring the meaning of words also has important implications for acquiring syntax. MIT linguist and cognitive scientist Steven Pinker, whose important article with Paul Bloom on language evolution was mentioned earlier in this chapter, has pointed out that many puzzling exceptions to some basic syntactic patterns in English appear to render their being learned by children very difficult if not impossible. Consider the following sentences (an asterisk precedes words and sentences that are un-grammatical in English):

(1)   Beth sold the cookies to Eric.

(2)   Beth sold Eric the cookies.

(3)   Beth pulled the cookies to Eric.

(4)   *Beth pulled Eric the cookies.

From the first two sentences, it is clear that a speaker of English can use one of two different grammatical structures for sentences containing both a direct object (*the cookies*) and indirect object (*Eric*). We can put the direct object after the verb followed by *to* and the indirect object. Or we can drop the *to* and switch the positions of the two objects. But notice that although the second structure seems to work fine for the verb *sold*, it does not sound right to most speakers of English for the verb *pulled* as used in (4), despite the fact that both verbs behave similarly in (1) and (3).

Let's consider a few more sentences to show that this is not an isolated example.

(5)   Christopher kicked Erin.

(6)   Erin was kicked by Christopher.

(7)   Christopher resembled Erin.

(8)   *Erin was resembled by Christopher.

Here we have examples of the active and passive voices. In the active voice construction of sentence (5), the doer of the action is before the verb and the recipient after the verb. But in the passive voice construction of (6), the recipient is before the verb (to which *was* has been added) and the doer is now after the verb (after which *by* has been added). Note that countless verbs could be substituted into sentences (5) and (6) and yield grammatical sentences, such as *loved*, *heard*, *kissed*, *believed*, and *served*. But for some reason the passive construction using *resembled* in (8) is clearly not an acceptable English sentence.

Now if children learned language by simply listening to and memorizing sentences, and if they never said a sentence that they hadn't already heard, these inconsistencies would pose no problem. But countless studies and observations reveal that children are *not* conservative in their language learning. We already noted their use of past tense forms they could not have heard from an adult, such as *cutted, *drawed, and *breaked. Some additional examples of children's creativity in attempting to figure out and apply rules of English grammar are:

(9)   *How was it shoelaced?

(10)   *Jay said me no.

(11)   *I'm just gonna fall this on her.

(12)   *I'm gonna pour it with water.[48]

So instead of being linguistically conservative, children are quite creative speakers in venturing beyond the words and sentences they hear others produce. Now if it is true (as Pinker believes) that children receive no useful information from adults concerning which sentences they produce are ungrammatical,[49] then the fact that we do not produce such ungrammatical sentences as adults is one of the most interesting dilemmas of human learning. If children obtain no information concerning the grammaticality of their sentences, how are they able to eliminate the ungrammatical ones? This is often referred to as the problem of the "learnability" of language.

Pinker attempted to provide a solution to this problem by showing that the syntactic exceptions of the types shown above are not arbitrary but depend on often quite subtle differences in the meanings of the verbs and the meanings of syntactic constructions. For example, the so-called dative indirect object can be used when the verb of the sentence indicates a change of possession, as in (2) *Beth sold Eric the cookies*, but not when only motion is implied as in (4) *Beth pulled Eric the cookies*. Also, if the verb implies acting upon an object, then a passive form is acceptable, as in *Adam was hit by Anne* and (6), but usually not otherwise as in *Money is lacked by Matilda* and (8).

So by Pinker's account, learning the meanings of words is essential to producing syntactic sentences. So how does he suggest that these meanings are learned? To quote, with a bit of added commentary in brackets: "What we need to show is that the child is capable of entertaining as a hypothesis any possible verb meaning [that is, consider any of a large number of possible variations], and that he or she is capable of eliminating any incorrect hypotheses [and consequently selecting the better ones] as a result of observing how the verb is used across situations."[50] Pinker sounds even more Darwinian and selectionist when discussing the learning of morphemes (the meaningful entities that make up words) in stating that "as the child continues to work on that morpheme over a large set of sentences, all incorrect hypotheses will be discarded at some point or another, any correct hypothesis will be hypothesized sooner or later . . . and only the correct ones will survive in the limit."[51]

For the "unlearning" (elimination) of overgeneralized verb forms such as *drawed* and *hitted*, Pinker (as well as several other researchers) invokes what is called the uniqueness principle by which the child expects that there cannot be two ways of expressing exactly the same meaning. So when the child hears an adult say *Nicholas drew a nice picture* and it is clear that the adult is referring to the past act of drawing, the child will understand that *drew* has the same meaning as *drawed*, and he will eventually replace the latter with the former.[52]

Although Pinker would almost certainly not characterize his theory in this way, he nonetheless has proposed a procedure by which children are able to generate guesses about the structure of the language they hear and then eliminate the incorrect ones without the benefit of having adults indicate which utterances are incorrect, that is, without access to negative evidence. To pull this off, children must be sensitive to some very subtle semantic distinctions among verbs, which is in itself quite remarkable. It leads Pinker to speculate that such "lexicosemantic" concepts (which appear to pertain only to language learning and use) must be innate and part of a separate language component of the mind having little to do with other cognitive abilities. But his overall theory of language learning can nonetheless be understood as selectionist in that innate knowledge, arising from natural selection *among* humans during biological evolution, interacts with selectionist cognitive processes of hypothesis formation and elimination (selection *within* humans) to arrive at adult language competence.

Whereas Pinker's theory of language learning can be construed as implicitly selectionist, the one proposed by American psycholinguists Elizabeth Bates and Brian MacWhinney, called the competition model, is explicitly so. According to the competition model, the child's learning of word meanings and grammar has three necessary stages:

First, the child develops a function to express. We will call this *functional acquisition*. Then the child makes a first stab at a way of mapping the function into a form. We will call this *jumping in*. Then a period of *competition* ensues during which the range of the form is narrowed or widened.[53]

Let us take a brief look at each of these three stages. *Functional acquisition* has to do with the child's assignment of meaning to the objects, actions, and relationships around her. These meaning functions must be developed before she can understand language referring to them, and before she can use language herself to express them.

The child then has to associate the sounds of the language she hears to these meaning functions. Since at the early stages of language acquisition she can do no better than make a guess as to the meaning of a word, phrase, or sentence, this stage is referred to as *jumping in*. Such initial guesses may be quite wide of the mark, but as the child learns more and more word meanings, this knowledge can be used to help discover the meanings of new words in much the same way that the final words of a crossword puzzle are usually easier to identify than the first ones attempted. These jump-ins are, of course, nothing but preliminary guesses as to the meaning of a word (or grammatical form) and are the necessary source of variation for subsequent selection.

Selection is accomplished by a process of *competition* in which words and grammatical forms compete for meanings based on the assumption that two different forms must have different meanings (as in Pinker's uniqueness principle), and if no difference can be found, one of them is wrong. For example, the child may hear the words *plate* and *saucer* referring to what initially appear to be the same type of object. These two words will then compete for these two meanings until, after several presentations and perhaps some correction ("That's not a plate, it's a saucer"), *plate* wins out for round, thin objects on which food is placed, and *saucer* wins out for round, thin objects on which cups are placed. MacWhinney likens this to the competition of two species for the same environment resulting in each species establishing its niche in that part of the environment for which it is best adapted.[54] In other situations, one of the forms may be eliminated entirely (become extinct) as when *goed and *cutted are eventually replaced by *went* and *cut*.

Such competition is not limited to word meanings. Different languages use different ways of expressing grammatical relations, and the child must learn these syntactic rules as well. In English, the order of words in a sentence is the primary determinant in assigning the roles of subject (almost always placed before the verb) and object (usually after the verb), so that in the sentence *The food ate the dog*, the food would normally be understood as the entity doing the eating despite the fact that this makes little real-world sense (perhaps the dog fell into a vat of highly acidic hot sauce?). But the same sentence structure in certain other languages (such as Spanish in *La comida comió el perro*) would be immediately understood as the more sensible "the dog ate the food." Bates and MacWhinney and their associates in their extensive research on both children learning their first language, and students and adults learning second languages, indicated that learners rely on certain cues to resolve the various competing words, meanings, and syntactic forms.

It should be clear from even this brief description that the competition model is selectionist. Although Bates and MacWhinney do not use that word or its variants, the process of competition they propose is clearly one through which selection (and elimination) takes place. Children make guesses concerning the meanings and forms of the language they hear and eventually fine-tune these guesses by cumulative selection as words and forms compete for various meanings and functions. The theory is informed by both linguistics and psychology, and it draws support from a considerable body of research. And it does not assume that the child possesses an extraordinary store of detailed innate linguistic knowledge. The selectionism is clear when MacWhinney states that "the underlying idea in the Competition Model is that mental processing is competitive in the classical Darwinian sense."[55]

A Near-Common Denominator—Selection

We have now considered several theories of language learning that differ in a number of respects. One useful way of comparing and contrasting them is to consider the equation: Innate knowledge × experience × learning = language knowledge.

This equation states that the child's ability to use language is the result of the interaction of the child's innate knowledge, experience, and learning. Innate knowledge is considered in the broad sense, including the brain and vocal tract structures shaped by biological evolution, in addition to any more general cognitive or specifically linguistic knowledge that could be considered to be a part of the child's biological endowment. Note that the interaction among innate knowledge, experience, and learning is considered to involve multiplication (not addition), since this recognizes that if any one of the three factors did not exist (were zero), the child's language knowledge would also not exist (would also be zero).

Using this formula, it can be seen that the various theories of child language acquisition discussed thus far differ in the importance they ascribe to the three factors in a more or less compensatory fashion. Skinner's and other behaviorists' accounts of language learning emphasize the role of experience and learning and downplay innate knowledge.

Chomsky and Fodor, in contrast, minimize the role of both experience (recall Chomsky's argument of the poverty of the stimulus) and learning (Fodor believing that any form of learning is impossible) while emphasizing the importance of innate knowledge. Indeed, the standard practice of most linguists today is to put as much as they possibly can in the innate knowledge factor, which they refer to as "universal grammar," while still allowing children to be able to learn different languages, that is, while providing the minimum necessary role for experience, since it is clear that children speak the same language they hear. It should also be noted that linguists' conception of language knowledge is much richer and sophisticated than that conceived of by Skinner.

Pinker appears to place more emphasis on experience than do Chomsky and Fodor, but he nonetheless maintains one aspect of the poverty of the stimulus in his belief that negative evidence, which would allow the child to reject easily mistaken hypotheses about the language being acquired, is not available to the child. Pinker also emphasizes the role of innate knowledge, since his theory depends on the child being able to classify verbs in quite subtle and sophisticated ways using categories and concepts he believes to be quite specific to the domain of language.

Bates and MacWhinney's competition model of language learning contrasts with Chomsky's and Pinker's views and brings us back closer to

Skinner's in minimizing the role of innate knowledge and providing a larger role for learning. Bates and MacWhinney also place more importance on experience than either Chomsky or Pinker, believing that children have access to negative evidence about language as provided by the corrections and repetitions of their parents and possibly other adults.

But although these theorists differ considerably in the relative importance ascribed to these three elements, they appear to be much more in agreement (than perhaps they would care to admit) in their perspectives on the role of selection in language learning in the factors they consider most important. As we saw in chapter 7, Skinner has on many occasions drawn a parallel between operant conditioning and natural selection,[56] and the selectionist nature of the competition model has been noted.

Pinker, although appearing somewhat reluctant to recognize the necessary variation component in language acquisition, suggests that he sees all learning as a selectionist process in noting that "despite all its complex guises, learning can always be analyzed as a set of 'hypotheses' the organism is capable of entertaining and of a 'confirmation function' by which the environmental input tells the organism which one to keep."[57] His contribution to our appreciation of the role of among-organism selection in the evolution of language was recognized in the first section of this chapter.[58]

The one notable exception is Chomsky, who not only sees no place for selection in language learning, but rejects a Darwinian account of the evolution of language itself. It is tempting to speculate that this latter stance is related to his innatist beliefs concerning language acquisition, since if he were to admit that language gradually evolved along with our species through natural selection among humans, he might have to confront the possibility that language knowledge could also emerge through within-human selection processes in the growing mind of the child.

But regardless of the role that innate linguistic knowledge may play, it can only go so far. It cannot provide the child with knowledge of the specific sounds used in her mother tongue, the meanings of the words she hears, or knowledge of how each grammatical form maps to a function. Additional knowledge must somehow be developed as she adapts her developing linguistic system to her linguistic environment and communicative needs. The form that this additional constructed knowledge can take may be strongly biased and constrained by already achieved innate knowledge about human language. However, the variation and selection of linguistic hypotheses cannot be completely eliminated from the adaptive process of language acquisition. Wherever it may be that knowledge obtained through biological evolution leaves off, we can expect a selectionist process to take over in the generation of varied hypotheses concerning the meanings and functions of words and grammatical structures coupled with the elimination and fine-tuning of those hypotheses found by the child to be inadequate in some way. And although few linguists or language acquisition researchers now describe and model child language learning as a Darwinian process, the success that selectionist models of learning have in other areas (to be described in chapters 13 and 14) seems certain eventually to provide new insights into one of the most remarkable feats of human learning.[59]

The Use of Language

Let us now finally turn to a brief consideration of how language functions, enabling us to communicate our thoughts and intentions to others. Much could be said about this topic considering the prodigious amounts of relevant research conducted by psychologists, linguists, psycholinguists, and educators. I will make no attempt to review all this research here, but will rather provide a concise argument, using a few examples, that language use also involves a Darwinian process of variation and selection.

Once two or more people have acquired the same meanings for words (semantics) and knowledge of the same rules for combining words to express meaning (syntax), it might seem that language could be used to transmit meaning from one to another. Surely, if we understand the words *dog*, *cat*, *bit*, and *the*, and know that in English the normal ordering of sentences is subject-verb-object, my declaration of "The dog bit the cat" should provide the obvious information to you. But in practice things are not so straightforward, the reason being that words themselves do not carry meaning. Rather, they can only elicit meanings that already exist in the brain of the listener. And the meaning that they elicit depends on the listener's relevant experiences and the context in which the words are used. For example, consider the following five sentences.

(1)   Where did you put the newspaper?

(2)   There was an interesting story in the newspaper yesterday.

(3)   The newspaper is going on strike.

(4)   Workers are demonstrating outside the newspaper.

(5)   The newspaper is experiencing financial difficulties.[60]

It is readily apparent that the word *newspaper* has a quite different meaning in each of these sentences. In (1) a particular physical copy of the newspaper is intended. In (2) the word refers to all copies of a particular edition of the newspaper. It refers to employees in (3), and in (4) to the building where the publication is produced. Finally, in (5) *newspaper* means the institution that publishes the newspaper.

But it is not just the meaning of individual words that depends on context and the experiences and imagination of the listener, but larger stretches of words as well. To demonstrate this, consider the following passage:

The procedure is actually quite simple. First you arrange things into different groups depending on their makeup. Of course, one pile may be sufficient depending on how much there is to do. If you have to go somewhere else due to lack of facilities that is the next step, otherwise you are pretty well set. It is important not to overdo any particular endeavor. That is, it is better to do too few things at once than too many. In the short run this may not seem important, but complications from doing too many can easily arise. A mistake can be expensive. The manipulation of the appropriate mechanisms should be self-explanatory, and we need not dwell on it here. At first the whole procedure will seem complicated. Soon, however, it will become just another facet of life. It is difficult to foresee any end to the necessity for this task in the immediate future, but then one never can tell.[61]

I would venture to guess that you did not have difficulty understanding the meaning of any of the individual words in this passage. But I would also venture to guess that the passage as a whole probably did not make much sense to you when you first read it. But if I now inform you that the passage has something to do with washing clothes (and assuming that you have had some experience in washing clothes), reading it again will likely be quite a different experience as it will now elicit meanings it did not before. This is because I helped you to constrain your hypothesis about what the passage is about. But again, the meaning must actually be created by you and is not transmitted by the words, phrases, or sentences of the passage.

These examples are meant to demonstrate that language comprehension is not a matter of receiving meaning from a speaker or writer, but rather is an active process of *constructing* meaning as the listener or reader attempts to make the words, intentions, and context of the situation fit. As child language researcher Gordon Wells put it:

When I communicate with other people, whether it be to inform, request, or persuade, what I have in mind is an idea—an event, action, or outcome—that I intend they should understand. However, this idea arises from my mental model of the world, which is itself the product of my unique personal biography. Nobody else has exactly the same mental model of the world, since nobody else has had exactly the same experience. It follows, therefore, that nobody can have exactly the same ideas I have.

What all this leads to is a recognition that one never *knows* what other people mean by what they say or write. One can only make an informed guess, taking into account all the cues that are available: from the communication context, from one's own relevant experience, and from the actual linguistic signal. To put it differently, I cannot know what idea is in your mind as you speak or write. I can only know what ideas I would have had in mind if I had produced the same lexico-grammatical sequence as I believe you to have produced in the context that I think you think we currently share.[62]

So according to Wells (and he is certainly not alone in his interpretation), understanding language involves making informed guesses about the intent of the speaker or writer. Some guesses will be wrong and will be quickly eliminated. Others will be wrong but not so easily eliminated, resulting in misunderstanding, which is most likely when the individuals involved are from different cultures, age groups, sexes, or social classes.[63] Of course, we expect that most guesses will be quite close to the speaker's or writer's intended meaning. And that they usually are in normal conversation with our family (except very young children), friends, and work associates is what makes it appear as if using language does involve the transmission of meaning from one person to another. But this is an illusion, which is quickly revealed when we experience difficulties in communication, and when we recognize that language can at best elicit and help select meanings that already exist in the listener's or reader's head. This selectionist view of language use has important implications for understanding the process of education, to which we turn next.

12

Education
The Provision and Transmission of Truth,
or the Selectionist Growth of Fallible Knowledge?

Providence

But referring now to all things which we understand, we consult, not the speaker who utters words, but the guardian of truth within the mind itself, because we have perhaps been reminded by words to do so. Moreover, He who is consulted teaches; for He who is said to reside in the interior man is Christ, that is, the unchangeable excellence of God and His everlasting wisdom, which every rational soul does indeed consult.

—St. Augustine[1]

Instruction

Instead of paper, we have pupils, pupils whose minds have to be impressed with the symbols of knowledge. Instead of type, we have classbooks and the rest of the apparatus devised to facilitate the operation of teaching. The ink is replaced by the voice of the masters, since it is this that conveys information to the minds of the listener, while the press is school discipline, which keeps the pupils up to their work and compels them to learn.

—John Comenius[2]

Selection

Let me first briefly summarize the critical approach [to education]. It is based on evolutionary epistemology, which claims that we never receive knowledge, but rather create it; we create it by modifying the knowledge we already have; and we modify our existing knowledge only when we uncover inadequacies in it that we had not recognized heretofore. Accepting this as an explanation of how knowledge grows, I have suggested that teachers construe their roles as facilitators of the growth of their students' knowledge.

—Henry Perkinson[3]

The important roles that tradition, technology, and science play in the survival and proliferation of our species and the quality of our life make it imperative that new generations both acquire this cultural knowledge and be able to revise and improve it in response to changing physical and social environments. As the scope and complexity of human knowledge have increased, just about every human society has instituted formal methods of education to facilitate the acquisition of knowledge by its children.

Some isolated communities still do not make formal education a requirement for children, relying still on more informal means of cultural continuation such as apprenticeships. Universal formal education is the stated goal, however, of all countries. In the developing world the goal of universal primary education is to provide formal schooling for at least five or six years to allow all children to become literate in either their native language or the official non-native language of the country (for example, English or French in most of sub-Saharan Africa). In the more industrialized countries children are usually required by law to attend school until age 15 or 16, and to become a doctor, lawyer, or university professor in any country may require formal education until age 25 to 30.

Considerable financial resources are devoted to formal education, especially in the industrialized countries, which have great need for skilled workers, technicians, and professionals, including, of course, teachers. Since citizens contribute a significant portion of their paycheck in the form of taxes to public education, it should not be surprising to find considerable controversy over the schools' effectiveness, methods, and curricula. Nonetheless, it is clear that despite its problems, education does increase the fit of students' knowledge and skills to the physical and social environments in which they will live as adults. Students learn to read and write. They learn how to use mathematics to solve problems ranging from the mundane (balancing checkbooks and filing tax returns) to the creative (designing new automobiles, sending space probes to the planets and beyond). They gain knowledge of other times and places in geography and social studies classes. They learn additional languages to facilitate travel and to participate in our new world economy, which requires knowledge of English but also often Spanish, French, Russian, Arabic, Japanese, or Chinese. And they develop expressive abilities in dance, music, and art classes (often in special schools and programs), and sports and recreational skills in physical education classes.

The purpose of this chapter is to explain the puzzles of fit that are the consequences of education. Some of the discussion may overlap and draw on previous ones, but its focus will be on formal education and past and present views of how education results in new and improved knowledge and skills.

Education as Providence

Perhaps the most well-known geometry lesson of all time is the one presented by Socrates to Meno's slave boy. Although we addressed the Meno dilemma in chapter 6, it will be useful to start our discussion of education by taking a closer look at that interaction between teacher and student.

The problem Socrates describes concerns the area of a square. After establishing that the boy understands that a two-by-two square contains an area of four square units, he asks the boy how long each side would have to be for the square to contain eight square units. Since eight square units is twice as much as four, the boy quickly (and erroneously) concludes that a four-by-four square would contain the desired area.

But when Socrates sketches this four-by-four square, it becomes apparent that such a square contains not eight but rather sixteen square units. The boy, then realizing that the desired eight-square-unit figure must be larger than two by two but smaller than four by four, guesses that the answer is three. But when asked to reconsider this answer, the boy realizes that it is also in error, since such a square would have an area of nine square units, not eight.

At this point the boy admits that he does not know the answer, and Socrates points out to Meno, who is observing the interaction with keen interest, that in his admitted ignorance and confusion the boy is actually better off than he was before when he believed he knew the answer but in fact did not. With further careful questioning by Socrates, the boy finally discovers the correct answer, that is, that a square made of sides equal in length to the diagonal of a two-by-two square (equal to the square root of eight) has an area equal to the desired eight square units.

Since Meno states that the boy had never been instructed in geometry, Socrates concludes that the knowledge the boy finally demonstrated in solving the problem could not have been learned earlier in his life. And since Socrates did not instruct him, but only questioned him, he did not give the boy his new knowledge. Socrates therefore argues that the knowledge the boy used to solve the problem must have been possessed by the boy all along and was therefore simply *recollected* during their encounter from knowledge provided by an immortal soul. That is, inquiry is the recollection of knowledge that we already have—a providential view of knowledge that leads to a providential view of education as well.

Such a view of education did not die with Socrates and Plato. St. Augustine (354–430), the intellectual father of Christianity, fashioned a philosophy in which God is the answer to all questions. God made the universe, was responsible for everything that happened in it, and was the source of all the knowledge that humankind was able to acquire about it. In his *De Magistro* ("concerning the teacher"), Augustine adopted the same basic conclusion as Socrates concerning teaching, but instead of attributing all knowledge to immortal personal souls, he acknowledged God as the source of all knowledge, "the unchangeable excellence of God and His everlasting wisdom, which every rational soul does indeed consult."[4]

Therefore according to Augustine, teachers cannot transmit knowledge to their students by instruction, since teachers can only utter words. If the words are already familiar, students can learn nothing from them. And if the words are unfamiliar, the students can still learn nothing from them:

For it is the truest reasoning and most correctly said that when words are uttered we either know already what they signify or we do not know; if we know, then we remember rather than learn, but if we do not know, then we do not even remember, though perhaps we are prompted to ask.[5]

The noninstructionist argument concerning words that Augustine makes is remarkably similar to the one at the end of the previous chapter that language use does not involve the transmission of meaning or concepts or information from one person to another. To make Augustine's point clear and relevant to education today, imagine that you are a physics teacher attempting to teach a student the concept of acceleration. If the student already has a good idea of what velocity is and understands what it means for a quantity to change over time, you may explain acceleration as the change in velocity over time, and the student may now have a basic idea of what acceleration is. This is because she already knew what velocity and change over time mean, so putting these two familiar concepts together yielded a new one. But if the student has no idea of what velocity is or doesn't understand what is meant by a changing quantity, a teacher cannot do much to teach the concept. Nonetheless, by demonstrating acceleration and allowing both teacher and student to ask and answer questions, the concept can be acquired. Since it is not clear how such new knowledge is acquired, Augustine can only imagine that it is provided by God. For him, therefore, teachers can only hope to guide students to knowledge that is provided by God as divine illumination.

Education as Instruction

But a quite different view of education was to accompany the new philosophy brought about by the beginning of modern science in sixteenth-century Europe. Exemplified in the writings of Sir Francis Bacon, it rejected God, religious dogma, and the classic philosophical texts as the sole sources of knowledge. Instead of looking within oneself for God's revealed truth, or studying books and listening to the words of the teacher, Bacon insisted that nature be studied directly, for it was only in understanding nature and thereby advancing science that the human condition could be improved. According to Bacon, if you want to know how many teeth your horse has, you must look into its mouth and count them, not waste time reading what scholars and philosophers have to say about the matter. This new philosophy also stressed the importance of practical knowledge—knowledge that could be put to use for growing crops, building factories, and conquering the seas and the unknown lands that lay beyond the horizon.

This change in perspective concerning knowledge and science was to have a significant impact on education. Whereas before education had been considered a sort of initiation into the culture of the society or a process of discovering the truth within provided by God, it now began to be regarded as a process of instruction. Direct and careful observation of the world would allow knowledge to grow by the transmission of information through the senses to the mind. Where direct experience could not be easily had, the knowledge already gained by teachers (who now became instructors) would be transmitted into the minds of students through their spoken and written words. To facilitate the latter, textbooks came to replace classic texts. These textbooks were designed to present knowledge to students in the most effective manner. Subject matter was broken down into small, easily digested components and then carefully organized and sequenced to facilitate learning. Teachers and textbooks thus became transmission relay stations whose job was to reflect knowledge of the world into the minds of students.

This instructionist view of education had, and still has, important consequences for all aspects of education. First of all, if education is the transmission of knowledge from the teacher (or textbook) to the student, then the knowledge possessed by the teacher (or textbook) must be accurate. Instruction makes no sense if what is to be taught is not true. Teachers and textbook authors must therefore be (or at least pretend to be) unquestioned authorities on the subjects being taught and written about.

Second, an instruction-oriented view of education tends to put the blame on the student for failure to learn, since if the information to be transmitted is presented clearly and accurately, such failure must be due to problems on the receiving end. Inattentivenss, laziness, and lack of motivation are therefore often met with punishment and ridicule in an effort to make students pay attention and do their work.

Third, if education is the transmission of knowledge from instructor and textbook to student, then the usual test of its success is whether students can reproduce the transmitted information in spoken or written form. This encourages the memorization of what is "known" to be "true" and the use of standardized, objective tests as indicators of academic success.

In a word, a transmission-based, instructionist perspective on education is authoritarian. As New York University educational historian Henry Perkinson observed:

> . . . the transmission metaphor [for education] persisted through the eighteenth, nineteenth, and into the twentieth century. It persists down to the present, where many educators still remain caught in its spell. Believing that we inductively learn from experience, they strive valiantly to *transmit* knowledge to their pupils. Inevitably, this leads teachers to authoritarianism; the attempt to secure greater control over the educational process. Take for example the subject matter: teachers who seek to transmit knowledge attempt to control the subject matter by packaging it into a transmittable curriculum. . . . Take as another example of authoritarianism, the treatment of pupils: teachers who seek to transmit knowledge to students attempt to control them; they "prepare" them, "stimulate" them, "motivate" them, "get them to pay attention," "get them moving." All these tactics are attempts to control pupils so that teachers can more expeditiously and efficaciously transmit knowledge.[6]

Whereas the consequences of an instructionist view of education may be problematic, its greatest problem is belief in the transmission of knowledge itself. David Hume identified this problem in the eighteenth century with his critique of induction. Recall from chapter 6 that he concluded that we can never justify knowledge by observation, since to be useful, our knowledge (for example, objects fall to the earth) must go beyond what we have personally experienced (*some* objects fall to the earth at *certain* times and places). But although Hume admitted the logical impossibility of the environment being able to instruct the mind by way of the senses, he did not reject a psychological theory of induction. That is, he concluded that learning from observation was logically unjustifiable; however, this is what we actually do, since we believe that repeated observations of a phenomenon indicate a general law of nature. So it is logically invalid to assert that the sun will rise tomorrow since it has risen every morning up to and including today, but we still believe that we know that the sun will rise tomorrow based on our prior repeated observations of mornings and sunrises.[7]

If Hume is correct in asserting that no amount or type of experience can provide us with certain knowledge, yet we believe that experience leads to knowledge, this puts education in a very curious situation. For it means that although the knowledge possessed by teachers and contained in textbooks is almost certain to be mistaken, students will tend to believe (and will be encouraged to believe) that it is unquestionably accurate. From this perspective, education is not a process by which students acquire and improve their knowledge, but is a type of indoctrination in which students are encouraged and compelled to accept as true the inevitable errors of their teachers and textbooks.

But is it actually the case that we acquire what must be uncertain knowledge from our sensory experiences of the world? That is, does induction work psychologically even though it is a logically invalid process? To answer this question, we must consider again the work of Karl Popper who was also discussed in chapter 6. The belief in the transmission of knowledge from sense experience to mind is part of what Popper referred to as both the common-sense theory of knowledge and the bucket theory of mind; he convincingly argued that this view of learning is inaccurate from both logical and psychological perspectives. Hume believed that we are conditioned by habit to believe that repeated observations lead to useful generalizations. Popper noted that learning cannot proceed in this manner *since to recognize that an observation is in some sense a repetition of a previous one requires knowledge that cannot be obtained by observation alone.* This is surely a difficult idea to understand, but it is essential to grasp if we are to appreciate how knowledge cannot be a product of instruction either from environment to person or from one person to another.

In an attempt to make this important point clear, let us consider one of philosophy's favorite animals, the swan. According to Hume and the common sense or bucket theory of knowledge, repeated observations of white swans will lead to the idea that all swans are white regardless of the fact that such a conclusion is logically indefensible; there may be many nonwhite swans that simply have not yet been observed. But a problem immediately surfaces if we carefully consider such repeated observations, for it turns out that no observation is ever likely to be an exact repetition of a previous one. The swan we see today is not exactly the same as the one we saw yesterday. Even if we encounter the same swan, it will be not adopt exactly the same posture or movements it had yesterday, its feathers will likely be either cleaner or dirtier, and the light reflecting off the bird will not be of exactly the same hue and intensity. Thus to observe similarity one must have some prior idea concerning the way in which similarity will be observed. As Perkinson paraphrases Popper's argument:

A repeated observation, say, of a white swan, presupposes that the observer perceives the second observed swan as similar to the first, and to do this, the observer must have an expectation, a theory, about the two observations that make the first observation similar to the second. In other words, the theory "all swans are white" could not have been the result of conditioning, the outcome of repeated observations of white swans, since in order to experience a repetition, one must already have a theory that all swans are white.[8]

Turning to Popper's own words:

It is therefore impossible to explain anticipations, or expectations, as resulting from many repetitions as suggested by Hume. For even the first repetition-for-us must be based upon similarity-for-us, and, therefore, upon expectations—precisely the kind of thing we wished to explain.[9]

In other words, if the observation of similarity depends on an *expectation* of similarity, we obviously cannot use repeated observations to explain the origin of these initial expectations. And since much if not all knowledge can be considered a type of expectation (knowledge that unsupported objects fall to the earth will lead a mountain climber to expect to fall if he loses his grip on the rock face), repeated observations without prior expectations cannot in themselves lead to new and better knowledge. This analysis suggests that the knowledge acquired in educational settings is not acquired by transmission from teacher and textbook to student, and in this sense the teacher cannot literally instruct the student.

Hume's arguments against the logic of induction and Popper's argument against the psychology of induction would probably be considered irrelevant philosophical nonsense by most educators, particularly since our everyday experiences suggest that such transmission of knowledge from teacher to student takes place routinely. So although these ideas have had a considerable impact on philosophy, particularly the philosophy of science, we should not expect educators to be much influenced by them, especially since they are usually more concerned with the practical difficulties of education than with the philosophical ones.

So an instructionist conception of education remains very much alive and influential today. Nonetheless, the twentieth century has seen the emergence of educational theories that have rejected the instructionist view of education. This has come about as more and more psychologists and educators dismiss the conception of students as passive buckets into which knowledge is poured by teachers and textbooks, and replace it with a view of students as active creators of their own knowledge.

Education as Darwinian Selection

Piaget and Montessori
Jean Piaget's theory of cognitive development (touched on in chapter 9) has had perhaps the most impact on this change of perspective. Piaget showed that children see and understand important aspects of the world in ways very different from those of adults despite the fact that both are exposed to the same world. If knowledge results from taking in information, how is it that children perceive things so differently? Why does a two-year-old call a butterfly a bird? Why does a four-year-old insist that spreading out eight coins on a table results in more coins than when the same coins were grouped closer together? And why does a five-year-old "explain" that the sun does not fall down because it is yellow, or that the sun pushes the clouds across the sky? The fact that children have had fewer and less varied experiences than adults might explain why they may lack certain types of knowledge or have less knowledge. But it cannot explain why their understandings are *qualitatively* different from those of adults, and how they

come up with such extraordinary and creative explanations that cannot be traced to actual experiences.

According to Piaget, these differences arise because children do not passively receive knowledge from their environment but rather make it themselves. From this perspective, knowledge resulting from the transmission of information from environment (or teacher) to student is replaced by the concept of *growth*. Children grow in knowledge because they construct it, often by recombining what they already know in new ways and testing it, and the environment, whether it be the physical environment of hard knocks or mommy's verbal response, provides feedback concerning the adequacy of their constructed knowledge. Thus the role of the environment for Piaget is not that of a provider or transmitter or instructor of knowledge, but rather as the selector of the knowledge created by the child.

Piaget's own words (translated from French) indicate the degree to which he rejected a transmission view of education and advocated a constructive one:

I'm not an educator; I have no advice to give. Education is an arena of its own and educators must find the appropriate methods, but what I've found in my research seems to speak in favor of an active methodology in teaching. Children should be able to do their own experimenting and their own research. Teachers, of course, can guide them by providing appropriate materials, but the essential thing is that in order for a child to understand something, he must construct it for himself; he must reinvent it.[10]

Renowned Italian educator Maria Montessori (1870–1952), unlike Piaget, had little time or inclination for psychological theory or experimental research. Instead she developed

the most successful method of education in the twentieth century, perhaps the most successful method in the history of education. In Montessori schools, children of three years of age learn to dust, to dry, to set the table, to serve at table, to wash dishes. At the same time, they learn to take care of themselves: they wash, bathe, dress and undress themselves, they arrange their clothes in their locker or in a drawer, tie their shoes, comb their hair, and so on. By four and a half years of age, they learn how to write and how to read and can do basic arithmetic calculations.[11]

Montessori did not see the role of the teacher as an instructor or transmitter of knowledge. Instead, in a Montessori classroom the teacher tends to stay in the background, acting as a kind of valet who creates an interesting and challenging environment. The child's own natural curiosity and desire to master the environment results in learning, learning that depends on attempting new skills, making errors, and learning from mistakes. As for Piaget, the learning environment, including the teacher, acts not as a source of knowledge but rather as a selector of the knowledge constructed by the child.

Learning from Mistakes

In his book that reviews twentieth-century educational theory, Perkinson contends that the general approach of Piaget and Montessori to education (as well as aspects of those of B. F. Skinner, A. S. Neill, and Carl Rogers) is essentially Darwinian, and he contrasts this selectionist approach to a transmission-instructionist conception with respect to the process of education, and the roles of the teacher, the subject matter, and the student. In contrast to a view of education as a process of transmission, selection sees education as "a procedure of (Darwinian) growth; trial-and-error elimination; the continuous modification of existing knowledge."

From a transmission perspective "the teacher prepares the student, prepares the subject matter, and transmits (instructs, matches) the subject to the student in the form of lessons that the student learns," whereas from a selectionist perspective "the teacher creates an educative environment— an environment that is free, responsive, and supportive—wherein the student can improve (modify) his present knowledge through trial-and-error elimination."

A transmission view sees subject matter as "what is transmitted" but a selectionist perspective considers it "an agenda that specifies what aspects of the students' present knowledge are to be improved" and as that which "evokes the students' present knowledge and tests it (reveals the inadequacies in that present knowledge)."

Finally, transmission conceives of the student as "a learner, a more or less passive receptor who needs to be controlled and motivated" while selection sees the student as "a fallible, active creator of knowledge who seeks order" so that "when he discerns contradictions (errors, mistakes, inadequacies) in his present knowledge, he will modify that present knowledge."[12]

Perkinson does not pretend that Piaget, Montessori, Skinner, Neill, or Rogers would agree with his Darwinian, selectionist reinterpretation of their work. Indeed, we saw in chapter 9 how Piaget explicitly rejected

Darwinian theory and in chapter 7 how Skinner believed cultural evolution (which includes education) to depend on Lamarckian transmission. Nonetheless, all of these individuals (with the possible exception of Skinner) considered education to be a process of creative growth that results in the individual becoming better adapted to the environment and therefore better able to control aspects of the environment. Since they rejected both providential and instructionist explanations of education, their views lead quite naturally to a selectionist conception, even if they did not (again with the exception of Skinner) explicitly make Darwinian or selectionist arguments.

The selectionist orientation can perhaps be best appreciated by assessing the role of error. If this perspective views education as involving the fallible creation of knowledge, then educational theories and practices must consider error to be an essential part of educational growth. That is because the process creates fit not by the clairvoyant, advance fashioning of adapted thoughts or skills, but rather by the blind production of variations and the subsequent hindsighted selection of the thoughts and skills that better fit the needs and purposes of the learner. It is therefore not surprising that educational theories that reject transmission emphasize the positive role of error, since a selectionist process inevitably results in more errors (unfit variations to be eliminated) than successes (fit variations).

In Piaget's theory, the child experiences cognitive growth by realizing the mistakes inherent in his view of the world. It is a mistake to believe that there are now more coins on the table simply because they have been spread out to take up more space, and the child will develop more advanced modes of perception and thinking when he realizes these mistakes. For Piaget, cognitive development proceeds as the child creates new and better ways of interacting with his environment. But since these new ways of interacting are not determined or instructed by the environment, it is inevitable that the child will make many mistakes along the way. Thus, error and its elimination can be considered to be the basis of cognitive growth and education.

Montessori also stresses the central role of errors in education:

Supposing we study the phenomenon of error in itself; it becomes apparent that everyone makes mistakes. This is one of life's realities, and to admit it is already to have taken a great step forward. If we are to tread the narrow path of truth and keep our hold on reality, we have to agree that all of us can err; otherwise, we should all be perfect. So, it is well to cultivate a friendly feeling toward error, to treat it as a companion inseparable from our lives, as something having a purpose, which it truly has. . . . Whichever way we look, a certain "Mr. Error" is always present! If we seek perfection, we must pay attention to our own defects, for it is only by correcting these that we can improve ourselves.[13]

The essential role of error (that is, unfit variations) in a selectionist view provides the most striking contrast to a transmission or instructionist view with respect to the role of the environment. Since from an instructionist perspective the purpose of education is to transmit knowledge to the student accurately and efficiently, any error on the part of the student is an indication that something has gone wrong in the transmission process, usually with the student considered at fault for inattentiveness or laziness. In contrast, for selectionism, the educative environment reveals errors of behavior or thinking to the student and is responsive to the student's attempts to revise his behavior or thinking for the better. Thus, a teacher's primary responsibility is not to transmit knowledge, but to *assist the student in discovering the ways in which his current knowledge is inadequate*. But since such revelations of inadequacy can be quite threatening to a student's self-esteem, the teacher must provide an environment that is supportive of the student's attempt to better his knowledge. This environment should also be free, so that the student will be not be prevented from attempting bold new solutions.[14] In short, students should be eager to encounter their mistakes and will, it is hoped, find themselves in an environment that encourages them to revise their thinking and actions to arrive at better solutions to their problems.

The Dangers of Instruction

This century has seen the development and spread of selectionist views of education; however, many if not most educators retain an instructionist approach. That is because education as usually practiced admittedly looks like transmission. And learning obviously does take place in classrooms where teachers believe and act as if they are in the business of transmitting knowledge.

According to Perkinson, the learning that takes place in such environments is not due to the transmission of knowledge that is attempted (which is both logically and psychologically impossible) but rather despite such efforts. Even a transmission-oriented classroom is free to some extent, and effective teachers usually do provide a supportive environment for their students (although this support may be limited to the "good" students). Furthermore, the teachers provide critical feedback to their students in the form of question-and-answer sessions, discussions, quizzes, and test results, which reveal errors in the students' knowledge and lead them to modify what they know. Thus at least some learning does take place for at least some students, but Perkinson points out some serious disadvantages of a transmission approach. As already mentioned, these classrooms tend to be authoritarian and coercive. This leads to one of three possible reactions.

First, there are those pupils who withdraw, either from fear or from resentment of the coercion. They do not participate in the trial-and-error elimination and so do not improve those skills and understandings of concern to the teachers. The teacher classifies them as the stupid ones.

Second, there are those pupils for whom schooling becomes a game—the game of finding out what the teacher *wants* and then fabricating those skills or understandings. These are the hipsters, those who create pseudo-knowledge, knowledge created especially for the teacher, which, in the course of events, usually disappears—after the test.

The third group are the true believers. These are the pupils who have undergone intellectual socialization. They regard the teacher (or the textbook or the experts in the field) as final authorities, and they modify their own knowledge into accord with whatever pronouncements the authorities promulgate.[15]

Many educators and parents might at first consider this last possibility as a positive outcome. It must be realized, however, that such a student would be unable or at least reluctant to revise and improve his or her knowledge or skills when they were found to be inadequate.

Selectionist Teaching

What are the alternatives? Is it possible to move educational practice away from a coercive, transmission orientation? It certainly is, and teachers influenced by Piaget, Montessori, and other selectionist-oriented psychologists and educators have shown how it is possible. Perkinson has offered suggestions to educators to facilitate students' intellectual growth within the existing arrangements of most schools:

1. It is possible to *present* the subject matter rather than try to transmit it.

2. It is possible to invite students to *encounter* the subject matter critically rather than try to get them to accept it.

3. It is possible to view these critical encounters as a selection procedure of *trial-and-error elimination* wherein knowledge grows.

4. Regardless of institutional constraints, teachers can facilitate this growth by construing their role to be that of creating a classroom environment that is more free,
more responsive, and more supportive: a place where students can more readily
learn from their mistakes.

5. Finally, it is possible, in the schools as they presently are, for teachers to reconceptualize the aim of schooling as an attempt to develop concerned critics who can
and will facilitate the growth of our culture.[16]

Education as the Reorganization of Perceptual Control Systems

Although Perkinson provides convincing arguments for a selectionist view
of education, he does not attempt to describe the specific mechanisms by
which learning takes place, or examine teaching and learning as purposeful activities. To address these aspects, we will now consider educational
growth from the perspective of perceptual control theory as described in
chapter 8.

It will be recalled that perceptual control theory sees adapted (that is, fit)
behavior as allowing an organism to control some aspect of its environment. And since it is only through perception that an organism can know
anything of its environment, adapted behavior is in effect the control of perception. This view contrasts sharply with all other psychological theories
that consider an organism's perception to be in control of its behavior. For
perceptual control theory, learning is the reorganization of an organism's
control systems that allow it to control perceptual variables it could not
previously control, a reorganization resulting from a Darwinian process of
cumulative blind variation and selective retention.

To apply the idea of reorganization to education, let us use the example
of a person learning to swim. In its most rudimentary form, being able to
swim can be defined as staying alive in water that is deeper than one is tall,
that is, being able to tread water. One way to "teach" a nonswimmer to
swim is to throw the person into a body of deep water (we could call this
the immersion method). This will likely create error, since the student will
have difficulty keeping her head above the water.[17] This perceived error in a
crucial variable will trigger reorganization so that the student will immediately begin to move her arms and legs vigorously in random patterns to find
some way to maintain her ability to breathe. If she finds a behavioral
pattern (actually a perceptual-behavioral control loop) that allows her to

breathe, if even only a few gasps before she disappears below the surface again, the randomness of the movements will decline until she is able to keep her head above water continuously, at which point we would say that she has learned to swim. In effect, the student has now gained control over a variable that she could not control previously, and so by our definition learning has taken place.

Since the student did not initially know how to swim, her initial movements were of necessity blind attempts to do so. But although she did not know how to keep her mouth above water, she could perceive how successful she was in her attempts (getting her eyes above the water is better than below, but not quite good enough). This then provides a criterion for selection among the various behavioral patterns attempted, and allows the student to learn from her mistakes, eliminating patterns that did not succeed in getting her head above the surface and retaining those that did.

It is easy to imagine that the learner would be very highly motivated, since failure to learn to swim would result in death from drowning. According to perceptual control theory, motivation simply refers to error (that is, a difference between a perception and the reference level for that perception) that results in action to eliminate the error (see figure 8.1). Motivation is therefore considered to be *internal* to the student, since the reference level of the controlled variable is determined by the student, not by the environment.

We must point out, however, that the immersion method of swimming instruction may well fail for any particular student, since there is no guarantee that the student will come up with an effective control system for treading water within the few minutes available before lack of oxygen leads to unconsciousness and death. Clearly, a less drastic approach is called for.

This method could be improved in a number of ways. First, we could simply allow more time for learning to take place. This could be accomplished by having the student practice at the edge of a swimming pool so that she could reach out and hold onto the edge of the pool when she felt herself going under water. Or she could practice in water that was only neck deep so that she could simply stand in the water at any time to breathe. Given more time to try out new patterns of movement and eliminate those that are ineffective, the likelihood of successful learning would increase.

Another approach would be to attempt to accelerate the learning process with verbal instructions ("move your hands horizontally in the water from your sides to the front and back again"), demonstrating a model for imitation, or a combination of the two ("do it like this"). Such instruction might be useful in *constraining* the student's attempts; for example, she would not now attempt vertical movements of her hands. But no matter how effective, it could not *transmit* the skill to be learned from teacher to student. Even if the teacher provides a model, the student must still learn on her own how to imitate it. The perceptions the student has of the teacher demonstrating the technique are very different from the perceptions she will have when she is able to perform the technique successfully herself (watching someone else swim is a very different experience than that of actually swimming oneself). Models and instruction can provide useful information in the form of constraints on what not to try, but they cannot provide explicit instructions concerning exactly what to do.

In addition to giving the student more time to learn and offering constraints in the form of models and verbal instruction, the teacher can provide easier access to the knowledge or skill by suggesting a series of less-demanding intermediate goals. One way is to break down the skill into a number of subskills and make opportunities for them to be acquired. The swimming teacher could have the student stand in shoulder-depth water and make horizontal movements with her arms until she feels an upward force lifting her weight from her legs. After she masters this, the student could hold onto a float and kick her legs until she feels herself rising from the water. After practicing the arm and leg movements separately, she could attempt to combine them, first in shoulder-depth and then in deeper water.

Breaking down a complex problem into easier subproblems facilitates learning since the probability of finding a solution to each subproblem is higher than that of finding a solution to the more complex problem. Success in learning to make effective arm movements alone in swimming is more likely than success in learning to make both arm and leg movements together.[18] A selectionist-reorganization view of learning sees the teacher as constantly aware of the student's current abilities and continually imposing upon her tasks that are just a bit beyond these abilities. Assuming that the student wants to be able to gain control over this new situation, reorganization will take place until she achieves control, at which time new demands are imposed (after learning to tread water, the breaststroke is attempted; after addition is learned, subtraction is introduced).

Such a view of learning is consistent with Russian psychologist Lev Vygotsky's (1896–1934) concept of the "zone of proximal development" in which the student tries and eventually successfully masters new problems that are beyond her independent capabilities but can be learned with the assistance of a teacher.[19] Note that the teacher is not a transmitter or instructor of information or knowledge, but rather one who provides support to the student and arranges the learning environment in such a way that she is continuously challenged by problems that are just a bit beyond her current competence. In other words, the teacher arranges the environment so that the student is continually encountering error, but error that is not too large, so that any reorganizing efforts are more likely to be successful and set the stage for the next introduction of error. This view is also consistent with the idea now popular in education that a successful teacher provides educational "scaffolds" for students. These are platforms that provide support in breaking down complex physical and cognitive problems into more easily mastered subproblems.[20] All this is applicable to the physical skill of learning to swim, as well as other more cognitive skills such as learning mathematics, developing reading skills, and learning to write.[21]

But how can we account for knowledge that is acquired without any essential accompanying behavior, as when a university student is expected to learn by listening to lectures and studying textbooks? How can perceptual control theory help us to understand how this is possible?

Here we will have to consider not just one control system as in figure 8.1, but see the person as made up of a complex hierarchy of control systems as described in chapter 8 and illustrated in figure 8.2. It will be recalled that this hierarchy has two principal features. First, it is a hierarchy of *perception*. At the bottom level, perception is limited to perceived *intensity* of stimulation of the sense organs—sounds can be loud or soft, lights can be bright or dim. But as we move up the hierarchy, more and more complex perceptions are possible. Certain combinations of intensities give rise to particular *sensations*, for example, the taste of orange juice or the color red, and certain combinations of sensations result in perceptions of *configurations*, such as those involved in seeing and recognizing an apple or a pencil. Perceived changes in configurations result in the perception of *transitions*, as when a baseball batter senses the approach of a baseball. This combination of lower-level perceptions into more complex, higher-level ones continues in Powers's current model to include a total of 11 perceptual levels ending in what he refers to as a *system concept*.

The second major characteristic of this hierarchy is that control systems are organized so that the reference level of any lower-level control system is given by the output of the next higher-order system. So this is a hierarchy of goals as well as perception. This means that to answer the question of why a person is controlling a certain variable, we have to consider the reference level of the system above it. Why, you ask me, am I now opening my car door and getting behind the steering wheel? Because I am going to drive to Peoria. Why drive to Peoria? To visit a sick friend in the hospital there. Why visit a sick friend in Peoria? Because he is a good friend and I feel I should visit good friends who are ill. Why do you feel you should visit good old sick friends? Because I consider myself to be a kind and compassionate person. Why do you consider yourself to be kind and compassionate? Because that is the type of person I want to be. Why be this kind of person? I am not sure. When one reaches a high enough level, it becomes difficult to answer any more "why" questions. Nonetheless, a hierarchy of control systems in which higher-level systems pass down reference levels (goals) to lower systems makes sense out of much of human behavior as well as our perceptions of our own behavior by which higher goals (being a kind and considerate person) influence behavior by setting lower-level goals (getting in my car and contracting my muscles in such a way as to drive to Peoria). And it provides a very useful framework for understanding behavior as being purposeful at many different levels.

But note that in this hierarchy only the very bottom level interacts with the physical environment outside of the person. My driving to Peoria, although dependent on higher-order goals, can be accomplished only by the integrated contraction of dozens of muscles in my feet, legs, hands, arms, and neck. That is the only way that I can actually perceive myself driving to and eventually arriving in Peoria. However, I am able to plan and imagine the drive without moving a single muscle. I can imagine seeing the highway entrance ramp approach through my windshield, accelerating onto the highway that will bring me there, hearing the whine of the motor as it revs up in third gear, and feeling the stickshift in my hand and accelerator and clutch pedals beneath my feet. And again, all this without moving a muscle.

One way of explaining how we can plan and imagine certain experiences is to suppose that "imagination connections" can exist in the control system hierarchy between the outputs at a given level and the input at the same level. When these are activated, as indicated in figure 12.1, lower-level systems are in effect bypassed so that perceptions can be made to match reference levels *without having to act on the environment at all*. Powers provides an example from playing chess:

Suppose one is trying to find a good program—say a program for dealing with the next few moves in a chess game. "If he moves his knight *there*, I'll move my rook *here*, but if he moves his knight the other way, I'll take it with the pawn." Of course in chess one is not permitted to move the pieces freely, nor would a good chess player (as I imagine him) babble to himself like this. Instead, he would simply *imagine* the relationships, not actually making the memory-derived images into active reference signals, but looking at them as if they had been accomplished. That is how a reference signal would look via the imagination connection—as if the lower-order systems had acted instantly and perfectly to make perception match the reference image.[22]

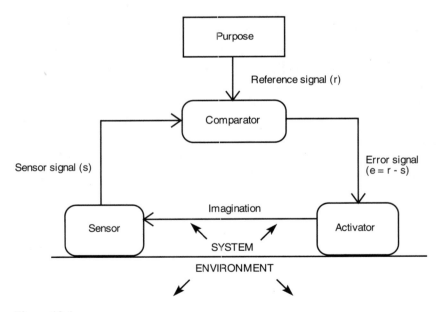

Figure 12.1
Imagination connection in a basic control system (after McClelland, 1991).

From this example we can recall that this hypothesized process of imagination is essentially what we considered in chapter 9 as *thinking*. But such thinking does not necessarily yield the instantly perfect world to which Powers refers. Indeed, the power of thinking lies in its ability to reveal problems that would not have become apparent otherwise—realizing, for instance, that taking the knight with my pawn would expose my queen to danger from my opponent's bishop, and that therefore I had better come up with another plan. Seen in this light, thinking becomes the vicarious trial-and-error-elimination process that was presented in chapter 9, equivalent to Campbell's visually and mnemonically supported thought.

Thinking, then, becomes a way of discovering problems and attempting to solve them without overt behavior. If the problem involves variables for which control systems are already in place (for example, "what is 2 plus 2?"), it can be solved quickly and routinely. But if it does not permit the routine application of already existing control systems (for example, "what is the relationship between the roots of two polynomials whose coefficients are reversed?"[23]), reorganization as a form of blind variation and selective retention involving visually and/or mnemonically supported thought will be called on to solve the problem (with, of course, no guarantee that it will be solved).

We can thus conceptualize the type of silent, covert learning that takes place in lecture and study halls as a type of internal problem solving that involves the reorganization of existing control systems. If I am taking a course on molecular evolution and want a good grade, I will have to understand how DNA replicates and how mutations can occur during the replication process. Right now, I perceive myself as not being able to do this, and so there is error between a goal and a perception. Reorganization must therefore take place so that I can reduce this error. Attending lectures and studying my textbook will facilitate this reorganization, not by transmitting to me the required information, but by enhancing the production of new ideas concerning DNA replication and mutation (blind variation) and helping me to eliminate my wrong ideas and retain the better ones (selective retention).

Thus, a perceptual control theory view of thinking and the reorganization that must take place during the acquisition of new knowledge is consistent with a selectionist view of knowledge processes in general and with education in particular. I should not give the impression, however, that such a view is widely known and accepted in psychological and education circles. On the contrary, perceptual control theory is undoubtedly one of the best-kept secrets in psychology and education, since these fields continue to see behavior and thinking as determined by a person's environment and have no adequate explanation of how they are functions of the internal goals of the individual; that is, behavior and thinking are *purposeful* processes. Nonetheless, the selectionist perspective on learning offered by the theory is consistent with both Campbell's insight into the ubiquity of cumulative blind variation and selective retention in all knowledge processes, as well as twentieth-century theories that emphasize knowledge growth and reject the idea of education as the transmission of knowledge from teacher and textbook to student.

IV

The Use of Selection

13

Evolutionary Computing

Selection Within Silicon

When man wanted to fly, he first turned to a natural example—the bird—to develop his early notions of how to accomplish this difficult task. Notable failures by Daedalus and numerous bird-like contraptions (ornithopters) at first pointed in the wrong direction, but eventually, persistence and the abstraction of the appropriate knowledge (lift over an airfoil) resulted in successful glider and powered flight. In contrast to this example, isn't it peculiar that when man has tried to build machines to think, learn, and adapt he has ignored and largely continues to ignore one of nature's most powerful examples of adaptation, genetics and natural selection?
—David Goldberg[1]

Over the last several decades the digital computer has become an important tool in a growing number of human activities. From writing school reports and managing household budgets to performing complex numerical analyses and simulating astrophysical events, the computer quickly becomes indispensable to anyone who takes the trouble to learn to use one. Its ability to store, manipulate, and analyze large amounts of data and provide stunning visual displays thereof have made it particularly useful in all fields of scientific inquiry.

Among its many scientific achievements, the computer has helped us to further our knowledge of biological evolution and the cumulative blind variation and selection processes on which adaptive evolution depends. Perhaps even more significant is that computers are now being used to solve extremely complex problems in many areas of science, engineering, and mathematics, generating solutions not imagined by the scientists who posed the problems and translated them into computer-readable form.

In this chapter we will survey some exciting new developments at the intersection of computers and evolution. In doing so, we will gain a better appreciation both of the power of the evolutionary process and how this power can be harnessed to solve complex problems—not over the centuries and millennia of geological time as in biological evolution, but over much briefer periods of time.

Recreating the Process of Evolution

Darwinian evolution remains the accepted scientific explanation for the origin and design of all the life forms on our planet; however, the slowness of biological evolution makes it impossible to observe in the way that other scientific phenomena can be studied. Solar eclipses can be observed from start to finish; cannonballs can be dropped from towers and their descents timed; and the entire life cycles of fruit flies, snakes, and rabbits can be studied. But biological evolution takes place over such long periods of time that its effects cannot readily be observed during any one human lifetime.[2] The large time scale of organic evolutionary change is no doubt a major reason for its rejection and misunderstanding by so many otherwise educated and well-informed individuals.

But computers have now made it possible to model the processes involved in biological evolution and to do this quickly enough so that they can be observed and studied over periods of time as short as hours, minutes, and even seconds. One of the first to create such simulations was Oxford evolutionary zoologist Richard Dawkins who reported his results in *The Blind Watchmaker*, his popular and colorful introduction to evolution.[3]

Dawkins's computer program is made up of two parts.[4] The first program, called DEVELOPMENT, produces on the computer screen a set of treelike line drawings called biomorphs, each representing an organism. The generation of each biomorph is based on a genome comprising nine genes, each of which can take a value from -9 to +9. For example, one gene controls width, so that a low value for this gene would result in a narrow biomorph and a higher value would produce a wider one. The remaining eight genes control other aspects of the biomorphs, such as height and the number of branchings added to the central stem. By specifying values for each of the genes, DEVELOPMENT produces the corresponding biomorph using the gene values as a sort of genetic recipe, not unlike the way in which real genes direct the development of living organisms.

But since evolution as we know it cannot occur without reproduction, this is the function of the second part of the program. REPRODUCTION takes the genes of the original single parent and passes them down to a set of 14 progeny. Although the reproduction is asexual, a rather high rate of random mutation is used so that each offspring differs from the parent on one of the nine genes. This provides the blind variation on which selection can operate.

Selection is determined by the eye of the program user. After each new litter of biomorphs is generated, the operator may select any one of the 14 mutant children to be the parent for the next generation. The criteria for this are completely up to the user. Dawkins provides a lively account of what happened when he decided to select cumulatively those biomorphs that most resembled an insect:

> When I wrote the program, I never thought it would evolve anything more than a variety of tree-like shapes. . . . Nothing in my biologist's intuition, nothing in my 20 years' experience of programming computers, and nothing in my wildest dreams, prepared me for what actually emerged on the screen. I can't remember exactly when in the sequence it first began to dawn on me that an evolved resemblance to something like an insect was possible. With wild surmise, I began to breed, generation after generation, from whichever child looked most like an insect. My incredulity grew in parallel with the evolving resemblance. . . . Admittedly they have eight legs like a spider, instead of six like an insect, but even so! I still cannot conceal from you my feeling of exultation as I first watched these exquisite creatures emerging before my eyes. I distinctly heard the triumphal opening chords of *Also sprach Zarathustra* (the 2001 theme) in my mind. I couldn't eat, and that night "my" insects swarmed behind my eyelids as I tried to sleep.[5]

Dawkins's program, although different in many ways from genuine biological evolution, dramatically demonstrates the potential of cumulative blind variation and selection. Each generation of 14 progeny is created from blind, random mutations of the parent's set of genes. Over this the human operator has no control. The operator does not create these forms, but rather they emerge from the blind variation algorithm of the computer program as part of the more than 322 billion combinations of nine genes, each having 19 different possible values. But by cumulatively selecting for certain characteristics, one can gradually discover biomorphs that resemble bats, lunar landers, foxes, scorpions, airplanes, or any of countless other possibilities (it took Dawkins only 29 generations to evolve his insect from a single point). And these myriad forms can be produced by a set of only nine genes, many orders of magnitude fewer than the millions that make up the genome of even relatively simple organisms.

As impressive as Dawkins's biomorph program is, it is quite unlike genuine biological evolution in that a human being is the sole agent of selection. This unnatural selection-by-humans has been used for hundreds of years in the selective breeding of food crops, flowers, and domesticated animals. In natural selection it is the organism's own success at surviving and reproducing that determines which variations will survive and which others will feel the blow of Darwin's hammer. And so Thomas Ray of the University of Delaware became obsessed with the idea of creating a computer environment that could simulate evolution free of any active human involvement in the selection process.[6]

In the artificial computer environment Ray calls Tierra, organisms are modeled as strings of computer code that compete with each other for memory space and processing time. Organisms that are successful in finding matching bits of code in their environment are able to reproduce, and those that are unsuccessful are eventually eliminated. Genetic variation is provided through processes designed to mimic the genetic mutations of real organisms caused by cosmic radiation and errors of replication.

Ray completed his programming of Tierra just as 1990 began, and on the night of January 3 introduced a single self-replicating organism into his silicon-based environment, or soup, as he called it. He thought at the time that this original ancestor would provide only a preliminary test of his simulator, and that it would likely take years of additional programming before a sustainable process resembling real evolution would emerge. But his rather pessimistic expectations were not to be borne out, as he recounts that "I never had to write another creature."[7]

Ray's original creature, called the Ancestor, consisted of computer code including 80 instructions. As it reproduced, its clones began to fill up the available memory space while the Ancestor and its oldest progeny began to die off. But then mutants containing 79 instructions appeared and successfully competed with the original clones. Before too long a new mutant arrived on the scene, but it had only 45 instructions, too few, Ray reckoned, for it to replicate on its own. And yet there it was, successfully competing with the much more complex creatures. Apparently this new organism was a type of parasite that had evolved the ability to borrow the parts of the necessary replication code from the more complex organisms. But as these parasites grew in numbers, they began to crowd out the host organisms on which their own reproduction depended, and so they began to die off as well. This decline in the numbers of parasites led in turn to an increase in the population of host creatures, causing an oscillatory cycle between the two types of organisms in the same way that cycles of growth and decline between hosts and parasites or predators and prey are observed in nature.

But more than just a simple moving pendulum between hosts and parasites was taking place. The hosts began to evolve characteristics that would make them resistant to the parasites, and the parasites found ways of circumventing these new defense systems. An ever-escalating evolutionary arms race was in progress, the very phenomenon that is believed to have provided the springboard for the increasing adaptive complexity of living organisms over evolutionary time.[8]

Ray also found other types of evolutionary processes occurring, such as forms of cooperation among highly related organisms, cheating, and sexual reproduction. Although he had designed both the Tierran environment and its first inhabitant, his role now switched to that of observer once this Ancestor began to reproduce, mutate, and evolve. But instead of observing only the end products of evolution as he had done during his many years of field work in the rain forests of Costa Rica, Ray was able to observe the process of evolution itself. And the new field of artificial life was finally on its way to becoming recognized by the scientific community as an important new arena for research.[9]

The work of Dawkins and Ray on simulated evolution are just two examples of the many investigations of artificial life that have been carried out.[10] Their work was among the first to show how computers could be used to simulate the same evolutionary processes responsible for all of earth's life forms. For those skeptical that the processes of cumulative blind variation and selection can produce the complex, adapted structures that make up living organisms, research in artificial evolution reports striking evidence consistent with the selectionist explanation proposed by Darwin for the origin and adaptive evolution of species over a century ago. As McGill University biologist Graham Bell noted, "Many people doubt that the theory of evolution is logically possible. . . . Now, one can simply point to the output of Ray's programs; they are the ultimate demonstration of the logical coherence of evolution by selection."[11] And although neither Dawkins nor Ray initially doubted the power of natural selection, even they were surprised to discover how easy it was to make evolution happen on a computer once the basic processes of variation, selection, and reproduction were modeled as programs.

Dawkins's and Ray's work also demonstrated the creative nature of evolution. Even though they were responsible for designing the computer environment, algorithms, and original digital organisms, the organisms that later evolved and the complex interactions among them were not explicitly programmed. Rather, they emerged in bottom-up fashion as the creative products of the evolutionary process itself. And if artificial evolution could be used to create complex line drawings and digital creatures able to survive and reproduce in a challenging silicon environment, it seemed reasonable to expect that the same process could be harnessed to solve practical, real-world problems, and perhaps to create intelligence itself.

The Computer *Can* Know More than the Programmer

Attempts to use evolution-based techniques to find solutions to novel problems and to develop artificial intelligence actually predate by quite some time the work of Dawkins and Ray. In 1966 Lawrence Fogel, Alvin Owens, and Michael Walsh published a slim volume in which they demonstrated how evolutionary processes implemented on a computer could be used to find solutions to problems involving prediction, diagnosis, pattern recognition, and control system design.

In their prediction experiments, these researchers started out by creating a single parent "machine" out of computer code. This machine was made up of a set of rules relating input conditions to output conditions, which was used to make the desired prediction. The prediction would be made and evaluated, after which the parent machine would produce an offspring with a slight, random mutation. The offspring's prediction would then be evaluated and compared with that of the parent's. If it was better, the offspring would be mutated to form the next machine; if the parent's prediction was better, the offspring would be discarded and the parent would be permitted to reproduce again. This process would continue until the predictions were within a desired range of error, or until some predetermined time or computing limit was reached.

This method can perhaps be better understood using the analogy of preparing a stew. You start with a recipe, perhaps one found in a cookbook, or provided by your mother or other respected culinary artist. After making up a pot of stew, you keep it warm while you make up another pot, randomly changing some part of the recipe, perhaps adding more salt or cooking it 20 minutes longer. You then compare the taste of the two pots of stew and discard the one (and recipe) you judge inferior. If the original stew was retained, another mutant stew would be prepared, tasted, and evaluated, but if the second stew was preferred it would be mutated again and compared with the new result. By repeating this procedure many times, the stew should continue to get better until any further change would only make it worse (according to your taste, that is).

Of course, for someone with knowledge of foods and cooking techniques, the outcome of a recipe will itself suggest certain changes—if the food is undercooked, cook it longer; if it is too spicy, cut back on the chili peppers. But in Fogel's method, as in biological evolution, mutations are completely blind in that the shortcomings of the current computer code provide no information about how the code should be changed to make it perform better. So for the cooking analogy to hold, we have to imagine a pure novice with absolutely no knowledge of cooking.

But this procedure is in one respect quite different from biological evolution. Whereas natural selection typically operates on a large population of organisms that vary in many different ways, this method compared only one parent and a single mutated offspring at each step. Nonetheless, this approach to evolutionary computing used a (rather limited) source of essential blind variation that, when coupled with cumulative selection, led to new solutions that had not been foreseen or in any way explicitly programmed into the simulation. And although their results were quite modest and did not have a major impact at the time on the field of artificial intelligence, Fogel and his associates were among the first to demonstrate a new, evolution-based way of using computers and to recognize its potential almost 30 years ago:

Computer technology is now entering a new phase, one in which it will no longer be necessary to specify exactly how the problem is to be solved. Instead, it will only be necessary to provide an exact statement of the problem in terms of a "goal" and the "allowable expenditure," in order to allow the evolution of a best program by the available computation facility. . . . The old saw "the computer never knows more than the programmer" is simply no longer true.[12]

Another pioneer in the application of evolutionary ideas to computing was John Holland of the University of Michigan. During the 1940s he was involved in the development of IBM's first commercial electronic calculator, the 701, and was intrigued by the problem of getting computers to learn in a bottom-up fashion from their environment. After completing his doctoral dissertation at the Massachusetts Institute of Technology in the late 1950s on the design of a parallel-processing computer (and thereby earning the first American Ph.D. in computer science), he discovered R. A. Fisher's landmark work on biological evolution published in 1930, entitled *The Genetical Theory of Natural Selection*. This book was the first attempt to provide a mathematical account of evolutionary theory, and in it Holland saw the potential of using a form of artificial evolution to enable computers to learn, adapt, and develop intelligence on their own.

Holland's intimate knowledge of computers and his doctoral research on the design of a multiprocessor computer led him beyond the asexual mutations and parent-offspring comparisons used by Fogel's group. Holland realized that in biological evolution many processes were occurring simultaneously. Natural evolution works using large populations of organisms. Each organism interacts with its environment, and those that are most successful in surviving and reproducing leave behind the most offspring for the next generation. In addition, most multicellular organisms use sexual reproduction in which genetic material from two parents is shuffled into new combinations involving a genetic crossover process and results in offspring that are both like and yet different from either parent. By adhering closely to these biological principles, Holland and his students developed what is now known as the genetic algorithm.

A genetic algorithm begins by generating a random population of binary strings, that is, a series of zeros and ones of a certain length such as 0100101110010. Each of these strings is analogous to an organism's genome and represents a tentative solution to the problem at hand. The problem may be as trivial as finding the value of the square of 13 (in which case the string would simply be treated as a binary number). Typically, however, it is a much more complex problem for which no solution or analytic technique is known to exist, such as finding the shortest path connecting 20 cities or maintaining a constant pressure in a natural gas pipeline network subject to varying user loads and occasional leaks. Each string is a type of shorthand for a computer program that can be run to provide a trial solution. These string-generated programs are then evaluated, with those most successful (for example, the top 10%) retained and the others eliminated. The selected strings are then allowed to pair off and "mate" with each other, each mating resulting in two offspring. For example, imagine two relatively successful parent strings 10000 and 01011. If the position between the second and third bits had been randomly chosen as the crossover point, the two parent strings would be cut in this position, yielding 10-000 and 01-011. By combining the first part of the first parent with the second part of the second parent, one offspring would be 10011; and combining the second part of the first parent with the first part of the second parent would generate the other offspring 01000. Occasionally, a bit or two is randomly flipped to simulate mutation,[13] and each string of each new generation is again evaluated. This process is repeated again and again, selecting the best strings and allowing them to mate and reproduce until a certain generation limit or error criterion is reached.[14]

Holland was betting that the processes of cumulative blind variation and selection that over many millions of years led to remarkable instances of biological adaptation and knowledge could be exploited on the computer to evolve useful algorithms using the computer's speed to compress hundreds of generations into mere seconds. It turns out that he won his bet, ultimately finding genetic algorithms capable of quickly and reliably evolving solutions to problems of daunting complexity. Scientists and engineers the world over are now using these algorithms to solve problems in many different areas. General Electric applied them to design gas and steam turbines and to make the jet engines of the new Boeing 777 more efficient. The *Faceprints* system developed at New Mexico State University allows a witness of a crime to evolve an image of a suspect's face using a system not unlike Dawkins's blind watchmaker program described earlier. The Prediction Company in Santa Fe, New Mexico, with the support of a U.S.-based affiliate of the Swiss Bank Corporation, uses genetic algorithms to make predictions useful for currency trading. Genetic algorithms have also helped design fiberoptic telecommunication networks, detect enemy targets in infrared images, improve mining operations, and facilitate geophysical surveys for oil exploration.[15] Even personal computer spreadsheet users can now find solutions to financial problems with genetic algorithms.[16]

But the evolution of evolutionary computation techniques was not to stop there. John Koza, a former student of Holland working at Stanford University, saw limitations in representing computer programs as unidimensional strings of zeros and ones. Since most complex computer programs are organized in a hierarchical fashion with higher-order routines making use of lower-order subroutines, he looked for a way of applying cumulative blind variation and selection directly to hierarchically structured computer programs. He did this with a computer language called LISP, which is structured in such a way that it lends itself particularly well to the crossover technique of sexual reproduction.

Each LISP program is structured as a hierarchical tree composed of mathematical and logical operators and data. In the technique he named genetic programming, Koza would first generate a random population of such program trees, including the operators he believed could help solve a particular problem. Programs represented by the trees would then be run and their results evaluated. Most of these programs, since they were randomly generated, provided very poor solutions, but at least a few would always be better than others. As in the genetic algorithm technique, programs that achieved the best results would be mated, and offspring programs would be produced by swapping randomly chosen branches—one each from each of the two parent trees—to produce two new program trees that were both similar to and yet different from their parents (figure 13.1). These new program trees would be evaluated, and the process continued until a program providing a satisfactory solution was found.

Koza's 1992 book *Genetic Programming: On the Programming of Computers by Means of Natural Selection* and companion videotape offer a veritable tour de force of genetic programming applications.[17] These include, among others, the solution of algebraic equations, image compression, robot arm control, animal foraging behavior, broom balancing, game-playing strategies, and backing up a truck to a loading dock. In applying genetic programming to the problem of finding the relationship between the distance of a planet from the sun and the time taken by the planet to complete one revolution around the sun, Johannes Kepler's third law of planetary motion was found, which relates the cube of the distance to the square of the time. Particularly intriguing was that on its way to this conclusion, the technique found a less accurate solution to this relationship that was the

Parent Computer Programs

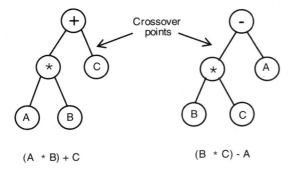

(A * B) + C   (B * C) - A

Offspring Computer Programs

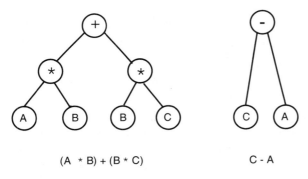

(A * B) + (B * C)   C - A

Figure 13.1
Sexual reproduction in genetic programming.

same as the one Kepler first published in 1608, 10 years before he discovered his error and published the correct version of his third law.[18] We can only wonder whether Kepler would have come up with the correct version years earlier if he had had access to genetic programming and let the program run beyond the intial solution.

In offering these many examples, Koza demonstrated that valuable computer programs can be developed by generating a random population of programs and then using a technique modeled directly after natural selection that takes the best programs and allows them to create new progeny sexually. Whereas such a technique would be hopelessly slow and ineffective if done manually, the recruitment of high-speed computers for this task has made it not only feasible but highly practical. Indeed, Koza showed that useful programs can be developed in this manner after as few as 19 generations. Among them are those that can generate crawling and walking behavior in simulated insects, perform aspects of natural language processing, make optimal bids and offers in commodity trading, perform financial analyses, control robots, generate art, and even produce bebop jazz melodies.[19]

Evolutionary computing techniques may turn out to be one of the most important developments in computer science of the second half of the twentieth century. The techniques are still quite young, but they are already beginning to find important commercial applications, and a number of regular scientific meetings, new journals, and electronic forums are available for disseminating the findings of research in this area.[20]

But the evolutionary approach to computing still encounters much resistance. The idea of using populations of randomly generated digital strings or LISP trees with trial-and-error elimination to write a computer program is so radically different from traditional approaches to software development that such resistance is perhaps not surprising. As science writer Steven Levy recounts from an interview with Koza:

Many traditional programmers, Koza explains, deplored the fact that a degree of chance was involved, both in the initial randomization and the probabilistic choices of strings that survived to the next generation. "Some people just broil at the idea of any algorithm working that way," says Koza. "And then you've got complaints that its sloppy. Because the first thing that comes to your mind when you hear about crossover is, 'Oh, I can think of a case where crossover won't work.' Of course you can think of a case. The key to genetic algorithms [as well as natural selection] is the population, that thousands of things are being tried at once, most of which don't work out—but the idea is to find a few that do."[21]

Indeed, many of the criticisms are not unlike the arguments marshalled against biological evolution itself. How can a complex, adapted system, be it a living organism or a computer program, possibly emerge from chance and randomness? How can such a wasteful procedure, in which almost all of the blindly generated entities in the initial and subsequent populations are eliminated, be effective in solving a complex problem, be it one imposed by a researcher on a computer program or one imposed by nature on a species? The power of nonrandom cumulative selection, of course, is the answer to these questions, and it works on the computer even better (or at least a lot more quickly) than it does in nature. And so the power of selection has begun to change the role of the computer from that of a recipient of the human programmer's knowledge to that of a generator of new knowledge.

Computers as Wind Tunnels: Simulations and Virtual Reality

Captain Emerson was definitely on edge as he lined up the Boeing 747 jumbo jet on his final approach to runway 27. He had spent thousands of hours in the air as an airline pilot, but this would be his first attempt to land the world's largest commercial passenger plane. In many ways, the Boeing 747 handled much like the much smaller 737 he knew so well. But the view of the runway from the cockpit situated four stories above the landing gear was quite unlike that of any plane he had flown. Believing that he still had several meters of altitude left when in reality he did not, the jumbo jet hit the runway hard, with a lot more force than even the monstrous landing gear of the 747 could absorb. Not only would this be the captain's first landing of a 747, it would also be his first *crash* landing.

Fortunately, however, Captain Emerson's error resulted in no injuries and no damage to the aircraft. For no actual aircraft or flying was involved. Rather, the captain had been training on a $16 million full-flight simulator, a piece of equipment that in some respects is actually more complex than the aircraft whose behavior it mimics. And so Emerson was able to learn from his error with no risk to life, limb, or equipment. He might well make a few more rough landings on the simulator, but none would be as rough as the first, and most would be better than the one before. When his turn would finally come to pilot a real 747, his first landing would seem just like one more of the many he had already performed in simulation.

The widespread use of flight simulators for pilot training is just one example of how computers are being applied to provide learning experiences and research opportunities that would otherwise be prohibitively expensive, time consuming, or risky. Although a certified flight simulator can cost many millions of dollars, many inexpensive simulation-based programs, including less sophisticated versions of flight simulators, are now available for use on personal computers. For example, the popular SimCity program created for IBM-compatible and Macintosh computers allows the user to design, build, maintain, and operate a simulated city complete with buildings, transportation systems, power-generating plants, police and fire departments, and occasional disasters such as fires, earthquakes, airplane crashes, nuclear power plant meltdowns, and even monster attacks. Another program called SimLife allows the user to design a planet, populate it with certain organisms, and allow simulated evolution to take place.

But specially designed simulation software is really not necessary to exploit selection on a computer to solve problems. Computer spreadsheets allow accountants and budget planners to investigate the consequences of various financial actions. Income tax preparation software permits taxpayers quickly and easily to investigate the consequences of various deductions and filing options, such as whether a married couple should file joint or separate returns. Drawing and design programs allow architects, engineers, and artists to try out their ideas for skyscrapers, bridges, lawnmowers, coffee makers, and sculptures without lifting a T square, hammer, screwdriver, or chisel. Even the ubiquitous word processor can be seen as a simulation of a typewriter and sheet of paper. Instead of making marks on paper, the word processor allows the writer to put easily changed characters on computer displays and magnetic disks, making revising and editing easier to accomplish than working directly on paper. More sophisticated examples are the simulations of mammalian nervous systems (to be considered below) and of new computer architectures using existing computers.

Although these examples may initially appear quite diverse, they are all alike insofar as the computer is used to simulate an environment that is more conducive to repeated cycles of trial-and-error elimination than the actual environment. Computer-aided variations can be generated more easily, more quickly, and more cheaply, thereby increasing the chances and reducing the cost of finding one that provides a good fit to a problem. In this way, the computer may be seen as a type of digital wind tunnel. We saw in chapter 10 how Wilbur and Orville Wright tested many of their ideas for aircraft design using quickly produced, inexpensive models in a wind tunnel, while their French competitors proceeded directly to full working implementations of their designs. By simulating the conditions of flight using models of propellers and wing shapes, the Wright brothers solved the problem of powered flight relatively quickly, and the French went from one crash to another. As aeronautical historian Walter Vincenti noted, "use of vicarious trial, both experimental and analytical, was a strength of the Wright brothers in comparison with their French contemporaries."[22] In much the same way, the computer now provides a means by which home-work assignments, hair styles, public transportation systems, and new supercomputer architectures (to name only a few examples) can be pro-posed, tested, and refined before producing a printed copy or real working model or prototype.

This vicarious variation and selection has much in common with and extends the use of the human brain to propose and test solutions to prob-lems, as discussed in chapter 9. We saw there that important functions of thinking are generating, testing, and selecting thought trials. Instead of hav-ing repeatedly to rearrange all the furniture in his living room until an acceptable new arrangement is found, the customer in the piano showroom can mentally generate, evaluate, reject, and finally select a new furniture arrangement that will accommodate his new piano. But although the brain has quite good verbal, visual, and auditory pattern-generating skills, it is less adept at performing certain complex mathematical computations and logical reasoning. So by including these mathematical and logical aspects in a simulated computer environment, a quite powerful combination of human intellectual and machine computational resources is formed for finding solutions to difficult problems.

The ultimate development of this brain-computer collaboration can be seen in the creation of what is known as virtual reality. The goal of devel-opers of virtual reality systems goes far beyond the small video displays and anemic loudspeakers found on most computers by making it possible for users to experience a simulated environment in much the same way that they experience the real world. Instead of seeing a new design for a home or office building on blueprints, architect and client can instead don dis-play goggles, step onto a steerable treadmill, and walk through a virtual building. An ultrasound-generated image placed between physician and patient allows a surgeon to see inside the patient to guide the initial critical moves of the scalpel. Molecular engineers can now touch and manipulate atoms the size of tennis balls to create new molecular structures that may have important scientific, engineering, and medical uses. Even virtual wind tunnels have been developed that allow aerospace and automotive engineers to enter the airstream virtually with a new vehicle design to examine from several perspectives how air flows past it at different speeds.[23]

Daunting technological challenges must still be overcome to make virtual reality a useful and accessible tool,[24] but virtual environments appear to be the ultimate step in facilitating variation and selection processes for generating new knowledge to solve new problems. A virtual reality-based architect can eliminate walls and add windows with a sweep of the hand to determine whether the change should be kept or undone. The wings of a new jet fighter can be modified instantly to see if some better combination of stability and maneuverability can be achieved. And the surgeon can rule out certain interventions without having first to subject the patient to exploratory surgery. Virtual reality is still in its infancy, but it seems inevitable it will play an increasing role in furthering our knowledge and technology in many fields due to the way that such virtual environments can greatly facilitate and accelerate the generation, variation, evaluation, and selection of potential solutions to complex problems.

Neural Networks

The relatively new field of artificial intelligence involves the creation of computer hardware and software designed to mimic and in some ways to surpass the perceptual, thinking, and reasoning powers of the human brain. Although many observers of this field have been disappointed with the results to date, there has been considerable progress in work that involves computer modeling and simulation of the workings of the brain and nervous system. Based on our current understanding of the brain as consisting of interconnected networks of billions of relatively simple units (neurons), developers of neural networks create computer simulations of neurons (sometimes referred to as neurodes) and connect them up in various ways to form networks that can learn and act intelligently in some way.[25] Such networks are able to recognize human speech and written letters, play games such as backgammon, and determine whether sonar echoes have been reflected from undersea rocks or mines.[26]

One of the most widely used neural network architectures is known as the backpropagation network (figure 13.2). This is typically a three-layer

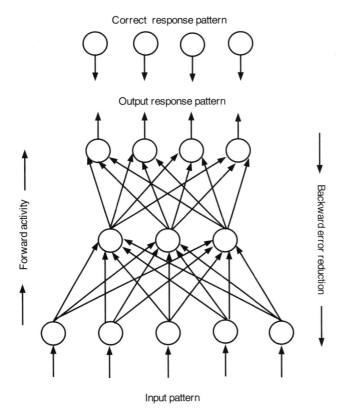

Figure 13.2
Backpropagation neural network (after Caudill & Butler, 1992).

network with a set of input neurodes connected through a set of middle-level neurodes to a set of output neurodes. It can produce certain desired responses (output patterns) when presented with certain input patterns. For example, the input patterns might correspond to received sonar reflections of the type used by submarine operators, and the desired output would be a pattern indicating the presence of either a rock or a mine, providing in effect an automatic and accurate classification of the object being sensed.

For a neural network to behave in this way, a particular pattern of connection strengths (analogous to the strengths of synaptic connections between neurons) must first be found between each neurode and the neurodes in neighboring levels. With the connection strengths among the neurodes initially set randomly, a pattern is presented to the input level that causes the network to produce an output pattern that will be interpreted as either "rock" or "mine." Since the initial connection weights are set at random, the initial outputs are most likely to be wrong. However, a training procedure is used whereby each output pattern is compared with the desired (correct) output pattern, and changes to the connection strengths are made according to a mathematical formula that changes the connection strengths of those neurodes that most contributed to the error. This use of error to go backward into the network to change the connection weights gives the backpropagation network its name. After many iterations of this procedure—providing an input, observing the output, noting the discrepancy between the actual output and desired output, and changing the synaptic weights among the neurodes to reduce the error—the network may learn to classify correctly not only input patterns on which it was trained, but also new ones it has not experienced before. In addition to classifying sonar signals, such networks can remove noise from medical instrument readings, produce synthesized speech, recognize written characters, and play various board games.[27]

It is of particular interest to note that no trial-and-error and no blind variation and selection are involved in the actual training and functioning of a backpropagation neural network. Instead, during training the difference between the network's actual output and the desired output is used to modify synaptic weights in a deterministic manner calculated to reduce error the next time that input pattern is encountered. This is the first clear example we have seen in this book of adaptive change resulting from an instructionist process—the instruction provided by that part of the computer program that trains the network.

But this instructionist approach to training a neural network has certain noteworthy limitations. First, the designer must initially decide on the number and organization of the neurodes in the network. Particularly important is the number of middle-level neurodes—too few, and the network will not be able to distinguish between subtle but important differences in input patterns; too many, and the system may not be able to generalize what it has learned to new input patterns on which it was not trained. And since there is no way to know beforehand the optimum number of such units, this initial design is actually the result of trial and error on the part of the network developer.

Second, the correct corresponding output, that is, the proper classification for each encountered input pattern, *must already be known* to train the network. A sonar-detecting backpropagation network cannot be used to distinguish rocks from mines if it is not already known which signals indicate rocks and which indicate mines. In other words, the instructionist learning procedure used by backpropagation neural networks can work only if the instructor already knows all the right answers for the training inputs. So this process cannot be used to discover new knowledge or develop new skills, but is a way of transferring knowledge from one source to another.

Third, backpropagation neural networks cannot always be trusted to find just the right combination of connection weights to make correct classifications because of the possibility of being trapped in local minima. The instruction procedure is one in which error is continually reduced, like walking down a hill, but there is no guarantee that continuous walking downhill will take you to the bottom of the hill. Instead, you may find yourself in a valley or small depression part way down the hill that requires you to climb up again before you can complete your descent. Similarly, if a backpropagation neural network finds itself in a local minimum of error, it may be unable to climb out and find an adequate solution that minimizes the errors of its classifications. In such cases, one may have to start over again with a new set of initial random connection weights, making this a selectionist procedure of blind variation and selection. One might also use other procedures such as adding "momentum" to the learning procedure, analogous to the way a skier can ski uphill for a while after having attained enough speed first going downhill, or adding noise to the procedure to escape such traps, again a form of variation and selection.

Finally, for those looking to research on neural networks to further our understanding of how real neurons work in real brains, backpropagation neural networks are not biologically plausible. Unlike the neurodes and their interconnections in a backpropagation network that conduct signals in both directions (one way for responding, the other direction for learning), biological neural pathways conduct signals in one direction only. Also, much if not all human learning occurs without the direct instruction and the patient training of the type backpropagation networks require. So although these networks can be useful in quite a number of interesting applications, their instructionist nature does impose limitations on both their adaptive flexibility and their applicability to biological neural systems.[28]

Given the limitations of instructionst training, it should not be surprising that selectionist approaches to designing neural networks have also been developed.[29] Some of the first were developed by Andrew Barto and his colleagues at the University of Massachusetts in the early 1980s. These researchers demonstrated that neural networks could learn to solve difficult problems by a selectionist process referred to as reinforcement learning as opposed to the supervised learning characteristic of backpropagation and other instructionist networks.[30] For example, one system learned to balance a pole hinged to a movable cart,[31] and others solved spatial learning problems.[32]

Another type of selectionist neural network is known as a competitive filter network. Like the typical backpropagation network, it consists of three groups of interconnected neurodes—the input layer, the middle or competitive layer, and the output layer. The network learns to make useful output responses to input patterns, but without the help of an instructor. An example is learning to recognize spoken words. For such a task the input layer encodes the sounds of the detected word and sends the resulting pattern of activity on to the middle layer. Since each neurode in the middle layer is connected to each and every neurode in the input layer, every middle-layer neurode is activated to some extent by the input pattern. But each middle-layer neurode has a different pattern of (initially random) weights that mediates the effect of the input pattern. Those having the pattern that best matches the input pattern are the most active.

In addition, connections among the middle-layer neurodes are arranged so that a highly active neurode excites neighboring neurodes while inhibiting those neurodes farther away. The net result of all this simulated neural activity is *competition* among the middle-layer neurodes, with the best-fitting ones sending their signals on to the output layer and allowed to adjust their weights so that they match the input signal even more closely than they did previously. After many trials, such a network can learn to categorize input patterns into meaningful and useful categories so that one and only one middle-layer neurode will respond to all or nearly all pronunciations of a specific word. Teuvo Kohonen, who did extensive work with such competitive filter networks, used such a system to create one of the first voice typewriters.[33]

It is not difficult to see the evolutionary processes of variation, selection based on competition, and retention operating in such a network on the connection weights between neurodes. The initial distribution of weights for the middle-layer neurodes is usually determined randomly. Then competition occurs among these neurodes, with the most active one winning and selected to be modified to match the original input pattern even better. But because this modification is directed toward a better matching of the just-received input pattern, there is no guarantee that this modification of weights will be retained, as more input patterns are received and further rounds of activation, competition, selection, and modification of connection weights take place.

Another class of neural networks known as adaptive resonance networks also use processes of variation, competition, and selection. For them,

> . . . the basic mode of operation is one of hypothesis testing. The input pattern is passed to the upper layer, which attempts to recognize it. The upper layer makes a guess about the category this bottom-up pattern belongs in and sends it, in the guise of the top-down pattern, to the lower layer. The result is then compared to the original pattern; if the guess is correct . . . the bottom-up trial pattern and the top-down guess mutually reinforce each other and all is well. If the guess is incorrect . . . the upper layer will make another guess. . . . thus, the upper layer forms a "hypothesis" of the correct category for each input pattern; this hypothesis is then tested by sending it back down to the lower layer to see if a correct match has been made. A good match results in a validated hypothesis; a poor match results in a new hypothesis.[34]

Gerald Edelman, whose neural Darwinism was discussed in chapter 5, has also been actively involved with his associates in applying Darwinian selection to neural networks to simulate the adaptive functioning of the brain.[35] Using what he calls synthetic neural modeling, he and his associates have created a series of neural simulations named after Darwin that explore and demonstrate the principles of neuronal selection. Darwin III is a computer simulation of a sessile (seated) creature possessing a head with a movable seeing eye and a jointed arm that has both touch and kinesthetic sensors.[36] Inhabiting a computer world of simulated stationary and moving objects of various shapes and textures, Darwin III learns to grasp certain objects and repel others. Although certain initial biases are built into Darwin III, for example, a preference to grasp round, smooth objects and reject square, bumpy ones, the actual perceptual categories and behaviors are not programmed. Rather, Darwin III learns to fixate visually and track objects, distinguish different types of objects, and grasp some while rejecting others through Darwinian selection of connection weights between the neurodes, in much the same way it appears that the human brain is able to learn through the variation and selection of synaptic connections among its neurons. With the recent creation of Darwin IV, Edelman and his associates have moved beyond computer simulations and created a working robot that performs the tasks of its predecessors in a real-world environment of physical objects.[37]

Before leaving our discussion of neural networks, let us return briefly to the instructionist networks described at the beginning of this section. It will be recalled that these systems of simulated neurons can indeed grow in adapted complexity through an instructionist process by which synaptic weights are modified in the proper directions without variation and selection. This being the case, however, we must wonder how it is that these network architectures and training procedures came to be, being themselves examples of adapted complexity and thereby posing additional puzzles of fit. Unless they were discovered in some novel manner completely unlike that responsible for other scientific and technological innovations, we must suspect that they owe their design to repeated cycles of blind variation and selection, not in the working of the networks themselves but in repeated overt and vicarious (cognitive) generation of the networks and their subsequent testing by scientists. And since, as we saw earlier in this chapter, computers can now simulate such selection processes, we should not be too surprised to learn that recent work in genetic programming has shown that the computer itself can be used to evolve neural networks to solve problems and perform various tasks.[38]

It can be argued that the digital computer is the most important technological tool of the twentieth century, but it is quite often viewed as a sophisticated electronic combination of file cabinet, calculator, and typewriter. Most computers are indeed still used in these ways. But we are now beginning to see exciting new applications that go far beyond their standard office and scientific functions. Computers have made it possible to model the evolutionary process itself and thereby convincingly demonstrate the cumulative power of the combined processes of blind variation and selective retention. They can be used to simulate real-world environments, allowing scientists and engineers to test many trial solutions rapidly, economically, and safely. They are being programmed to invent, test, and improve their own solutions to exceedingly complex problems using the techniques of genetic algorithms and genetic programming.

One cannot help but wonder where all this will ultimately lead. Will future computers and programs and perhaps even robots be bred in ways analogous to the ways that chickens, ornamental flowers, and corn are bred today, only much faster? Will programs and the machines they run on begin to evolve intelligence and understanding that will eventually approach and perhaps even surpass that of their human designers? Or will certain undiscovered limits to the evolutionary potential of computer hardware and software prevent them from achieving anything near the adapted complexity that organic evolution has achieved over the last four billion years? These are questions that only time can answer. But it now seems almost a certainty that as computers themselves continue their own technological evolution, thereby acquiring ever-increasing memory, processing power, speed, data storage, and audiovisual capabilities, the evolutionary processes first discovered among carbon-based life forms will be used increasingly within silicon-based machines.

14

The Artificial Selection of
Organisms and Molecules

It is wonderful what the principle of selection by man, that is the picking out of individuals with any desired quality, and breeding from them, and again picking out, can do. Even breeders have been astounded at their own results. . . . Man, by his power of accumulating variations, adapts living beings to his wants—may be said to make the wool of one sheep good for carpets, of another for cloth, &c.
—Charles Darwin[1]

Ever since the appearance of the first life forms on our planet, organisms have influenced each other in their evolution and resulting adaptations. They compete for sunlight, food, shelter, and mates, with the most successful passing down the accumulated knowledge of their genomes to the next generation. Species involved in parasite-host or predator-prey relationships may evolve sophisticated offensive and defensive equipment and behaviors in a continuing evolutionary arms race. Other species have come to depend on each other for survival as symbiotic relationships evolved, such as when flowering plants offer sweet nectar to insects in return for the insects' dissemination of the plants' pollen.

Although relative latecomers to this scene, humans driven by their continuing and increasing need for food, shelter, clothing, fuel, beauty, and amusement, have had a particularly dramatic effect on the evolution of some species. In this chapter we will see how that influence predates by many thousands of years our knowledge of evolution, and how our current understanding of natural selection coupled with advances in biotechnology have provided us with unprecedented power to create new and useful organisms and molecules.

The Human Selection of Plants and Animals

For most of its existence, our species lived much like other mammals, roaming its habitat and obtaining food by hunting, scavenging, and gathering edible leaves, nuts, seeds, fruits, and tubers. But this nomadic way of life began to change as humans discovered the advantages of taking more direct control over the growth and breeding of plants and animals. The first domestication of food plants probably took place between 10,000 and 13,000 B.C. in southeast Asia where crops such as rice and beans were planted and harvested. Two other sites where agriculture appears to have developed independently were the Fertile Crescent, which includes parts of present-day Iraq, Iran, Turkey, Lebanon, Israel, and Jordan, and south central Mexico. In the Fertile Crescent animals were also domesticated, such as the horse, donkey, camel, and sheep.[2]

This change in occupation from hunting and gathering to cultivating food crops and raising livestock had important consequences for human cultural evolution. In contrast to moving from one temporary camp to another, agriculture permitted the establishment of permanent villages and ultimately cities, city-states, and empires. Increasing agricultural productivity freed a significant portion of the population from the demands of growing food and allowed them to take up scientific, technological, religious, and artistic endeavors. But in addition to these sweeping and comparatively rapid cultural changes, another slower one was taking place—we were beginning to direct the evolution of increasing numbers of plant and animal species.

No understanding of evolution or genetics was necessary to begin this transformation of useful organisms; neither was a conscious and purposeful selection of plants and animals. That is because the practice of agriculture itself necessarily imposes new environmental conditions on plants and animals. Crops are planted at certain times of the year using specific methods of cultivation. As in natural selection, plants that by chance are better suited to these human-made conditions grow well and produce more seeds for the next planting. For example, most wild plants produce seeds that fall to the ground and are dispersed at maturity, but a few hold on to their seeds, facilitating their gathering. By selecting and then sowing more seeds from the latter, a selection pressure was created toward the evolution of plants with so-called non-shattering seeds that were easier to harvest.

But although such early artificial selection may have well been accomplished unintentionally, even these prehistoric peoples no doubt noticed that, among living plants and animals, like begets like. Horses give birth to horses, not ducks. From the seeds of a fig tree more fig trees sprout, not palm trees. And human couples reproduce children who bear an obvious resemblance to their parents. So as these people chose seeds to plant the next season's crop from the largest and most productive stems of wheat, barley, corn, or rice, and allowed the largest, strongest, best-tasting, or gentlest horses, cattle, or camels to mate, they were effectively exploiting the mechanisms of cumulative blind variation and selection to produce plants and livestock that were better and better adapted to human requirements.

Indeed, it was the observation that domesticated plants and animals changed under the selection pressures imposed by humans that led Charles Darwin to his theory of natural selection. The first chapter of the *Origin* is entitled "Variation Under Domestication," and here Darwin presents many examples of how domesticated plants and animals changed over time to become better adapted to human needs. He explains these changes as "man's power of accumulative selection: nature gives successive variations; man adds them up in certain directions useful to him. In this sense he may be said to make for himself useful breeds."[3] When he realized that such selection pressures also exist in nature without meddling by human agents, his theory was born.

The theory made explicit the principles of evolution that agriculturists the world over had been unwittingly using for millennia. For some obscure reason nature serves up variations, and only certain ones are selected by humans for breeding the next generation. These ancient farmers had no way to control the amount or direction of these naturally occurring variations, but by eliminating the undesirable and breeding from the desirable, they could control just about any observable or measurable characteristic of plant or animal.

The amazing success of early plant breeders was demonstrated by Native Americans who over the course of 4000 years transformed a stingy grass into one of the world's most productive food crops. When Europeans first set foot on the shores of the New World

Adapted maize [corn] cultivars extended from the southern part of South America to the north shore of the St. Lawrence River; from sea level to elevations of 3,355 m (11,000 ft.). Types included flint, flour, pod, and popcorn as well as red, blue, black, yellow, white, and variegated kernels. There is no doubt about the competence in plant breeding of the American Indians, as it took the development of F1 hybrid maize of modern genetics to exceed the performance capability of Indian maize. It is understandable that European settlers grew to respect the maize crop and its developers, for maize is said to be the greatest gift from the Indians.[4]

But although Native Americans and other early agriculturists were successful in breeding plants and animals suited to their conditions and needs, it was not until the insights of Darwin into evolution and those of Gregor Mendel concerning the genetic basis of inheritance that plant and animal breeding could begin to be put on a firm scientific foundation. This foundation permitted rapid advances in breeding techniques beginning in the 1900s.

Since plant and animal breeding is a form of adaptive biological evolution, it depends on the three components of variation, selection, and reproduction. Advances therefore involve one or more of these components. By increasing variation, the probability of finding a variant with desired characteristics is increased. Breeders do this with techniques such as raising large numbers of plants or animals, cross-breeding different varieties to produce new hybrids, and using irradiation and chemicals to increase the occurrence of mutations.

Other techniques that improve the accuracy and efficiency of selection include procedures involving physical and chemical comparisons coupled with statistical and screening methods to eliminate unsuitable specimens quickly. For example, to develop a new variety of wheat that is resistant to high levels of salinity, high levels of salt are applied to the soil. This will kill almost all of the affected wheat plants, but almost certainly a few plants will survive. These survivors can then be used to breed a line of salt-tolerant plants. Screening (the breeder's term for selecting certain individuals from a large population of specimens) is now possible in the test tube where it is referred to as in vitro selection. Individual cells from plants may be subjected to certain chemical toxins. Those that are able to grow despite the presence of a toxin are mutants that are naturally resistant. Using this method, corn with 10 to 100 times more resistance to certain herbicides has been developed.[5]

Advances in molecular biology have also had an impact on breeding. Once a particular gene has been identified that is associated with a desired characteristic, breeders no longer have to wait until the plant or animal matures to make their selection. They can do it by examining the genes (or inserted gene markers) of the cells of immature plants and animals, thus cutting in half the time normally required to determine whether a newly bred plant or animal has the desired genetic characteristic. With this technique, genes that cause corn to produce kernels with high oil content have been identified, and this corn is now being cross-bred with other varieties having other desirable characteristics such as high yield or standability. Genetic screening allows the rapid selection of only those plants that retain the gene for high oil content.

Other methods have been developed to enhance reproduction. The realization that each cell of a plant or animal contains all the genes necessary for the production of a new, identical organism led to attempts to reproduce organisms using single cells. This asexual propagation of normally sexually reproducing plants and animals is called cloning, and it permits useful varieties to be preserved with little or no genetic variation over many generations. Where sexual reproduction cannot be avoided (at this writing, no mammal clones have yet been produced), artificial insemination permits a bull with desirable genes to impregnate many more cows than he could ever inseminate using traditional bovine love-making techniques. And the in vitro fertilization of eggs from a desired cow with the sperm of a desired bull and their transplantation into surrogate mother cows for gestation and birth allow a single particularly desirable couple to produce a large herd of animals, all of them siblings.

Genetic Engineering

These twentieth-century advances that permit more control over the variation, selection, and reproduction of living organisms greatly facilitated the breeding of domesticated plants and animals. However, none of these methods makes it possible to produce the desired variations directly. Variation can be increased; more accurate and quicker selection methods can be employed; and reproduction of desired varieties can be increased. But the breeder is still limited to selecting among those chance variations that the organisms themselves provide.

To gain more direct control over desired characteristics, breeders had to await developments of the second half of this century in the relatively new field of molecular biology. After Watson and Crick's discovery of the structure of the DNA molecule in 1953 and the subsequent breaking of the genetic code by which sequences of nucleotides orchestrate the construction of proteins, it was only a matter of time before scientists developed methods to reach deep into the center of living cells and manipulate their genes.

Many different techniques are now employed in genetic engineering, which is also known as gene splicing and recombinant DNA technology.[6] They all involve manipulating an organism's genes. One technique involves introducing a gene from a plant or animal cell into a bacterium. Many bacteria have tiny rings of DNA, known as plasmids, that are more accessible and easier to manipulate than the more tightly packed genes in the chromosomes. When a gene that produces a desired protein is removed from a plant or animal cell and spliced into a bacterium's plasmid, the gene will continue to produce its designated protein product in its new bacterial setting. Provided with a nutrient-rich environment, the bacterium quickly grows and divides, producing more cells that also contain the foreign gene and that therefore also produce the desired protein, which can then be harvested from these bacterial chemical factories. With this technique, bacteria have been genetically engineered to produce insulin, human growth hormone, and the anticancer drug interferon, among thousands of other important substances.

But what about introducing foreign genes into the chromosomes of plants and animals to alter their characteristics and to create brand new organisms, such as the fire-breathing chimera of Greek mythology, which combined the head of a lion, the body of a goat, and the tail of a snake? Unfortunately, manipulating chromosomal DNA is a more difficult affair. For certain plants, the aid of *Agrobacterium tumefaciens* is enlisted. This soil bacterium has been doing genetic engineering of its own for millions of years by inserting genes from its plasmids into the chromosomes of plant cells. The foreign DNA causes the plant cells to form tumors that produce unusual compounds (opines) that serve as food for the bacteria. Thus, this bacterium is a naturally occurring genetic engineer that has evolved the ability to alter plant cells to provide food for itself. This ability has been exploited by human genetic engineers who can now splice desired genes (for example, one that provides resistance to a pathogenic virus) into *Agro-bacterium's* plasmid, which inserts it into a plant cell's chromosomal DNA. The targeted plant cell is then coaxed into developing into a complete fertile plant that will pass on the engineered DNA and accompanying viral resistance to its progeny.

For manipulating the genes of animals and those plants for which this technique will not work, such as the major food crops of wheat, rice, and corn, a messenger virus is used, or the desired gene is directly injected into the nucleus of a plant cell or into a one-cell animal embryo. In the case of animals, the genetically altered embryos are implanted into the uteri of surrogate mothers, and the resulting transgenic animals can be used to breed more progeny by traditional methods, with each offspring containing the altered gene. The first transgenic animal was produced in this manner in 1981 when a rabbit gene was inserted into a mouse embryo.

Although genetic engineering is a very young field, it has already had some impact on food production and promises to have much more in the future. About half the cheese produced in the United States uses an enzyme created by bacteria containing a cow gene. Similarly, recombinant bovine somatotropin (rBST) is grown in bacteria, isolated, and injected into dairy cows, increasing milk production up to 20%. The Flavr Savr tomato, genetically engineered to stay fresh longer before spoiling, made its appearance in American supermarkets in May 1994. And extensive research is under way by firms such as Calgene, Monsanto, and DuPont to genetically engineer crops resistant to herbicides, harmful insects, viruses, bacteria, and fungi.

Whereas genetic engineering now provides means for directly manipulating genes, these techniques have not eliminated the use of cumulative trial-and-error research from the design of new and valuable organisms. First of all, to know which gene to insert into an organism, the gene's function must be known. One way to determine the function of a particular gene is to expose a large number of organisms to a mutagen, select those that differ in some interesting way from normal organisms, and then establish through DNA sequencing which gene was mutated. If, for example, a mutated bacterium is unable to replicate its DNA, and the genes that were changed by mutation can be determined, it will be known that these genes must in some way be involved in DNA replication. This method allows no control over what mutations will result (thus they remain blind variations), and organisms are consequently selected on the basis of some particular characteristics. This technique can be effective for large populations of single-cell organisms whose genomes are relatively small (*E. coli* contains about 3000 genes in its genome), but it is less so for larger organisms with larger genomes (the common fruit fly has 20,000 genes and the mouse about 200,000). For these organisms, researchers traditionally relied on the study of naturally occurring mutations, but a new technique allows them to manipulate the genes of their choice.

Currently applied to mice, with which we share more than 90 percent of our genes, targeted gene replacement (also called homologous recombination), developed by Italian-born Mario Capecchi at the University of Utah School of Medicine, makes it possible to create mice with a mutation in any gene of interest.[7] But this relatively precise genetic control is still largely a hit-and-miss process since it is successful in only a very small percentage of treated cells. It is much more likely that the introduced gene will either not be incorporated into the cell's chromosomes, or that it will be, but at the wrong location. By incorporating two additional marker genes into the foreign gene, one that provides resistance to a certain drug and the other that provides sensitivity to another drug, it is possible to screen the treated cells using the two drugs and easily find the one cell in a million whose genome had been altered in the desired way. Only these cells are selected and injected into a mouse embryo, eventually leading to transgenic mice with the precise desired mutation. The mice are examined for physical or behavioral abnormalities that provide clues to the function of the altered gene. So once again we see that a selectionist screening procedure is necessary to separate successfully genetically engineered cells from the others. Similar screening methods employing antibody or enzyme detection are applied in many of the genetic engineering techniques described earlier.[8]

But even the identification of the function of individual genes does not permit one easily to create new organisms with desired characteristics for the simple reason that many of the most important traits are controlled by combinations of hundreds or perhaps even thousands of genes. It is therefore necessary to test various combinations of genes to determine which one provides the desired characteristic, such as early maturity, high oil content, height, or drought resistance, while maintaining other beneficial traits. It is for this reason that despite the rapid progress in genetic engineering, much more research must be done before plants can be grown that combine the advantages of both corn and soybeans, or animals are created that both give milk like cows and grow wool like sheep.

The Evolution of Drugs and Methods of Drug Design

Regardless of its remarkable achievements, the continued presence of a plethora of human illnesses is a constant reminder that biological evolution does not fashion perfectly adapted organisms. Indeed, we now understand infectious diseases as Darwinian competition between us and various pathogenic parasites, bacteria, and viruses. Many of these organisms have evolved remarkably effective methods for infecting human hosts, and have certain important advantages over us, such as much larger population sizes, prolific reproduction, and high rates of mutation. But although vastly outnumbered, we can and do fight back against our microbial adversaries using our intelligence and the knowledge and technology that it has generated.

The discovery, refinement, and administration of drugs is our most important weapon in this continuing battle. Drugs have been used since ancient times, with the Greeks and Romans prescribing opium to relieve pain, the Egyptians taking castor oil for worms, the Chinese consuming liver to treat anemia, and twelfth-century Arabs ingesting sponges (which have a high iodine content) to treat goiter. But since these early users of drugs had virtually no understanding of chemistry and physiology, these and other agents could have been developed only by the crudest trial-and-error methods. Perhaps healers noted that an individual with a certain malady recovered after ingesting a certain food or substance, and tried it on another person suffering from the same illness. Or they may have observed animals eating certain plants when they appeared to be suffering from ill health. Such methods probably led to the discovery of certain useful drugs such as quinine to treat malaria and the aspirin-like substance in the bark of willow trees to treat pain and fever. However, the lack of a systematic approach to drug research, coupled with ignorance of the structure and functioning of the body, inevitably resulted in many ineffective and harmful drug treatments.

The Beginning of Scientific Drug Development

Important advances in drug development began about 1800, although they too often resulted from serendipity rather than systematic scientific research. In the 1790s English physician Edward Jenner heard that dairymaids who contracted the relatively mild cowpox disease seldom suffered from the much more serious and often fatal smallpox. Consequently, he suspected that cowpox somehow provided immunity against smallpox, and developed a smallpox vaccine from the pus of cowpox sores. In the 1870s French chemist Louis Pasteur observed that chickens inadvertently injected with weakened cholera bacteria developed immunity to this disease; he later developed vaccines against anthrax and rabies. And in 1928, Scottish bacteriologist Sir Alexander Fleming observed that a mold that accidentally contaminated a culture of *Staphylococcus* appeared to stop the bacterium's growth, and thereby discovered penicillin.[9]

The development of these and other modern drugs was greatly facilitated by knowledge of the causes of disease. It was only through his awareness that infectious diseases were caused by microorganisms that Fleming was able to recognize the importance of the antibiotic produced by the penicillium mold. Recent discoveries of the key role of enzymes and various receptors in cellular activity permitted important advances in the development of a wide range of new drugs.

Finding Drugs by Large-Scale Random Screening

Drugs exert their effect by providing molecules that are able to fit and attach themselves to other molecules in the body, not unlike the way that an antibody fits an antigen, as discussed in chapter 4. For example, angiotensin-converting enzyme (ACE) acts as a catalyst to convert angiotensin I to angiotensin II, the latter being a vasoconstrictor, which reduces the diameter of blood vessels, thereby increasing blood pressure. The drug lisinopril has a distinctive molecular shape that tightly fits and effectively plugs up the active site of ACE, thereby inhibiting its activity in converting angiotensin I to the vasoconstricting angiotensin II. The net effect of this drug is therefore to reduce levels of angiotensin II, resulting in lowered blood pressure in hypertensive patients.

From the molecular perspective, finding an effective drug can be likened to finding a key for a lock. In what is now referred to as the classic method of modern drug design, as many as tens of thousands of synthetic and natural substances may be randomly tested in groups of 50 to 100 as potential keys to a desired lock such as ACE. If it is found that one of these substances binds with the target site, further screening is done to isolate the particular keylike compound, which is referred to as the lead compound or molecule.

Once a lead compound is found, much more work remains to be done in the form of additional modifications and more tests. Variations of the lead molecule are then synthesized and tested. Some of these variations may be less effective than the original molecule, and others may be much more potent. An example of the latter is etorphine, a variant of morphine that is 1000 times more potent than the morphine. But this is only the beginning. The ability of the molecule and its variations to reach the target site (its bioavailability) must be tested using living cells and tissue cultures. Then the toxicity of the drug and its side effects must be determined in trials with animals and humans. The object of all these modifications and testing is to find the compound that is most effective while having the least toxicity and fewest unwanted side effects. Modern drug development involves many sophisticated techniques for zeroing in on such compounds,[10] and the entire process is clearly one of a rather lengthy cumulative-variation-and-selection procedure. Because of the large number of compounds to be screened, followed by required rigorous testing on cells, tissues, animals, and finally humans, it may take more than 10 years of intensive research effort and many millions of dollars to develop a new drug and bring it to your medicine cabinet. This provides some explanation for the high price of many drugs.

Nonetheless, the initial in vitro chemical screening of large numbers of molecules is much more efficient (and safer) than having first to test all these compounds on animals or humans, particularly since pharmaceutical firms employ technology that largely automates the random screening process. Indeed, large-scale molecular screening can be seen as a type of vicarious blind variation and selection in which many types of molecules are tested, and only those that are found to have an affinity for the target enzyme or receptor are retained for additional testing and development.

But this method of molecular selection has an important limitation that becomes obvious when compared with the cumulative variation and selection of biological evolution (as well as the evolutionary computing techniques considered in the preceding chapter). Namely, only natural and synthetic compounds that are already on hand can be initially screened, as there is no way to create spontaneously new variations of successful molecules in the way that mutations and sexual genetic recombination provide new variations in biological evolution. In other words, the desired drug molecule must already be provided by the researchers, and no step in the selection process can automatically fine-tune those molecules that show some fit to the target molecule. Any subsequent fine tuning must be carried out painstakingly by chemists who attempt to determine the reasons for the selected molecule's success, and apply their knowledge of chemistry to improve it further through repeated rounds of variation and selective screening.

Structure-Based Rational Drug Design
It should also be noted that this classic method of drug development does not require initial knowledge of the actual structure of the interacting molecules. But since modern technologies involving X-ray crystallography and nuclear magnetic resonance spectroscopy can provide detailed information about the atomic structure of many molecules, this information is now being used in what is called rational drug design. If the classic approach to drug design can be likened to finding a key among tens of thousands of keys that will fit a particular lock, structure-based rational drug design is analogous to having information about the shape of the tumblers inside the lock. Since such knowledge would facilitate the making of a working key, information on the structure of a target site for a drug can facilitate the finding of an effective drug that will bind to the target.

But even with detailed structural information, designing an effective drug is not as straightforward as one might expect. A drug molecule cannot simply be ground into shape in the way that a key can be made from a blank. Instead it must be assembled from constituent atoms that will themselves fit together only in certain combinations and arrangements. To complicate matters further, a molecule's configuration may change dramatically when brought into proximity with another molecule, and atomic charges (which are not easily modeled) can affect the binding of one molecule to another. So whereas knowledge of the target's structure usefully constrains the number of candidate drug molecules, a considerable amount of trial and error is required to find a good fit. This search is now facilitated by three-dimensional computer displays of the target molecule that allow researchers to design molecules atom by atom to fit the simulated target site, in much the same way that wind tunnels and other computer-simulated environments facilitate the design of new products as noted previously. But since computer models are not perfect, promising compounds still have to be tried in a test tube and then on cells, tissues, live animals, and finally humans in clinical trials. Each stage in this long process represents yet another selective filter, with only those molecules that pass every test finally finding their way onto pharmacists' shelves. The cumulative variation and selection involved in this process is apparent in the description provided by three pioneers of structure-based drug design of their attempts to come up with a compound that would enhance the effect of certain anticancer and antiviral agents and be helpful in treating autoimmune disorders:

This iterative strategy—including repeated modeling, synthesis and structural analyses—led us to a handful of highly potent compounds that tested well in whole cells and in animals. Had a compound encountered difficulty in the cellular or animal tests (such as trouble passing through cell membranes), we would have revisited the computer to correct the deficiency. Then we would have cycled a modified drug through the circuit again.[11]

Recognizing both classic and structure-based drug design as types of evolutionary processes involving cumulative variation and selection helps us to understand the potential advantages of the latter approach over the former. In the classic method, thousands of compounds are first randomly tested, and the success of one compound provides little if any information on the potential usefulness of others still to be screened. Variation and selection occur to be sure, but there is no *cumulative* variation and selection in the initial screening procedure, so that every one of thousands of compounds must be tested. In the rational, structure-based approach, information concerning the structure of the target site is used to constrain the possible candidates. So instead of screening thousands of compounds, the researchers quoted above had to prepare only about 60 substances to find a promising drug, thereby saving considerable time, effort, and money. In addition, using the computer as a type of virtual test tube for testing drugs allows a vicarious means of variation and selection that can be more efficient and cost-effective than conducting all initial screenings chemically.

Directed Molecular Evolution: Selection in a Test Tube

But is it really necessary to know the structure of the target site to produce drugs efficiently? Biological evolution has come up with stunning achievements of design without any such knowledge at all. Of course, evolution has had over three and a half billion years. But if molecules themselves could somehow be bred using very large populations and rapid iterations of cumulative blind variation and selection, it should then be possible to direct the evolution of useful molecules for drugs and other purposes in much the same way that breeders traditionally directed the evolution of domesticated plants and animals.

The first demonstration that an evolutionary approach to molecular design was indeed possible was provided in the early 1960s by Sol Spiegelman of the University of Illinois at Urbana-Champaign.[12] Using a technique that causes strands of RNA (the molecular messenger of the genetic information archived in DNA) to replicate with a rather high error rate, Spiegelman began to breed a huge number of variations of a particular RNA molecule. For his selection criterion, he chose a rather simple one—speed of replication. Since he provided progressively shorter periods of time for replication, molecules that were the quickest at making copies of themselves were more likely to be selected as parents for the next generation of molecules. After 74 generations of this "serial transfer" experiment, Spiegelman had bred an RNA molecule that was 83% different from the original ancestor molecule and replicated itself 15 times faster. The artificial evolution of molecules in a test tube had been achieved.

Variations of this technique, referred to as directed molecular evolution, are beginning to be used to design molecules for drugs and other uses. Instead of selecting self-replicating molecules that replicate most quickly, a type of molecular obstacle course is set up that involves binding to a target molecule. A varied population of many millions of molecules is passed through a filtration column that contains the target molecule. Since this initial population is a random collection of molecules, only a very small percentage are likely to bind to the target molecule. But since the population is extremely large and varied, even an exceedingly small percentage of "hits" is virtually guaranteed to produce at least some molecules with an affinity for the target, and the vast majority that do not stick are simply washed down the drain (a watery version of Darwin's hammer). Only the relatively few molecules having enough affinity for the target molecule to adhere to it are retained, allowed to replicate (with a certain error rate to ensure additional variation), and passed through the filtration column once again—the process being repeated until a molecule that binds very tightly to the target is found. With this new, purely selectionist technique, a drug was developed that binds to the protein thrombin and consequently inhibits the formation of blood clots in patients who must be connected to heart-lung machines for surgery, or who must undergo blood dialysis because of kidney disease. Initial research in directed molecular evolution was restricted to RNA and DNA molecules, but work is currently under way to enable the same "irrational" technique of cumulative blind variation and selection to be applied to the evolutionary design of other types of molecules.

Although directed molecular evolution is still in its infancy, it has generated considerable excitement and activity in the biotechnology industry. One has only to consider the parallel with plant and animal breeding to understand why. As discussed earlier in this chapter, traditional selective breeding practices resulted in dramatic improvements in food crops and domesticated animals (at least from the perspective of human consumers). At best, such efforts may involve a population of thousands of plants that require months to mature and reproduce. In addition, screening (selection) may involve considerable time and effort, as when applying chemicals such as pesticides or salt, and having to wait and select the plants that are the least affected for the next round of breeding.

In contrast, breeding molecules using directed molecular evolution typically involves populations of 10^{13} to 10^{15} (ten million million to one thousand million million) molecules, each of which may take only an hour to reproduce. And selection can be as simple as passing the populations of molecules through a filtration column. Because of its promise, several companies such as Gilead, Ixsys, Nexagen, Osiris, Selectide, and Darwin Molecule are devoting their entire research program to directed molecular evolution, and Genentech, the grandfather of biotechnology firms, is also exploring this approach.[13] Researchers at Affymax even developed a test tube version of group sex in which *E. coli* genes from up to 10 bacteria are chopped up, randomly recombined, and reinserted into the bacteria, greatly increasing the odds that several favorable mutations will find their way into a few of the bacteria.[14]

It therefore seems quite likely that before long a new generation of powerful drugs and other substances will be available that evolved in the laboratory under the selective guidance of human researchers. But in contrast to the rational design of molecules, the researcher will (at least initially) have no real understanding of why a particular molecule is so successful at doing what it does, in the same way that traditional plant and animal breeders know nothing of the genes underlying the desirable characteristics that they select. It is of course likely that successfully evolved molecules will be analyzed and perhaps even improved by rational methods of design. But the potential power of cumulative blind variation and selection using large, heterogeneous populations of molecules is such that rational fine-tuning may not be necessary. Obviously impressed, Nobel prize-winning biochemist Manfred Eigen calls directed molecular evolution "the future of biotechnology."[15]

Three different technological revolutions have now taken place. In the nineteenth century the industrial revolution involved the exploitation of huge amounts of energy for manufacturing, agriculture, and transportation. The information revolution of the mid-twentieth century provided telecommunications and computer hardware and software to generate, analyze, manipulate, and transmit vast amounts of information. The most recent revolution, that involving biotechnology, now provides the means to manipulate the very core of living cells and direct the course of biological evolution for many species from bacteria to humans. It has also given us the ability to develop new drugs, vaccines, and gene therapies to fight disease and improve the length and quality of human life.

It was argued in chapter 10 that all technological development is dependent on cumulative blind variation and selection. But what is particularly intriguing about recent developments in the information and biotechnology domains is that these fields are now making explicit use of artificial evolution. The sequence of events leading to this development is noteworthy: evolution of the human brain by cumulative natural selection; the use of this brain's evolved capacity for thought (itself a form of vicarious cumulative variation and selection) to fashion an understanding of the nature and power of natural selection; and the application of this knowledge to solve problems by exploiting forms of artificial selection to direct evolution in agricultural plots, barns, computers, and test tubes.

The potential of these techniques for improving the human condition is immense and includes more productive crops and livestock, controlling and eliminating human disease, slowing the aging process, and engineering microbes able to transform the products of industrial pollution into harmless and even useful substances. But there is a potential darker side, such as crops that are able to produce their own pesticides that ultimately poison the birds and other animals that feed on them, the mutation of new supercrops into superweeds that are able to drive out native plants, and the release of novel pathogenic microbes into the environment against which humans and animals have no natural defenses. In addition, highly charged moral issues are raised when one begins to tinker with the human genome.

This third technological revolution brings with it the unprecedented ability to direct evolution itself. The human species has already had a tremendous impact on many life forms and physical features of the earth. Our harnessing of the very evolutionary process that created us will no doubt have a much greater impact.[16] Whether it will be positive or negative for the long-term survival of our and other species is, of course, the big question that only time and evolution itself can answer.

V

The Universality of Selection

15

From Providence
Through Instruction
to Selection
A Well-Traveled Road

Looking back into the history of biology, it appears that wherever a phenomenon resembles learning, an instructive theory was first proposed to account for the underlying mechanisms. In every case, this was later replaced by a selective theory. Thus the species were thought to have developed by learning or by adaptation of individuals to the environment, until Darwin showed this to have been a selective process. Resistance of bacteria to antibacterial agents was thought to be acquired by adaptation, until Luria and Delbrück showed the mechanism to be a selective one. Adaptive enzymes were shown by Monod and his school to be inducible enzymes arising through the selection of pre-existing genes. Finally, antibody formation that was thought to be based on instruction by the antigen is now found to result from the selection of already existing patterns. It thus remains to be asked if learning by the central nervous system might not also be a selective process; i.e., perhaps learning is not learning either.

—Niels Jerne[1]

It is nevertheless worth noting that in the history of ideas "instructive" hypotheses have most often preceded selective hypotheses. When Jean-Baptiste Lamarck tried to found his theory of "descendence" on a plausible biological mechanism, he proposed the "heredity of acquired characteristics," a tenet that advances in genetics would eventually destroy. One had to wait almost half a century before the idea of selection was proposed by Charles Darwin and Alfred Wallace and validated the principle, if not all the details of its application. In the same way the first theories about the production of antibodies were originally based on instructive models before selective mechanisms replaced them. It could conceivably be the same for theories of learning. To understand the reasons for this temporal succession, we must obviously examine the functioning of the scientist's brain. An instructive concept consists of only one step. It is the simplest possible approach. Moreover, whether we like it or not, it contains an "egocentric" component. "Nature directs forms" much as the sculptor models clay into a statue. . . . The concept of selection, on the other hand, implies further reflection. It involves two steps, and it satisfies the quest for a material mechanism totally devoid of "intentional" aspects. It is natural that this more complicated procedure, more difficult to execute, should have systematically appeared in second place throughout the history of scientific thought.

—Jean-Pierre Changeux[2]

. . . chance alone is at the source of every innovation, of all creation in the biosphere. Pure chance, absolutely free but blind, at the very root of the stupendous edifice of evolution: this central concept of modern biology is no longer one among other possible or even conceivable hypotheses. It is today the sole *conceivable hypothesis, the only one that squares with observed and tested fact. . . . There is no scientific concept, in any of the sciences, more destructive of anthropocentrism than this one, and no other so arouses an instinctive protest from the intensely teleonomic creatures that we are.*

—Jacques Monod[3]

Having now surveyed so many different fields of knowledge, it is finally time to entertain some general conclusions concerning their origins, insofar as such forms of knowledge are considered to be characterized by adapted complexity and constitute puzzles of fit of one system to another. Certainly, one conclusion should already be quite obvious, as it has provided the basic theme of this book—that selectionist theories have now been advanced in all these domains to account for the origin and growth of adapted complexity. But another similarity across many of these fields is also quite striking—remarkably similar patterns of evolution are evident in the growth of human knowledge itself.

Changing Perspectives on Fit

Biology provides one of the clearest examples of this progression of understanding. As we saw in chapter 2, the first reasoned explanation for the adapted complexity of living organisms was an extension of our everyday understanding of how the world works. As Reverend Paley so eloquently reasoned in *Natural Theology,* if we find a watch we naturally assume that the watch must have had a maker and the intricate design and function of the watch are due to the intelligence and skill of its creator. Similarly, Paley reasoned, an examination of the complexity in the functional design of living organisms leads to the conclusion that these life forms also had a creator, and that the adapted design and intelligence of these organisms is a reflection of the creator's knowledge, skill, and power. From this perspective, organisms, like watches, are considered to be the products and passive recipients of the knowledge provided by a supernatural entity in this providential theory of adapted complexity. Although modern biological science in its quest for naturalistic explanations no longer considers this to be a viable theory, it is probably still the most popularly held belief concerning the origin of the biosphere among nonscientists, even in a technologically advanced country such as the United States.

But it eventually became clear to some that living organisms were not the finished, final product of an omniscient and omnipotent creator. Instead, the accumulating fossil evidence indicated that the biosphere has undergone—and is still undergoing—dramatic changes, with some species dying out, new ones appearing, and others undergoing extensive modifications. Lamarck's explanation for these changes did not explicitly reject God's graces, but added another mechanism by which organisms would be modified advantageously by their interactions with the environment in ways that were passed down directly to their descendants.

Lamarck's theory of evolution had some important advantages over Paley's purely providential account of the design of life forms. It took into account the fact that species are not immutable but rather change over time. It also attempted to provide a naturalistic theory of evolution that did not rely exclusively on the omniscience and omnipotence of a supernatural being. But Lamarck's instructionist theory was inadequate to the task of explaining biological evolution, as it provided no explanation for either why the modifications undergone by organisms in their interaction with the environment should be adaptive (why should exercise make a muscle stronger or cold weather make a mammal's fur grow thicker?), or how these modifications could be transmitted to the next generation. The theory was an important step in moving from a supernatural to a naturalistic understanding of adaptive evolutionary change, but it did not go far enough in that it still required processes that were "smart," and the origin of this smartness was itself unaccounted for.

Natural selection provided an alternative explanation that removed all smartness from the process. Organisms varied, but why, Darwin did not pretend to know. By absolutely blind and ignorant luck, certain variations were better adapted to their environment, giving them a competitive edge over other organisms and allowing them to leave behind more offspring. These offspring, being more like their parents than not (again, Darwin did not know why), would share the traits and reproductive success of their parents, but with continued variations in their own offspring.

Darwin's account required no divine providence. Neither did it have magical instructions from the environment telling organisms how to adapt and how to pass on this information to their progeny. Darwin hit on a theory that, although incomplete because of his understandable ignorance of genetics and molecular biology, provided the first explanation for how adapted complexity could emerge on its own with no outside guiding hand or mysterious environmental instructions. But despite his insight that so stunned the world of science, he was unwilling to reject completely either the providentialism of Paley or the instructionism of Lamarck. As he stated in the first edition of the *Origin* and repeatedly emphasized in later editions, "I am convinced that natural selection has been the main but not the exclusive means of modification."[4] We will see later that he also did not reject a providential account of life's initial emergence.

And so it was left to the younger and more radical ultraselectionists, in particular August Weismann, to assert toward the end of the nineteenth century that natural selection was the sole process by which species grew in adapted complexity. And now more than 100 years later, this purely selectionist view of the emergence of design has so far withstood all challenges (including some quite recent ones to be considered in the next chapter) and continues to be the foundation for modern biology.

As recounted in chapter 4, a remarkably similar sequence of theory evolution, from providential through instructionist to selectionist, can be seen in the field of immunology. Ehrlich's original side-chain theory of antibody production was providential not in the sense of calling on God as an active creator of antibodies, but rather because it assumed that the information necessary to produce all possible antibodies was already contained in the genome.[5] But this theory was cast into doubt when it was realized that the immune system could produce antigens that were able to match and thereby recognize as foreign completely novel microscopic invaders.

As in biology, the first nonprovidential theory proposed for the functional design of antibodies was an instructionist one with a definite Lamarckian flavor. The template theory proposed that antibodies obtained information for their production from their environment, that is, from the antigens with which they came in contact. But after about 20 years it became apparent that the template theory could not account for certain key findings, such as that later-produced antigens were usually more effective in binding with an antigen than earlier ones.

To explain these findings, Jerne proposed a selectionist theory of anti-body production that in its basic conceptualization is still accepted today. The building plans for antibodies are not all specified in the genome. Nor are antibodies created from instructions or templates obligingly provided by the invading antigens. Instead, new antibodies having remarkable fit to a given antigen are produced through the random generation of a large population of varied antibodies, with those having the best fit selected for additional rounds of blind variation and selection.

So as in evolutionary biology, immunology proceeded through the same stages to explain its puzzle of fit. The formulation and acceptance of a selectionist theory of antibody production was of particular significance since it provided the first clear demonstration that the same basic process of cumulative blind variation and selection that occurs over eons of phylogenetic time among organisms also occurs within organisms during their much shorter life spans.

Although selectionist theories of the achievement of fit have become mainstays of evolutionary biology and immunology, they have not been well accepted into other fields seeking explanations for the puzzles of fit in animal and human learning, human thought, scientific progress, and cultural adaptation. In philosophy, the selectionist perspective is well represented by the work of Popper. But Popper's emphasis on human fallibilism and his evolutionary epistemology (which sees all knowledge growth as never-ending cycles of conjectures and refutations) is not very popular today. A minority of philosophers see his work as an important advance, but mainstream philosophy seems more attached to providential and instructionist theories of the origins of knowledge. An example of current providentialism is the linguistically influenced innatism of Chomsky and Fodor, and instructionism survives in the continuing efforts to wrestle with the problem of induction in attempts to explain how our limited sensory experiences can instruct us with justified, certainly true beliefs about the world.

Much the same state of affairs can be seen in the cognitive and social sciences. Despite the work of Campbell and a handful of others, the view that perception and cognition constitute forms of substitute trial-and-error-elimination is not widely held among psychologists and cognitive scientists. The same appears generally true of sociology, anthropology, and other fields concerned with understanding the dynamics of human culture and social institutions. So selectionist accounts of cognition and culture certainly exist, but mainstream currents in these fields still seem to vacillate between the providentialism of innatist theories (as in E. O. Wilson's sociobiology[6]) and the instructionism of much learning theory, which sees the environment in control of behavior.

Selection theory is doing somewhat better in the neurosciences, partly due to the influence of Changeux[7] and to Edelman's neural Darwinism. But there continues to be much resistance to selectionist theories of brain development and functioning, perhaps at least partly attributable to the difficulty of Edelman's writing[8] and possibly resentment against an outsider from the field of immunology attracting so much attention in the neurosciences.[9]

The largest successes of selection theory outside biology, immunology, and agriculture have been in the design of computer programs and molecules. But even here, particularly in computer science, much resistance remains. For many, resorting to the creative powers of ignorant, cumulative variation and selection as implemented in genetic algorithms and programming is an admission of programming incompetence, forcing one to rely on the computation muscle of fast, parallel processors to find a brute-force solution in the form of a hacked program that may do the job, but is inelegant and perhaps even incomprehensible to the programmer's eye.

The Rejection of Selection

Why are selectionist theories of design so little known and used in the scientific community outside of evolutionary biology and immunology to explain puzzles of fit and the growth of knowledge? There appear to be a number of reasons, some of which were mentioned in the discussion of biological evolution in chapter 2. But nowhere has more debate arisen than in the attempt to apply it to the growth of human knowledge in fields such as the philosophy and history of science, psychology, and cultural change. And so it is back to these fields we will turn to consider the arguments against a selectionist account of the ontogenetic growth of knowledge.

The Wastefulness and Improbability of Selection

Some of the arguments against a within-organism selectionist account of human knowledge growth are much the same as arguments that were (and still are) put forward against natural selection as an account of the adapted products of organic evolution. Some of them are primarily based on religious or aesthetic considerations. We saw in chapter 9 that Piaget rejected both Darwin's selectionist account of evolution and any application of selection theory to the cognitive domain, citing the "alarming waste" and the "fruitless trials" accepting such a theory would entail. Indeed, Piaget's concerns were warranted, even if his conclusions were not. The process of cumulative blind variation and selection is exceedingly wasteful since almost all variations, ignorant as blind probings must be, are dead ends (literally so in organic evolution). But this wastefulness can be understood as the unavoidable price that must be paid for a process from which new adaptations and knowledge can emerge without miraculous outside providentialist insights or mysterious instructionist guidance.

That we tend to forget or ignore the many failures in our attempts to better understand and control our surroundings and remember only the successes makes a selection theory epistemology appear unnecessary and unappealing. This is much the same as considering only the living and adapted end results of biological evolution and ignoring the countless unadaptive variations. But unlike biological evolution where the failed organisms and extinct species are usually well hidden from view, we can take notice of the fruitless trials of our experiments and conjectures.

An informative case from the history of technological innovation is Thomas Alva Edison's two-year attempt to find an appropriate substance for the filament in the first electric light bulb. After trying out dozens of substances, including red hair from a man's beard, Edison finally found success in December 1879 using carbonized sewing thread. His oft-quoted statement that "genius is 1% inspiration and 99% perspiration" reflects the long hours and countless failures that accompanied this and his many other inventions and technological advances.

But when we buy a new product, whether it be a new video camera or a more effective laundry detergent, we take no notice of the many failed attempts that preceded its development. We also know nothing of the countless would-be inventors and scientists who do not produce noteworthy breakthroughs. We usually consider only the successes and not the failures, making it appear as if science and technology progress through the sheer brilliance and insight of scientists and inventors rather than through painstaking trial and error and only occasional exhilaration of trial and success. Walter Vincenti, whose account of advances in aeronautical engineering was introduced in chapter 10, remarks:

> From outside or in retrospect, the entire process tends to seem more ordered and intentional—less blind—than it usually is. It is difficult to learn what goes on in even the conscious minds of others, and we all prefer to remember our rational achievements and forget the fumblings and ideas that didn't work out.[10]

Another argument against selection theory stresses the improbability of a blind, unintelligent process coming up with useful solutions to complex problems. This same argument was used repeatedly against natural selection in the evolution of species. It is of course true that any single blind change to a working system is almost certainly not going to make that system function better. A random wiring change inside the computer I am using to record these words is quite unlikely to make it perform better and much more likely to lead to a trip to the repair shop. However, iterative cycles of blind variation and selection based on large populations of such variations is quite another matter. The recent impressive successes of selectionist approaches to the design of both computer programs and molecules may convince some of the more open-minded skeptics of selection theory that such a process working cumulatively on populations can effectively tame the improbability of a blind process generating a fitter solution.

Examining Variation, Selection, and Transmission

But not all who argue against selectionist accounts of the growth of knowledge are offended by the wastefulness of variation and selection, nor do they all doubt its power, at least not in biological evolution. Rather, some insist that there are important differences between the biological evolution of adapted complexity and the growth of knowledge as manifested in both the progress of science and the cognitive development of the individual human brain. These arguments focus on the three principal themes of variation, selection, and transmission.

Probably the most frequently raised objection against the selectionist view of knowledge growth takes the form of an argument against the view that the variations are generated *blindly* as to their ultimate success. Other equally unflattering adjectives such as "unjustified," "unforesighted," "non-prescient," "undirected," "haphazard," "random," "groping," "stupid," and "dumb" have also been used to describe these necessary variations, but the word *blind* perhaps makes the essential point most clearly.

The argument against blindness in the variations of thought and theory leading to advances in human knowledge is that, although in biological evolution the generation of genetic variations may indeed be blind, the growth of human knowledge is consciously and purposely directed toward finding solutions to specific problems. As University of Waterloo (Canada) philosopher of science and cognitive scientist Paul Thagard proposed:

> Whereas genetic variation in organisms is not induced by the environmental conditions in which the individual is struggling to survive, scientific innovations are designed by their creators to solve recognized problems; they therefore are correlated with solutions to problems . . . Scientists also commonly seek new hypotheses that will correct error in their previous trials . . .[11]

But, we must ask, how does the fact that scientists have purposes (which few would doubt) provide emancipation from the necessity of blind variations in pushing back the frontiers of knowledge? The fact that a young scientist may be spending almost all of her waking hours in pursuit of room-temperature superconductivity does not, unfortunately, provide her with any clairvoyance as to the desired solution, if one does exist. Stating that these variations are "correlated with solutions to problems" begs the question as to how such prior guiding knowledge might have been achieved in the first place. Our scientist, unlike the process of organic evolution, most certainly does have a definite goal, and she generates methodological and theoretical variations in an attempt to accomplish this goal. But to the extent that new discoveries are made for which prior knowledge did not exist, this growth of scientific or technological knowledge is possible only by producing and testing new experimental variations *whose outcomes are unknown until tested.*[12] As Campbell put it, "rather than foresighted variation, hindsighted selection is the secret of rational innovation."[13]

But we must be careful to make clear what is meant by *blind* in this context. First, blindness does not imply that all variations are equally probable. For this reason, the word *random* is probably not a suitable descriptor since to some it may carry that connotation. Second, blindness does not mean that the process of producing variations of ideas, theories, and experiments for testing is necessarily unconstrained. Our superconductivity-seeking scientist is not likely to throw just anything into her concoction of chemicals, such as some of last night's leftover soup. Instead, she will rationally try out those substances in those proportions and under those conditions that, based on her knowledge of previous research and current theory, she believes have the greatest chance of success.

So it cannot be denied that previously achieved knowledge has an important role to play in constraining the variations to be investigated. Nonetheless, the new concoction is still a blind variation in the sense that the scientist does not know, and cannot know, if the resulting material will be an improvement over previous ones. It is in this important sense that the variation, although far from random and unconstrained, remains blind. The manner in which you grope about in a dark room to find the light switch changes significantly after making contact with the wall on which the switch is located. What were three-dimensional gropings now become two-dimensional ones. And as you encounter the molding along which you know the switch is located, your gropings become further constrained to just one dimension. But although they may become progressively and usefully constrained over time, an unavoidable blind component exists in your gropings until you actually find the switch. The same could be argued—although it is a much harder sell—about our use of vision to find objects and help us navigate around our environment (recall the discussion of the blind man in chapter 9).

To the extent, however, that constraints are effective in advancing knowledge (for example, whatever it is that prevents the scientist from adding the soup to her would-be superconducting material), they must be seen as additional puzzles of fit requiring explanation. And unless we are to return to providential or instructionist explanations for the existence of these adapted constraints, they can be explained only as the products of prior blind variation and selection.[14] As such, they may be quite well suited to guiding research into new, unexplored areas. But their fallibility must also be recognized since their use in finding answers to new problems may on occasion actually hinder progress rather than facilitate it. So, "it is not only the case that there is no prescience about which variations will lead to success, there is also no prescience about what part of the wisdom already achieved must be abandoned in order to go beyond it. In exploring new regions the cognitive constraint system is itself up for grabs."[15]

It has also been argued that the process of selection in the advancement of human knowledge is very different from natural selection in biological evolution. Again, Thagard contrasts humans as intentional agents in their role as selectors of theories in the growth of scientific knowledge with purposeless natural selection:

The differences between epistemological and biological selection arise from the fact that theory selection is performed by intentional agents working with a set of criteria, whereas natural selection is the result of different survival rates of the organism bearing adaptive genes.[16]

This certainly is a noteworthy difference, but we must again ask how it in any way invalidates a selectionist explanation of scientific achievement. In biological evolution, organisms that by the luck of their genome are better suited to their environment leave behind more progeny and therefore more copies of their genes than those less well adapted. It is this winnowing away of the less-fit organisms, not any foresightedness or clairvoyance on the part of the genetic variations, that is responsible for the fit of organism to environment. And different environments result in the selection of different adaptations, such as wings and lungs for air, and fins and gills for water.

Similarly, science progresses by the selection of theories that better fit the criteria used by scientists, such as explanatory power and parsimony. This is not to deny that certain practices and criteria used by scientists and communities of scientists may be irrelevant or even detrimental to the progress of science, such as the tendency to fund or follow a line a research due solely to the power, prestige, or popularity of its leading proponent. But insofar as science becomes progressively better at describing and explaining the objects, forces, and processes in the universe, it must be because the universe somehow interacts with the experiments and thoughts of scientists and thereby plays a role in determining which theories and hunches will be retained.

It has also been noted that biological evolution shows divergence leading to a great diversity of life forms, whereas science in marked contrast ultimately leads to convergence. The biosphere is rich in many strikingly different types of life forms, but physicists the world over use the same theories of relativity and quantum physics to account for and predict mechanical events. This difference has been taken by some as evidence that organic evolution and conceptual development must be fundamentally different. Thagard points out this difference by stating that survival of theories is the result of satisfaction of global criteria, criteria that apply over the whole range of science. But survival of genes is the result of satisfaction of local criteria, generated by a particular environment. Scientific communities are unlike natural environments in their ability to apply general standards.[17]

But just how are scientific communities able to apply these "general standards"? Is it not because the local criteria of modern scientists are much the same no matter where on the globe they may be located? The most obvious explanation for why scientific theories tend to converge is that they all share a very similar local environment. Light behaves very much the same in Sri Lanka as it does in Switzerland. So do falling bodies, chemical reactions, and cell division. Indeed, much of the technology and effort of scientific research is directed toward ensuring that experiments are conducted under highly controlled conditions that can be duplicated elsewhere with the same results. Thus a successful experiment that reveals a new regularity of nature should be replicable by other scientists with similar equipment anywhere in the world. In addition, the goals of scientists everywhere are much the same in their search for powerful yet simple theories with high explanatory and predictive powers. American biologist and philosopher David Hull addresses this issue in pointing out:

Conceptual evolution, especially in science, is both locally and globally progressive, not because scientists are conscious agents, not because they are striving to reach both local and global goals, but because these goals do exist. Eternal and immutable regularities exist out there in nature. If scientists did not strive to formulate laws of nature, they would discover them only by happy accident, but if these eternal, immutable regularities did not exist, any belief that a scientist might have that he or she had discovered one would be illusory.[18]

So regardless of the differences between biological evolution and the work of scientists, one can argue that scientific theories, like organisms, develop as they are edited by the selection pressures of their environments, which, although necessarily local, reflect both universal (as far as we know) regularities of nature and the shared practices, beliefs, and goals of modern earth-bound researchers. The unavoidable local nature of these apparently global criteria may someday be made quite clear when life is discovered on another planet that does not conform to terrestrial theories of life, or when it is revealed that the laws of physics in the vicinity of black holes have little resemblance to those that were formulated to account for phenomena closer to home.

It should also be noted that biological evolution, like science, shows convergence when similar problems are confronted by quite different organisms. The case of flight is perhaps the most striking example, with the asymmetrically curved wing having evolved independently in insects, reptiles (the extinct pterosaurs), flying fish, birds, and mammals (bats). The similar shapes of fish and marine mammals such as dolphins and whales is another example of convergent evolution. Indeed, the phenomenon often makes it difficult for biologists to disentangle the phylogenetic relationships among organisms based on physical appearance alone. Just because two organisms share a common feature, this does not necessarily mean that they are close to each other on the phylogenetic tree. Similarly, just because two scientists may come up with the same theory to solve some problem, it does not necessarily indicate that one of the scientists took the idea from the other. The independent discovery of natural selection to explain the origin of species by both Darwin and Wallace is a case in point.

So certain aspects of selection may at first appear different in biological evolution and scientific development, but its basic function of eliminating the less fit and retaining the fitter is actually very much the same. As Campbell put it so simply, "rather than foresighted variation, hindsighted selection is the secret of rational innovation."[19] And this "hindsighted selection" is as much a feature of scientific discovery as it is of organic evolution.

Differences in the transmission of accumulated knowledge between biological (genetic) and scientific systems have also been emphasized by opponents of selection theory epistemology. It has been suggested, even by such widely recognized experts in evolutionary theory as Gould[20] and Dawkins,[21] that the cultural transmission of ideas, including scientific ones, is a type of Lamarckian instructionist process, since the knowledge discovered by one individual can be promptly passed on to another. In this sense there appears to be a type of inheritance (or at least transmission) of acquired characteristics that simply does not seem possible in biological evolution.

However, such a view of the origin and progress of scientific and cultural knowledge encounters difficulties no less severe than Lamarck's account of organic evolution. Certainly, a strict Lamarckian interpretation is untenable since "in order for sociocultural evolution to be Lamarckian in a literal sense, the ideas that we acquire by interacting with our environment must somehow become programmed into our genes and then transmitted to subsequent generations."[22]

But even if a less literal interpretation of Lamarckism were applied to human knowledge, imposing problems remain. It was stated in chapters 11 and 12 that knowledge cannot simply be transmitted from one individual to another, either by language or any other means currently known. Instead, the process of understanding the ideas of another, whether expressed in oral or written language or other signs or gestures, requires the active generation of a variety of candidate ideas on the part of the "receiver," and the subsequent selection of the best ones. Virtually all modern theories of learning, education, and knowledge acquisition emphasize the active role of the learner in the construction of meaning, even if they do not explicitly embrace a selectionist account of communication.

So although scientific and cultural knowledge may appear to be transmitted from individual to individual and from generation to generation through a Lamarckian process, this turns out to be as unlikely as the inheritance of acquired characteristics during biological evolution. For such transmission to take place, it would have to be possible somehow for the knowledge contained in my brain to be transferred to yours in the same way that I can copy the computer file containing this chapter from my disk to yours. It is possible that such a technique could be developed in the future, perhaps by reading the pattern of synapses in one brain and then rewiring part of another brain to match this pattern. Until that time, however, we must tolerate the rather slow (although still many orders of magnitude faster than biological evolution) and inefficient necessity of recreating the knowledge of others using language and educational settings as facilitators of this constructive, selectionist process.

Other Criticisms

Many other criticisms have been made of the application of selection theory to the growth of human knowledge. It would be too tedious to consider all of them here, but many if not most of them share one or more of three characteristics.

The first is taking the process of biological evolution as the gold standard of selectionism and consequently treating any differences in the mechanics of variation, selection, and reproduction between that process and the growth of human knowledge as reason enough to discount a selectionist view of the latter. But as the previous chapters should have made clear, biological evolution is just one of many instances of cumulative blind variation and selection leading to the adaptation of one system to another. So although scientific theories, cultural practices, and genes may exist in very different forms and employ distinct modes of variation, selection, and replication, these and other superficial differences have in themselves little bearing on the argument that both thought and science make progress through a process of cumulative blind variation and hindsighted selection. It is for this reason that the terms "selection theory epistemology" and "selectionist epistemology" are preferable to the original "evolutionary epistemology" proposed by Campbell in his seminal 1974 chapter. Biological evolution, insofar as it leads to increases in adapted complexity, is a selectionist process. But not all selectionist processes have to mimic adaptive organic evolution in all of its biological details.

Second, many critics fail to take into account that selection theory is necessary only to explain the emergence of *new* fit, *new* adapted complexity, *new* knowledge, and not the routine application of old knowledge. We may not require a selectionist explanation for how a scientist, having acquired the accepted knowledge of his field, is able to apply this knowledge in a rather routine fashion to his work.[23] The use of current techniques for deciphering the genetic sequences in strands of DNA as is now being done as part of the human genome project may be an example of this type of research. As one defender of Campbell's selectionist dictum (which sees all instances of increase of fit resulting from blind variation and selective retention) pointed out:

> They [critics of Campbell] have typically not been sufficiently cognizant of the fact that the Dictum concerns the origins and the advancement of knowledge and not the utilization or extension of already acquired knowledge. It is certainly not an appropriate criticism of Campbell's Dictum to note how much previous knowledge is at work in one's daily activity or show how infrequent blind search seems to occur in most of our mundane efforts. The Dictum applies only to those instances when we are going beyond what we already know.[24]

Finally, a number of critics suggest that evolution itself has endowed our species with sensory and mental abilities that make it possible for us to acquire new knowledge without resorting to wasteful blind mental variation and selection. In essence, this argument contends that selection theory is appropriately applied to the phylogenetic development of human sensory and cognitive capabilities, but it need not and should not be applied to the ontogenetic acquisition of knowledge and skills developed during the life span of individuals.[25] In the same way that a girl does not have to figure out by trial and error how to grow breasts at puberty, this having been already figured out during the course of phylogenetic development of the human species and encoded in her genes, during the course of human evolution we acquired sensory organs and brain structures that allow us to acquire new knowledge without the need for further variation and selection.

But this argument that the growth of knowledge is able to bypass variation and selection in its basic mode of operation has at least three flaws. First, the achievements of biological evolution are due to what was selected in the past, and can provide no guarantee that these achievements will continue to be of use in the future. Fortunately, many aspects of our physical environment have remained relatively stable over long periods of time so that many of evolution's achievements (such as lungs, gills, wings, and feet) are still of use today. But the fact that extinction, not perpetual survival, is the usual fate of a species attests to the tentative nature of the knowledge achieved by a selectionist process. It therefore appears unlikely that natural selection could have provided us with innate abilities for acquiring all new knowledge, particularly in fields where our usual intuitions concerning space, time, and causality are violated, as in relativity theory and quantum mechanics.

Second, as presented in chapters 4 and 5, it has become increasingly clear that at least some (and perhaps all) aspects of ontogenetic development are dependent on selectionist processes. In the mammalian immune system, it is widely accepted that the fit of antibody to antigen is not solely due to past achievements of evolution, but requires an accompanying process of blind variation and selection during the lifetime of the animal. Selectionism is not quite as firmly established in the brain sciences, but thanks to increasing evidence, there is growing agreement that the fit of the brain to the needs of its owner is largely due to a continuous process of blind variation and selection of neurons and neuronal connections. Therefore, the argument that selectionist processes underlie only phylogenetic change resulting from among-organism selection and not ontogenetic development, which is now known to require within-organism selection, is no longer tenable.

Third, some critics have pointed to instructionist procedures by which new knowledge is attained as counterexamples to the conjecture being considered here that all knowledge has its roots in cumulative blind variation and selection. One example is Pavlovian conditioning as discussed in chapter 7. It will be recalled that in Pavlovian conditioning an animal learns to make an old response to a new stimulus, as when a dog began salivating on seeing a dark-colored liquid, after having had such a liquid containing acid placed its mouth. Such learning appears instructionist, since no obvious trial and error of responses is involved, and by pairing a neutral stimulus with an unconditional one that elicits a built-in response, the environment does appear to be instructing the organism to use the neutral stimulus to anticipate the unconditional one. It cannot be denied that such learning makes an animal's behavior better adapted to its environment, as it is now able to anticipate and therefore avoid dangerous situations (such as a bear fleeing at the sight of hunters) and seek out favorable ones (as in using the squeak of a mouse to locate food).[26]

But whereas such instructionist adaptation may occur and therefore pose a challenge to universal selection theory, it must be realized that such mechanisms of knowledge acquisition are quite limited in what they can accomplish. As already noted, Pavlovian conditioning does not account for the learning of new responses and the development of new perceptual abilities, but instead can account only for the association of old responses to new stimuli. And these new stimuli must be ones that immediately precede the unconditional stimuli. Stimuli that either follow or are presented long before the unconditional stimulus do not readily become conditional stimuli, at least not in nonhuman animals.

So Pavlovian conditioning may be an instructionist mechanism for attaching new meanings to old stimuli, but it cannot by itself provide innovative solutions to problems the way that natural selection and other selectionist processes have done and can do. Similarly, simulated nonstochastic neural networks may be capable of certain forms of learning without making use of cumulative blind variation and selection, but they are thereby condemned to sticking to their "innate" architecture and do not have the flexibility to reorganize themselves as demanded by novel problems in the way that evolutionary algorithms can. (Of course, it is the human researcher who changes them, using cumulative blind variation and selection, when it is clear that such networks have to be fine-tuned or more drastically overhauled.) Instructionist processes may play an important part in certain contexts for acquiring new knowledge, but they clearly cannot be the whole answer. And if they are effective, this in itself constitutes a puzzle of fit that must be ultimately accounted for, most likely by a selectionist explanation.

Finally, selection theory is not intended to account for changes that are not characterized by increases in adapted complexity. It is possible that a species may change over time in ways that are essentially neutral or perhaps even maladaptive. A decrease in the ozone layer may result in increased rates of radiation-induced mutations so that a species may no longer be able to preserve and pass down the accumulated wisdom of its genome. For various social, geographical, and economic reasons a society may lose adapted cultural and scientific knowledge; book burning and persecution of scientists occurred many times during periods of social conflict and upheaval. It is, of course, interesting to consider just how much evolutionary, cultural, scientific, and cognitive change is adaptive, and a comprehensive study of any of these fields will certainly have to go beyond selection theory. But to the extent that adaptive change has occurred, a selectionist perspective would see that cumulative blind variation and selection must be involved. To the extent that no change occurs or that change is not adaptive, other mechanisms (or factors interfering with cumulative blind variation and selection) must be involved. Selection theory is neither able nor intended to explain stasis or neutral or maladaptive developments, and to criticize it on this count indicates a misunderstanding of its intent.

But although it is neither able nor intended to explain neutral or maladaptive change, it can provide clues for understanding such processes, especially when adaptive change would be desired or normally expected. For example, despite continued attempts by breeders, no cattle have yet been developed that reliably produce significantly more female than male offspring.[27] Since these breeders have been successful in selecting for other traits such as increased milk production, selection theory suggests that there is no heritable genetic variation for the sex ratio of offspring. As we saw in chapter 10, cultures may fail to continue to adapt to changing environmental conditions. Selection theory would lead us to place the blame in such cases on lack of variation in cultural practices (perhaps no innovations are permitted by the society) or lack of information with which variations could be compared and selected.[28] And a school or classroom that is ineffective in fostering the growth of students' knowledge and skills would be suspected of not providing an educative environment that was both sufficiently free to allow students to generate and try out their variations, and responsive in giving sufficient feedback for students to select the better ones.

To return to what selection theory is intended to explain, it was observed that:

None of Campbell's critics have proposed rival models of how knowledge could have arisen out of ignorance, or how stupid processes could lead to intelligent adaptation. In rejecting Campbell's Dictum the critics have all noted that advances in knowledge are all based upon prior knowledge, but . . . this is largely an irrelevant criticism. In any situation in which the advances in knowledge are not wholly explicable in terms of previously attained knowledge, a BVSR [blind variation and selective retention] process must be at work. Unless of course, we really do live in a world in which prayer or meditation or passive induction can lead directly to new knowledge without any need for blind trials.[29]

So despite the many attacks on selection theory, no one has yet demonstrated how a process that completely circumvents blind variation and selection can generate new knowledge. This, of course, does not mean that such a demonstration may not be forthcoming. But until such time, selection theory appears to be the only ballgame in town.

The Innatist Misconstrual of Selection

Unfortunately, a rather serious problem has arisen from the use of within-organism selectionist explanations of fit. Some quite prominent scholars have used selectionism to advance innatist conclusions concerning the ontogenetic achievement of puzzles of fit that are inconsistent with selectionist processes and products. This misconstrual involves emphasizing the process of selection while ignoring or deemphasizing the generation of novel variants among which selection takes place.

The work of cognitive scientist Massimo Piattelli-Palmarini is one such example.[30] Piattelli-Palmarini, who has written on concept and language acquisition, makes many strong arguments against instructionist (which he calls "instructive") accounts of concept and language acquisition and for a selectionist (his "selective") one. He is also impressed by the selectionist functioning of the mammalian immune system and uses this (as I did in chapter 4) to demonstrate both that somatic, ontogenetic selection processes do take place, and how theories of immunology have progressed from instructionist ones to a selectionist process.

But Piattelli-Palmarini's perspective on within-organism selection is quite different from the one advanced here in that his requires all variations to be innately provided before selection can take place. For example, in his 1989 discussion of the selectionist functioning of the immune system he mentions only in passing the somatic (ontogenetic) genetic reshuffling and mutations that provide the necessary variation in antibodies. And despite the fact that these antibodies are not specified in the genome but rather emerge from the countless possibilities that the genome allows, he refers to this as "the innate repertoire of antibodies."[31] He thus leads the reader to surmise that all possible antibodies are innately specified and therefore provided before any selection takes place. In so doing, he essentially resurrects the long-dead, genetically providential germ-line theory of antibody production proposed by Ehrlich in 1900 (see chapter 4). Piattelli-Palmarini makes the same basic argument for concept and language acquisition in stating that "*Any* pattern is *already* in the actual repertoire of the organism."[32]

This innatist (and therefore genetically providential) perspective on selectionism ignores the fact that in the evolution of species and in the production of antibodies *unpredictable novelty emerges* from the blind recombination and mutation of DNA sequences in the genes regardless of the fact that it is always the same old four genetic building blocks that are reshuffled. Indeed, immunologists now make an important distinction between what they call innate immunity and adaptive immunity, with the latter clearly not innate.[33] Similarly, a selectionist account of human cognition should lead to an expectation of novelty in the emergence of human concepts, ideas, and problem solving, dependent on the blind variation and selection of neuronal connections as discussed in chapter 5. But for Piattelli-Palmarini (and also for Fodor, as observed in chapter 11), all such products of human cognition must be innately specified before they can be selected, despite the fact that such a conclusion is inconsistent with what is now known of the immune system, that is, all antibodies are not innately specified in the genome but rather exist only as potentialities, the majority of which remain unrealized. It is also inconsistent with what is known about biological evolution.

Compare a typical adult human with a typical adult mouse. I will take it as uncontroversial that the human possesses knowledge of the world, concepts, and language that the mouse does not. And yet remarkably "an estimated 99 percent or more of the genes in mice and humans are the same and serve the same purpose."[34] So mouse genes are very much like human genes, differing only in the sequence of relatively few nucleotide base pairs and how these are organized into chromosomes, with a meager 1% of genes that humans do not share with mice (and perhaps also differences in how the genes are organized) responsible for human cognition and language. Now by Piattelli-Palmarini's (and Fodor's[35]) reasoning, we would have to consider the mouse as having innate knowledge of human concepts and language (the genetic building blocks are all there; they just need to be rearranged a bit). Such a conclusion would of course be absurd. The confusion here appears to stem from not appreciating that the recombinatory shuffling of genes, as well as the recombination of synapses, computer algorithms, and parts of molecules, can result in the emergence of completely novel and unpredictable possibilities for adapted complexity.

Here are two more examples.[36] Shakespeare's play *Macbeth* is essentially a long, ordered string of about 35 typographical characters, including the 26 letters of the English alphabet (ignoring the difference between upper- and lower-case letters), a few punctuation marks, and spaces between words. Knowledge of these written characters, however, does not in itself provide any knowledge whatsoever about the play. Nor does knowledge of all the sounds of Russian provide any clues as to what Boris Yeltsin said yesterday. It should be clear that possessing the elements that make up a complex structure is not the same as knowing the complex structure itself.

Piattelli-Palmarini is not the only well-known scholar to misconstrue selectionism as requiring innately provided variations among which to choose. In a recent selectionist attempt to understand the workings of the human mind and brain, neurobiologist Michael Gazzaniga made the same innatist error in his book *Nature's Mind*. Although Gazzaniga, unlike Piattelli-Palmarini, does recognize that the generation of antibodies involves the somatic mutation of lymphocyte cells, the creative and variation-generating aspect of selection theory is not included in his view of human behavior and brain function. This oversight is made clear when he states "for the selectionist, the absolute truth is that all we do in life is discover what is already built into our brains"[37] and "selection theory is hard on the nature/nurture issue in arguing that all we are doing in life is catching up with what our brain already knows. We are discovering built-in capacities."[38] As one reviewer remarked, Gazzaniga's perspective on knowledge is not unlike Socrates' providentialist doctrine of recollection (or *anamnesis*; see chapter 6) in which "all knowledge is already in the mind, waiting to be remembered appropriately. Michael Gazzaniga's doctrine is remarkably similar, although it says, 'selected' rather than 'remembered'."[39]

A quick comparison of mouse and human (or amœba and mouse) makes it evident that the among-organism selection of biological evolution results in the emergence of new possibilities that were not present before, and that natural selection is therefore a creative process that constantly fashions innovative variations among which to select. We should therefore expect within-organism ontogenetic selection to do the same. Indeed, as we saw in chapter 5, the brain not only selects from preexisting neural circuits, but actually engineers new circuits by first adding new synaptic connections between neurons and then selecting some while eliminating others. And although the mechanisms for such within-organism variation and selection may be considered innate, its products cannot be regarded as such, since by the same reasoning, one would be obliged to conclude that the genetic information for the human body, brain, and cognitive abilities was already contained in the very first organism that used DNA for its genes.

Gazzaniga and Piattelli-Palmarini are right in contending that selection theory has important implications for the nature-nurture debate in psychology. But they err in their conclusions that the implications are necessarily innatist. If they were right, selection could operate only among already achieved variations, and that would mean that we could not have evolved to be what we are today.

As Immanuel Kant noted: "The Creation is never over. It had a beginning but it has no ending. Creation is always busy making new scenes, new things, and new Worlds."[40] In the same way that natural selection among organisms provides an explanation for the continuing adapted complexity and adaptive creativity of organic evolution, cognitive selection within the nervous system provides an explanation for the adapted complexity and continuing adaptive creativity of the human mind.

Universal Selection Theory

The Second Darwinian Revolution

. . . the growth of our knowledge is the result of a process closely resembling what Darwin called "natural selection"; that is, the natural selection of hypotheses . . .

—Karl Popper[1]

In the provocative essay "Universal Darwinism" (referred to at the end of chapter 2), Richard Dawkins maintains that if life were to be found elsewhere in the universe, we would have very good reasons to suspect that it had evolved as it did on Earth, that is, by natural selection. Let us recall that Dawkins's conclusion is based on the argument that the process of cumulative blind variation and selection is the only currently available scientific explanation that is in principle capable of explaining the emergence of the adapted complexity required for life.[2]

But Dawkins's argument is not limited to adaptive biological evolution. We saw in the preceding chapters how selectionist explanations for puzzles of fit themselves have evolved in many other disciplines that are also concerned with understanding various forms of knowledge growth involving the spontaneous emergence of fit between two or more interacting systems. Scottish philosopher and psychologist Alexander Bain may have been the first to apply a Darwinian perspective to human thought (see chapter 9), but it was not until 100 years after the publication of Darwin's *Origin* that Donald T. Campbell envisioned a comprehensive, all-embracing role for selection theory.[3] His selectionism emphasizes the psychological, scientific, and cultural growth of knowledge, but the recent selectionist discoveries in immunology and neurophysiology, and the applications of selection theory to the engineering of molecules and computer software finally have attracted the attention of the scientific community to the importance and potentially universal applicability of Darwin's selectionist insight.

Indeed, under certain conditions, the evolution of fit by way of cumulative blind variation and selection appears inevitable. Consider a population of self-replicating entities that vary in ways relevant to their reproductive success, and that inhabit an environment of limited space and resources that does not undergo large fluctuations from one generation of entities to the next. If these entities produce quite (but not always perfectly) accurate copies of themselves, after a few generations the winnowing effect of selection will be noticed as the population inexorably shifts toward a preponderance of new entities that better fit their environment. This, in a nutshell, is what the process of cumulative blind variation and selection is all about.

However, as we have now seen, these entities do not have to be restricted to living organisms and the genes they contain. They can be molecules, antibodies, neural synapses, behaviors, scientific theories, technological products, cultural beliefs, words, or computer programs. And selection does not have to be restricted to the natural and purposeless selection of Mother Nature, but may involve purposeful humans selecting for plants growing bigger tomatoes, cows giving more milk, scientific theories providing better predictions, automobile engines yielding greater efficiency, or molecules providing more powerful drugs. The robustness of the selection process was dramatically demonstrated by the findings of artificial life researchers such as Thomas Ray (see chapter 13) who were amazed at just how easy it is to get adaptive evolution happening on their computers. As long as the basic conditions of some (but not too much) variability, accurate (but not too accurate) replication, and a fairly stable environment prevail, the mechanistic, unforesighted evolution of fit appears inescapable.

The selectionist explanation for the emergence of adapted complexity and new knowledge may also appear inevitable on logical grounds. As Campbell put it, "increasing knowledge or adaptation of necessity involves exploring the unknown, going beyond existing knowledge and adaptive recipes. This of necessity involves unknowing, non-preadapted fumbling in the dark."[4] Of course to this we must add the selection and retention of those occasional results of this fumbling that are found by hindsight to provide a better solution to the problem at hand.

When stated this way, particularly using the phrase "of necessity," a selection theory epistemology may be inevitable by tautology. That is, it appears to be true simply by definition (as is the statement "a bachelor is an unmarried male") with no possibility of being disproved and consequently replaced with a better theory of knowledge growth.[5] A selectionist epistemology may be described tautologically; however, this does not mean either that it is false or that it cannot be stated in a more falsifiable way. For example, Campbell also made the bold claim that

A blind-variation-and-selective-retention process is fundamental to all inductive achievements, to all genuine increases in knowledge, to all increases in the fit of system to environment.[6]

And it was observed that

This statement is clearly not analytic.[7] One can easily imagine possible worlds in which genuine increases in knowledge are generated in some other way, by prayer for instance, or through the cultivation of omniscient meditative states. But we need not be so exotic in the search for alternative models of knowledge generation. In fact the received wisdom of Anglo-Saxon philosophy describes a world in which knowledge is generated otherwise. In that world knowledge is generated through direct passive-absorptive associations of ideas (passive induction). BVSR [blind variation and selective retention] as an analog for all knowledge generation was introduced specifically against such views, a point made most clearly by Popper.[8]

Thus, it should be possible to test the bold selectionist hypothesis of Campbell and discover—if they exist—nonselectionist processes underlying the emergence of adapted complexity and knowledge growth. This may not be easy in disciplines such as the history of science and cognitive psychology, but the new selectionist explanations for antibody production and brain development are based on empirical research findings, not on empty tautologies.

Campbell would thus have us believe that all knowledge, all problem solving, all skills, all adaptive physiological and neural changes, all useful cultural beliefs and practices, and all scientific and cultural progress have at their roots cumulative blind variation and selection—phylogenetic (among organisms), ontogenetic (within organisms), or both—of the same general type proposed by Darwin to account for the evolution of species. The findings, theories, and rationale for selectionist explanations of puzzles of fit reviewed in this book suggest that such a universal selection theory should now be taken seriously.

Such a theory has definite elegance as well as undeniable audacity. One supporter of Campbellian universal selection theory remarked that it

... points the way toward a unified theory of knowledge generation based upon mechanical processes. The natural process by which variation and selective survival designed the species may be seen at the root of all adaptive design and knowledge as a subspecies of adaptive design. The vastness of the conception, uniting knowledge and adaptation, biology and epistemology, and artificial and natural intelligence is what continues to tempt Campbell and others to insist on the blind variation and selective retention approach to knowledge. In all instances of fit to system it is (to paraphrase Campbell) hindsighted selection and not foresighted variation that is the key to adaptive advance.[9]

Selection as the Exploitation of Emergence

But perhaps Campbell goes too far. Is it really the case that only selectionist processes can create the puzzles of fit of adapted complexity?

As we saw in chapters 7 and 13, Pavlovian conditioning (possibly) and backpropagation neural networks (more clearly) provide examples of how adapted complexity may be achieved by instructionist processes. But we also noted that such processes are quite brittle in their operation since they cannot adapt to large, unpredictable changes in the context in which they work. So although Pavlovian conditioning may allow an animal to adapt behaviorally by in effect anticipating certain events, it cannot provide the animal with useful novel behaviors. And although a backpropagation neural network can adapt to fit the requirements of some very complex tasks, we have seen how it may instead find itself trapped at a seriously suboptimal location on its fitness landscape from which it cannot escape without randomly changing the starting weights of the neurodes (or making other structural modifications such as a change in the number of middle-level neurodes) and starting all over again, that is, resorting to blind variation and selection.

There are also mechanical examples of the achievment of fit by instruction. For example, it is possible to insert a brass key blank into a lock and use the markings left by the lock's interior to fashion a key to fit and operate the lock. Here the inside of the lock acts as a template that transmits information onto the key blank, which is then used to make the key fit the tumblers of the lock. The fact that a new automobile engine runs more

smoothly and efficiently after an initial break-in period is another example
of mechanical fit (involving the fit of pistons to cylinders and valve rods to
camshafts) resulting from an instructionist process whereby a component's
environment operates directly on it, and not by selection of previously generated variations, to cause it to fit better.

Let us consider two more examples of adapted complexity that appear to
have roots in instructionist mechanisms. On my computer I have a communications program called Mosaic. With Mosaic up and running, my computer becomes a marvelously well-adapted instrument for finding and
displaying information from the Internet in the form of text, images, and
sounds. Your computer cannot do likewise without this or similar software.
But if I take a diskette on which the program has been copied and insert it
into your computer, I can provide processing instructions (in the form of
software) that now enables your computer to do what mine can do (the version of Mosaic I use is available free, so this is legal).[10] So in effect your computer has achieved new adapted complexity with no variation and selection
on its (or our) part.

It is also possible to imagine how something similar may be possible in
the future with human brains. We described in chapter 5 that all we know
and can do appears dependent on patterns of interconnections among neurons. So if I could somehow determine how your brain is connected up and
reproduce that same pattern in my brain, I would know what you know
and be able to do what you can do. If you could speak both Chinese and
English and I only English, I would become bilingual as well. (I would hope
a way would be found so that only those connections underlying a particular desired competence would be reproduced in my brain, so I wouldn't
have to give up the knowledge, skills, and personality I already possess.)

In both cases, computers and brains, increases in adapted complexity
would result from an instructionist process. But if we look closer, we will
see that we have simply described a process of transmission of information
from one system to another, not the actual development of the knowledge.
Once a complex computer program has been developed through a
painstaking process of cumulative trial-and-bug elimination (as any programmer must admit), it can be transferred to and run on any compatible
computer. But no new adapted complexity is generated in this process. And,
of course, human knowledge has never been, and may never be, transmitted by having one brain directly instruct the synapses of another.

So we find ourselves considering selection once again. In its simplest form, selection can involve choosing among already provided alternatives with no way to modify further the given alternatives or create new ones. This is like approaching a locked room with 10 keys and not knowing which one will allow you to enter. If one key does not work, you can only choose another. And if none of them fits, you are out of luck. This simplest (and least powerful) selection process can be referred to as *nonconstructive* or *nongenerative selection*, since no new variations are generated and one is therefore limited to the variations already on hand. Ehrlich's original side-chain theory of antibody production (see chapter 4) was a nongenerative selection theory. And so appears Fodor's view of human hypothesis testing (with all hypotheses to be tested and consequently selected or rejected being innately provided), together with Piattelli-Palmarini's and Gazzaniga's application of selectionism to cognition, as discussed at the end of the previous chapter.

Much more powerful is *constructive* or *generative selection*, which involves the creation of new variations, whether they be organisms, antibodies, patterns of neural connections, behaviors, thoughts, concepts, or computer algorithms. By recombining elemental building blocks, the resulting variations have new, unpredictable properties that are not contained in any of the individual building blocks of which they are composed. In effect, variations with novel properties *emerge* from the recombination of the building blocks.

Complex three-dimensional configurations of proteins (their tertiary structures) emerge from sequences of animo acids, which in turn emerge from DNA sequences encoded in genes. And from proteins emerge organelles, from organelles emerge cells, from cells emerge organs, from organs emerge organ systems (such as the circulatory and nervous systems), from organ systems emerge organisms, and from organisms emerge societies and (for humans) social institutions. And from the simple yet marvelously coordinated activity of billions of individual neurons emerge human behavior, knowledge, and consciousness itself.

The emergence of new properties and complex systems is not restricted to the living world. In chemistry we see how combining sodium, an alkali metal that reacts violently with water, with chlorine, a deadly gas, produces sodium chloride, or ordinary table salt, that neither reacts violently with water nor is toxic. In computers, sequences of binary digits give rise to computer programs. The basic components of control systems, such as the cruise control device you may have in your car, are made up of rather simple input-output devices. Connecting them together in a special way (see chapter 8) gives the resulting control system the lifelike "willpower" to maintain a desired goal despite unpredictable disturbances.

This phenomenon of emergence is clearly very powerful stuff, and it provides the complexity without which life surely could not exist. But since the detailed properties of emergent variations—whether they be molecules, organelles, cells, organs, organ systems, organisms, societies, or social institutions—cannot initially be predicted from knowledge of their component building blocks, the only way that newly emerged entities can lead to adapted complexity is through a process of blind variation and hindsighted selection.

But we can make selection more powerful still. Not being content with a single-step selection process, we can instead take the best of the variations, vary them, and then select the best of the new generation, repeating the process over and over again. This, of course, is *constructive cumulative selection* (figure 16.1). This process of selecting and fine-tuning the occasional accidently useful emergent system turns out to be so powerful that we should not be surprised that the adaptive processes of biological evolution, antibody production,[11] learning, culture, and science all employ it, and

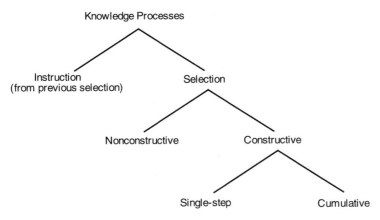

Figure 16.1
Types of knowledge processes: instruction and three types of selection.

that its power is now being explicitly exploited in the design of organisms, drugs, and computer software by one of evolution's most complex and adaptive creations—the human species.

So Campbell does go too far in his radical selectionism if he insists that all mechanisms leading to fit must be selectionist in their current operation. But he appears to be right on the mark if by new knowledge he means emergent knowledge. And, of course, any instructionist mechanism capable of generating new adapted complexity is itself an emerged adaptive system that must owe its own origin to the prior emergence and consequent selection of candidate instructionist systems.

Combining Conservative Hedging with Radical Gambling

Another useful perspective on selection can be had by considering the process of cumulative blind variation and selection as a *search-and-construction procedure* for finding solutions to novel problems.

Figure 16.2 provides representations of four different types of problem spaces or fitness landscapes. All four graphs represent functions with the value of y (height on the vertical dimension) determined by the value of x (lateral position on the horizontal dimension). Solving each problem requires finding the value of x that yields the highest (or at least close to the highest) value of y, starting out with no knowledge of the relationship between the two values, that is, no prior knowledge of the function as represented in the relevant graph.[12]

In the first landscape, we have essentially a problem of finding the needle in the haystack, since only a single value (or at most a very limited range of values) of x provides a nonzero value for y. An example of such a problem would be that of a thief trying to figure out the personal identification number (PIN) of a stolen bank card, where only one four-digit combination will work. Since this problem has but one solution, the best that the thief can do is to try each and every one of the 10,000 possible four-digit numbers between 0000 and 9999. But this could be considerably shortened if the result from each trial could somehow be used to provide a clue as to what number should be tried next. Unfortunately, the all-or-none nature of this problem means that the thief has no way to use the results of prior trials to zero in on the target, other than making sure not to try past failures a sec-

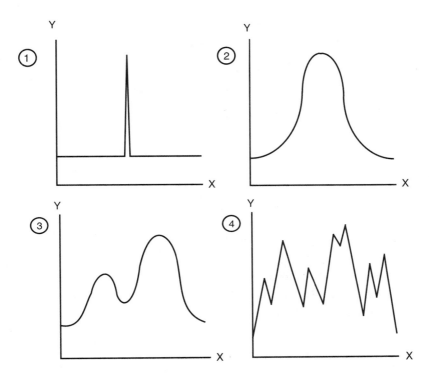

Figure 16.2
Four problem spaces (or fitness landscapes).

ond time. No gradual accumulation of knowledge is therefore possible, since each trial is no more or less likely than the previous one to provide the solution. So this is a form of noncumulative selection. Nonetheless, if the thief has enough time and patience, he should eventually succeed (although perhaps not before the card is reported stolen and canceled). Unfortunately, there might not be enough time left in the universe to find a solution using this method for problem spaces that are very large and multidimensional. So the only way a solution can be found is through a series of blind guesses and hindsighted selection. The essentially noncumulative character of the problem makes this at best a tedious and at worst an impossibly long process.

In the second panel we notice a broad peak to the solution, so that trials in the neighborhood of the peak provide better results than those farther away. An example of such a problem would be finding the amount of fertilizer that maximizes a certain crop yield, where too little fertilizer leads to diminished yields, as does too much. Now it is not necessary to try all possible quantities of fertilizer, since if it is found that more fertilizer is better, one will continue to try still more until the yield begins to drop again. Not much blind variation is involved in this problem, except perhaps for the guess of the initial quantity. Such single-peak problems can often be solved quite readily using what are called hill-climbing techniques or algebraic methods based on differential calculus, that is, solving for the x value whose associated y value has a slope of zero if the function relating y to x is known. Indeed, we discussed in chapter 13 and in the previous section that back-propagation neural networks achieve their fit by a type of instructionist hill-climbing technique. A variation and selection procedure can also work quite well, and much more quickly than in the previous problem, as long as new guesses about the value of x take advantage of the knowledge of the partial successes of previous trials.

Things become both more interesting and difficult when we consider a two-peak problem space as shown in the third panel. Now we have a local maximum (the lower peak), in addition to the global maximum representing the overall best solution, so we run the risk of getting stuck on the local maximum and never finding a better (or the best) solution. An example of such a problem would be finding the optimum dosage of a drug to treat hypertension, where a low dosage has some effect in lowering blood pressure, a high dosage has the greatest effect, and intermediate dosages have little effect. The only way to escape the trap of a local peak is to try some new, random values to see if they lead to a higher peak. Since it is not known where that peak may be located, one can only take a blind jump off the local peak and hope that it leads to a better solution. The cumulative-variation-and-selection approach to this problem would be to start with a wide-ranging population of values of x and allow them to recombine, reproduce, and mutate so that it is unlikely that the global peak would be missed.

But even this two-peak example is much simpler than many if not all of the real-world problems that we and other species encounter. The fourth panel represents a very complex problem space with many jagged peaks and valleys. On first consideration, finding the value of x that provides the best solution might seem as difficult as the first problem considered above. Regardless of the complexity of the landscape, constructive and cumulative blind variation and selection may be successful in finding a solution to this and similar problems if knowledge concerning the best solution of previous trials is applied to construct new variations. This is what is done in the biological world with sexual reproduction, and it is also used in genetic algorithms and genetic programming as described in chapter 13. By taking values that on previous trials provided the best solutions and using them to breed new values, patterns that may be too complex for the human eye to perceive can be exploited and a solution reached. By repeatedly breeding (with mutation or sexual recombination to ensure a continued source of new variations) from the best solutions found so far, knowledge obtained from past trials is preserved, and at the same time new blind variations are introduced in each generation that continue to grope for still better solutions.[13]

Representing problems as fitness landscapes can be useful for understanding how a selectionist procedure finds solutions. It can be somewhat misleading, however, giving the impression that the solutions to be searched all exist before the search takes place—in other words, that nonconstructive selection is involved. Instead, these solutions exist only as *potentialities*, and each one must first be constructed before it can be evaluated. This is obvious in biological evolution where pairs of sexually reproducing organisms whose offspring, no matter how many, represent only a minute fraction of the possible genetic recombinations of the parents' genomes. And when to this is added various possible mutations and other errors such as extra or missing chromosomes, the possibilities that can be constructed and evaluated by selection become infinite. The importance of the construction and consequent emergence of new forms to test and select led Henry Plotkin, an important advocate of universal selection theory, to use the phrase "generate-test-regenerate"[14] to describe what is referred to here as constructive cumulative selection.

Despite the very powerful generation-and-search procedures that constructive cumulative selection provides, it does have its limits. Notably, it cannot be guaranteed to find the optimum solution for every problem. But it has now been well demonstrated that many problems exist for which constructive cumulative selection can find, if not the best, at least useful solutions to a broad range of problems that are orders of magnitude more complex than the puny one-dimensional examples considered here, such as those related to the evolution of living organisms in complex and hostile environments, and finding better scientific theories to account for puzzling physical and biological phenomena. Constructive cumulative blind variation and selection can provide just the right blend of conservative old knowledge and risky innovation to push back the frontiers of knowledge. These two features as they relate to adaptive biological evolution (although they are relevant to all selectionist achievements of adapted complexity) are well characterized by Plotkin:

One is that it takes the logical form of induction, generalizing into the future what worked in the past. That is, the successful variants are fed back into the gene pool where they will be available for sampling by future organisms. This is the conservative, pragmatic part of the heuristic. The other is the generation of novel variants by chance processes. This is the radical, inventive component of the heuristic. It is nature's way of injecting new variants into the system in order, possibly, to make up for the deficiencies that may occur if what worked in the past no longer does so because the world has changed. When John Odling-Smee and I first wrote about this in the 1970s, we noted: "In effect the g-t-r [generate-test-regenerate] heuristic 'gambles' that the future will be the same as the past. At the same time it hedges its bets with aleatoric (chance) jumps, just in case it is not."[15]

Although it may be far from perfect, no other general-purpose construct-and-search procedure has yet shown itself to be as capable for such a broad range of problems, and none other is able to explain the remarkable achievements of fit we continually encounter in both natural and human-made environments.

Beyond Selection?

Universal selection theory draws heavily on biological evolution for its inspiration. Although the evolution of living forms is only one instance of a selectionist process resulting in adapted complexity, it provides the foundation and inspiration for all other selectionist theories of the emergence of fit. Biology has also lived longer with selectionist thinking than any other discipline. For these reasons, recent developments in biology that cast doubts on the fundamental role of natural selection in the emergence of the adapted complexity of living organisms are of considerable interest to those attempting to extend the selectionist perspective to other fields. If cumulative blind variation and selection is found to be lacking as an explanation for the emergence of design in organic evolution, an extension of selectionist principles to other achievements of adapted complexity would be suspect. We will therefore now confront five such would-be challengers to natural selection: punctuated equilibrium, directed mutation, exaptation, symbiosis, and self-organization.

Punctuated Equilibrium

According to classic Darwinian selection, biological evolution proceeds through the accumulation of very small changes over long periods of time. Gradual change is essential, since it is the only way that blind variation is likely to come up with improvements for selection. Whereas it is always possible that a large genetic change (or macromutation) may result in a fitter organism, for example, the transformation of an organism completely insensitive to light to one with a functioning eye in one generation, the laws of probability are almost certain to make large random changes less adaptive rather than more. As Dawkins explains:

To "tame" chance means to break down the very improbable into less improbable small components arranged in series. No matter how improbable it is that an X could have arisen from Y in a single step, it is always possible to conceive of a series of infinitesimal graded intermediates between them. However improbable a large-scale change may be, smaller changes are less improbable. And provided we postulate a sufficiently large series of sufficiently finely graded intermediates, we shall be able to derive anything from anything else, without astronomical improbabilities.[16]

The problem is, however, that the fossil record does not provide clear evidence for the gradual change of even one species into another. Darwin recognized the incompleteness of the fossil record, but believed that it was only a matter of time before these intermediate "missing links" would be found to provide hard evidence for the gradual emergence of new species over time. That these fossil gaps remain despite many new fossil finds has been taken by some as an indication that Darwin's emphasis on the gradualism of evolution was mistaken, and that evolution proceeds not by slow, gradual changes but rather by large and dramatic jumps, or saltations.

True saltationists are not easy to find among modern evolutionary biologists, since it is generally recognized that large, blind macromutations from parent to offspring are almost certain to be maladaptive.[17] But a well-known antigradualist perspective is present today in the theory of "punctuated equilibrium," developed by Gould and Eldredge.[18]

These researchers theorize that instead of continuous gradual change over time, the evolution of a species is marked by long periods of no or little change (stasis) interrupted occasionally by short periods of relatively rapid evolutionary change (punctuations). This may be a somewhat different picture of evolution than originally conceived by Darwin, but it is not inconsistent with the gradualism that is an essential part of natural selection. Although punctuated equilibrium describes relatively rapid change, this change still takes place over very long time periods, in the range of many thousands of years to much longer.

What is characteristic of punctuated equilibrium, then, is not the belief in adaptive macromutations arising in a single generation, but rather the long periods of stasis. But these periods need not be considered mysterious since they may simply be an indication that the species was already well adapted to its environment, and that the environment was not undergoing any rapid changes that would have created new selection pressures requiring new adaptations. So actually nothing in the theory of punctuated equilibrium is in any way fundamentally inconsistent with Darwin's conception of evolution.[19]

Directed Mutation

Similar compatibility is not the case, however, for another view of evolution that has attracted considerable interest and led to much recent controversy. In 1988 John Cairns, a well-respected molecular biologist and cancer researcher, published with two associates a paper in the prestigious British journal *Nature* that threatened to undermine the basic tenets of Darwinian evolution.[20]

Cairns and his colleagues claimed to have found evidence that *E. coli* was able somehow to direct its mutations to achieve adaptive changes when placed in a new, challenging environment. This research involved placing bacteria that could only metabolize glucose in an environment where only a foreign sugar (lactose) was available. Here the stressed bacteria continued to duplicate and, as would be expected, some of the descendants contained mutations that permitted them to metabolize the new sugar. This in itself is not surprising, since the genetic change necessary to transform an *E. coli* from a glucose- to a lactose-eating bacterium is quite small, and in a large colony it would be expected that at least some of the naturally occurring mutants would have stumbled on it by sheer blind chance. But these scientists reached the highly unorthodox conclusion that instead of being produced randomly, the bacteria were somehow able to produce the adaptive mutations at a much higher frequency than other, nonadaptive mutations. In other words, they believed that their studies provided evidence that "bacteria can choose which mutations they should produce" which would "provide a mechanism for the inheritance of acquired characteristics."[21]

As would be expected, these statements immediately elicited both considerable interest and controversy, since the central dogma of biology was being challenged, that is, that changes in the environment cannot direct (instruct) changes in the genome. Some researchers rejected this conclusion out of hand, but others were impressed enough to attempt to find possible mechanisms by which the environment could somehow instruct the genome to produce just the right mutations to allow it to digest the new sugar. Cairns himself proposed that environmental changes could affect changes in proteins that could consequently instruct the DNA to make certain adaptive changes in the genes, in flagrant violation of the central dogma.

However, it may well be that this and other explanations for directed or "instructed" mutation are not necessary after all. Australian microbiologist Donald MacPhee and his colleagues provided evidence that, when placed in a medium of lactose, the mutations produced by glucose-metabolizing *E. coli* are indeed produced blindly.[22] What seems to happen under the stressed condition of a glucose-poor environment is not a specific increase in the rate of adaptive mutations, but rather a general increase in the overall mutation rate due to inhibition of the mechanism that usually checks and repairs the genetic errors that arise during the normal functioning of the bacterium. So while mutations continue to be produced blindly, the higher rate of genetic change allows the bacteria to stumble on the adaptive genetic change more quickly than they would if left in their normal glucose-rich environment.

But let us continue to imagine for a moment that a bacterium was able to change just those genes regulating metabolism in just the right way to allow for the digestion of a foreign sugar. If this were the case, it would be yet another example of a puzzle of fit demonstrating that the bacterium had somehow acquired the ability to sense a new sugar in its environment and alter its genome to digest it. But then we would be led to ponder how this adapted complexity could have originated in the first place, with cumulative blind variation and selection as a prime candidate to explain the source of this remarkable ability that somehow permitted the bacterium to instruct its genome to make the required changes to digest the new, strange food that was being served.

Although no convincing evidence exists that adaptive changes in genes can be directed by the environment in a Lamarckian manner, the findings of Cairns and MacPhee and their respective colleagues are important. If organisms are able to increase their mutation rate in the presence of new environmental stresses but keep mutations in check when these stresses are absent, it would enable organisms to exert a certain degree of control over evolution that is absent from the classic neo-Darwinian perspective. Instead of producing mutations at a constant rate regardless of environmental conditions, organisms may produce more mutations and therefore more varied offspring just when such innovative variation is necessary to keep the species extant.[23]

This view ascribes to the evolutionary process decidedly more "intelligence" than does the neo-Darwinian perspective. It nonetheless preserves the required blindness of genetic variations. What is altered is only the rate of production of these variations. This sensitivity of mutation rate to environmental stress could simply be the result of a stress-related breakdown of genetic repair mechanisms. Or it could be the result of a more sophisticated active mechanism that itself had evolved by natural selection, since individuals that by chance produced more genetic variability under difficult environmental conditions would have been more likely to leave better adapted progeny than those insensitive to environmental stress.

The work of Cairns and MacPhee concerned the metabolism of different types of food. It is not difficult to imagine how other types of biological functions could also be involved, such as thermoregulation. For example, as temperatures dropped at the onset of an ice age, mammals would undergo stress as did Cairns's bacteria when placed in an environment where no useful food was available. This would lead to an increase in the mutation rate during reproduction, resulting in a second generation of animals with greater variation in the length and texture of their coats. Those particular descendants having, by chance, longer and thus warmer coats would suffer less from the cold environment, resulting in lower mutation rates and consequently less variation in the coats of their third generation, extra-hairy offspring. But those second-generation animals with short coats would maintain a higher rate of mutation, so that at least some of their offspring would likely have warmer coats then their parents.

This hypothesis has some interesting consequences. As in the ice-age example, by varying the mutation rate, a species would adapt more quickly to changing environmental conditions. It is also of interest to realize that such stress-dependent mutation rates would result in occasional short periods of relatively rapid (although still gradual) evolutionary change separated by longer periods of little or no change during periods of environmental stability. And this is exactly what Gould, Eldredge, and their associates refer to as punctuated equilibrium.

Exaptation

Another challenge to natural selection is posed by those advocating exaptation as a major mechanism of biological evolution. Although we briefly discussed exaptation in chapter 11, some additional remarks are appropriate here, as this perspective has received considerable attention as a potential rival to the selectionist account of adaptive evolution.

It will be recalled from chapters 5 and 11 that exaptation refers to the emergence of some feature of an organism that fits a current function, but did not originally evolve for this use. We considered how bird and insect wings, now used for flight, appear to have have evolved originally to aid in cooling and heating. Darwin's own example was how the lungs of terrestrial animals evolved from the swim bladder of fishes.[24] Exaptation also refers to the emergence of a current adaptedly complex characteristic that arose for no functional reason, probably as a nonselected correlate of some other adapted characteristic, but turns out later to be useful for some function.

The concept of exaptation is particularly valuable in understanding how the necessary gradualism of natural selection can account for the evolution of complex adaptations that would appear to be functionally useless in their incipient forms. Let us consider again the bird's wing. If wings evolved gradually over a long period of time, the first protowings would have been nothing but stubby protuberances from the backs of protobirds. Since they would have been ill suited for flight, the question arises as to why and how they would have begun to evolve in the first place.

Exaptation, by disentangling current use from the original selection pressures, makes this understandable. The protowings may not have been of much use for flight, but could have easily aided in eliminating excess body heat when stretched out in a shady breeze, and increasing warmth when extended to catch the rays of the morning sun. Selection for larger and larger wings for more effective thermoregulation would have also laid the groundwork for flying, although the first aeronautically useful wings probably mainly provided some protection from falling from trees or other high places. Once past a certain size, the switch in the primary role of wings from thermal equipment to flying equipment would have been made, and further refinements would have increased their fitness for flying.

But although some see exaptation as a challenge to natural selection and as a part of a "new theory of evolution,"[25] it actually provides no competition at all as an explanation of adapted complexity. For wings to have evolved as useful instruments for flight, they first had to evolve as useful instruments of thermoregulation. The only explanation for how this could have happened, aside from the astronomical improbability of a single lucky mutation, is by natural selection. In addition, further refinements were undoubtedly necessary to fashion wings for efficient flight. Once again, the gradual and cumulative selection of blind variations in wing shapes is the only process currently understood that can account for the evolution of such design. Therefore,

to identify a feature as an exaptation does not mean that feature "is not the product of natural selection"—only that it did not always serve its present function. Indeed, most human behaviors serving biological functions that enhance fitness should probably be regarded as exaptations, modified through natural selection for subsequent specialization. The manual dexterity enabling specialized human hand movements now used in food gathering, eating, caretaking, tool use, and communicational expressions derives from the ability of prosimian ancestors to grasp branches. Human emotional expressions evolved from ritualized facial and vocal displays in other ancestral species, display derived from behaviors that originally had no signaling function. . . . The use of stereotypical vocal patterns in human mothers' speech to infants is also undoubtedly an exaptation, with evolutionary origins in ancestral non-human primate vocalizations used for very different purposes. The claim that such human behaviors are exaptations rather than adaptations is a claim about origins, which does not require rejection of an adaptationist explanation for the current fit between these behaviors and the biological functions they serve.[26]

So whereas exaptation permits new, unanticipated twists and turns in the evolution of a species as new roles and consequently new selection pressures are found for old features, it relies on natural selection for all initial adaptations and consequent refinements of these adaptations for new functions. Exaptation may make it easier for natural selection to do its stuff, but it is most certainly not an alternative explanation for the adapted complexity of naturally occurring design.

Symbiosis

Other individuals interested in accounting for the products of evolution but who are not particularly enamored of Darwinian natural selection emphasize the role of cooperation among organisms. One particularly outspoken advocate of this view is American biologist Lynn Margulis. She has offered strong criticism of the Darwinian selectionist account of evolution, stating, "It's totally wrong. It's wrong like infectious medicine was wrong before Pasteur. It's wrong like phrenology is wrong. Every major tenet of it is *wrong*."[27]

Margulis does not doubt that natural selection occurs, but objects both to the gradualness of evolution and its emphasis on competition. In 1965 she proposed that eukaryotes, the nucleated cells that make up all organisms except bacteria, originated from the fusion of two different types of bacteria. Although her proposal was initially greeted with considerable skepticism, the theory of the symbiotic evolution of eukaryotes is now widely accepted among biologists.

But if symbiosis did occur, how does this make Darwinian theory "totally wrong"? To be sure, Darwin knew nothing of the genetic basis of life and so he could hardly have imagined that new organisms could emerge from the incorporation of the genetic material of one genome into the genome of another. Also, the cooperation between different species that characterizes evolutionary symbiosis does constrast with his very competitive view of evolution. But it must be kept in mind that the basic core of Darwinian theory is its reliance on hindsighted selection from a population of blindly constructed variations, and in this respect Margulis's symbiotic theory is Darwinian at its core. The variations available, however, are no longer limited to genetic mutations and recombinations within the gene pool of one species, but include genetic combinations across species. And these initial interspecific (between-species) genetic transfers must have occurred blindly, with probably almost all of them harmful to one or both organisms (as occurs when the genetic material of an influenza virus invades human cells) or of no consequence. So by pure chance it must have happened one lucky day that the accidental pooling of the genetic material or organelles of two types of unicellular organisms provided a new cell with survival and reproductive advantages over both of its constituents. But if one is to suggest that anything other than chance was involved in forming such symbiotic relationships, then one will be in the unenviable position of having to explain the origin of the knowledge required to circumvent chance.

Indeed, if such symbiotic evolutionary processes have in fact taken place, it would make Darwinian selectionism a more powerful, rather than a less powerful, explanation for adaptive evolution. Evolution would no longer be limited to fiddling with the genes of separate species by mutation and sexual recombination. Instead, a whole new world of possibilities becomes available in the mixing and matching of genes, organelles, and cells across species. A species would not have to evolve photosynthesis for itself if it could incorporate the photosynthetic know-how of another species, which is now believed to have occurred. But there still is no way that a primitive one-celled organism could possibly "know" which other type of cell or organism it should incorporate or borrow from to improve its chance of survival and reproduction. It could only await the blind forces of nature to bring a promising partner within reach, or await the invasion of an adventuresome virus bearing genetic material from a previously infected cell. Although the potential benefits of certain symbiotic unions would be great, it would not be possible to avoid the very same blind variation and selection that characterizes the natural selection that Margulis attacks so vigorously—at least not without the guidance of an intelligent and beneficent matchmaker working behind the scenes.

Self-Organization

But there may well be something more to adaptive evolution than natural selection after all. The second law of thermodynamics is the well-known law of increasing entropy, which states that in an isolated system (that is, a system that can neither gain nor lose energy or matter) we can expect order to decrease, energy to become less available, and a stable (and lifeless) equilibrium to be reached. But in an open system able to draw on sources of outside energy, the situation can be dramatically different. The evolution of life itself is the most striking example of a naturally occurring increase in complexity. But inanimate objects and systems can also demonstrate naturally emerging complexity in certain situations. Anyone who has marveled at the intricate symmetrical beauty of a snowflake, observed the coordinated ballet of grains of rice in a simmering pot of water, or encountered the organized fury of a tornado has noticed that complexity can also arise spontaneously in the inanimate world. And this spontaneous emergence of complexity, or self-organization as it is now usually called, has recently attracted the attention of a wide range of scientists, from physicists and biologists to cognitive scientists and economists.

That organized complexity can emerge spontaneously in inanimate systems may have far-reaching implications for understanding the origin of life and its continuing evolution. One of the major difficulties in coming up with a convincing nonmiraculous account is explaining how inanimate matter could have organized itself into the very first self-replicating life forms. The degree of complexity required for this first step has seemed to many biologists to be just too unlikely to be due to the random forces of nature. Darwin himself was reluctant to advance a nonmiraculous argument, and in the last paragraph of later editions of the *Origin* refers to the power of life "having been originally breathed by the Creator into a few forms or into one."

So if the blind laws of nature operating on inanimate entities could with high probability lead to the emergence of complex, self-organized molecules and networks of molecules, then the origin of life itself, as well as its continued evolution, becomes somewhat less of a mystery. This is the message that American biochemist and biophysicist Stuart Kauffman has been delivering for the past dozen or so years, and provides in detail in his influential book, *The Origins of Order*.[28]

It is now recognized that the laws of physics acting on nonliving entities can lead to spontaneous complexity, but nothing in these laws can guarantee *adapted* complexity of the type seen in living organisms, that is, the ubiquitous biological puzzles of fit. Of all the complex systems and structures that may self-organize due to the forces of nature, there can be no assurance that all or any of them will be of use for the survival and reproduction of living organisms. Selection, therefore, must choose among these various complex systems the ones with characteristics better suited to survival and reproduction, and eliminate others. As Kauffman remarked, "evolution is not just 'chance caught on the wing.' It is not just a tinkering of the ad hoc, of bricolage, or contraption. It is emergent order honored and honed by selection."[29]

The study of self-organizing systems is among the newest and most ambitious scientific ventures of the late twentieth century, and its discoveries may ultimately have a major impact on evolutionary theory and our understanding of the emergence of life itself. But from our present viewpoint it is difficult to see how self-organization could ever replace, as opposed to complement, natural selection. It may help to jump-start natural selection by blindly offering up a variety of already complex systems from which to choose. But it is only after-the-fact selection that can eliminate the nonviable complex systems and retain the viable ones.

To return one final time to the major theme of this book, selection theory has become an important part of many different disciplines either to explain puzzles of fit, as in the immune system, or to create them, as in genetic programming and directed molecular evolution. Where previous providential and instructionist theories of adaptive change and knowledge growth have been found to be inadequate, selection theory provides a truly naturalistic and nonmiraculous account of puzzles of fit, whether this fit occurred with or without the assistance of humankind. The evolution of theories in many different disciplines from providential through instructionist to selectionist is a provocative suggestion of the superiority of selectionism. And this recent movement in so many different fields of inquiry constitutes what may be considered a second Darwinian revolution.

In a number of respects this revolution is unlike the first. The first made a dramatic début on 24 November 1859 with Darwin's *The Origin of Species* selling out its entire first edition of 1250 copies the first day it was offered for sale. It involved a single discipline, biology, and a single question, the evolution of species. And Darwin himself (if not all of his fellow Darwinians) was insistent that processes other than natural selection—such as the instructionist Lamarckian processes of use and disuse—were involved in the emergence of adapted organic structures and behaviors.

In contrast, this second Darwinian revolution cannot be limited to any one significant event. Its roots lie even deeper than the theory of organic evolution. It involves parallel and simultaneous developments in many different disciplines, in much the same way that different species evolve simultaneously in conformance with the demands of their particular environments. The second revolution also appears to be moving in an exclusively selectionist direction in accounting for the emergence of truly novel knowledge. And although selectionism currently rules in evolutionary biology and immunology, it remains very much a minority viewpoint in the other fields surveyed in the preceding chapters.

But if the history of science has anything at all to tell us, it is that our current theories eventually become less than satisfactory. Theories that seem to be well founded and clear improvements over previous ones are eventually seen as inadequate and are replaced by newer, more encompassing perspectives in the way that Newtonian physics gave way to Einstein's relativity and Bohr's quantum mechanics. This continued development in science depends on the relentless criticism of both currently accepted and newly proposed theories in the form of continuing efforts to discover how they are inadequate and how they can be improved.

From this historical perspective, it would appear highly unlikely for selectionism to be the final explanation for the emergence of all puzzles of fit. Although it has already endured for a remarkably long period of time in evolutionary biology, we would normally expect it eventually to be surpassed. Newer and better theories might then explain everything in a given field that selection theory can, plus other phenomena that it cannot, even if at this moment it is exceedingly difficult to imagine what such a replacement would look like. That such a theory would have to be selected among competing ones (including selection theory itself) poses somewhat of a paradox to those who want to overturn selectionist accounts of knowledge growth.[30]

But since universal selection theory is itself a blind variation, although one that has so far resisted the arguments that have been fatal to many providential and instructionist accounts of adapted complexity, I must welcome the criticism and selection pressure that this book and its thesis will undoubtedly provoke. Such scrutiny is absolutely necessary to lead us to understandings that may go beyond selectionism, if indeed that is possible.

I do hope, however, that such criticism will not be based on faulty understandings of universal selection theory and the claims being made for it, and consequently result in a return to providential and instructionist theories that have been shown to be inadequate to the job of accounting for the emergence of adapted complexity.

In the meantime a strong case can be made that universal selection theory provides the best explanation for both naturally and artificially produced puzzles of fit. It relies on patient, iterative cycles of blind variation and selection that over the course of time can result in biological adaptations and new species, functional human cultures, technological breakthroughs, and scientific revolutions. And what is perhaps most appealing from the naturalist perspective of modern science, it provides this explanation without miracles—except for the illusory miracle of how such an inherently blind, stupid, wasteful, and sluggish process can be found at the very foundation of life, its marvelous design, and all the subsequent knowledge that life in its human form has generated.

Appendix
The Trouble With Miracles

Throughout this book miraculous accounts of the origin of adaptive complexity were discounted and natural, scientific ones advocated in their stead. But what exactly is the problem with miracles? Why should miracles and miraculous explanations be rejected in science in favor of nonmiraculous ones?

First, we have to define what we mean by *miracle*. In everyday usage, there seem to be two meanings. The first refers to an event that is considered to have a very low probability of occurrence, but that happens nonetheless. A man and his wife both winning the grand prize of a state lottery on successive weeks might be considered such an event. But, of course, there is really nothing miraculous about such happenings, since they are possible and their rarity is consistent with their low probability. We have also seen how the process of cumulative variation and selection can dramatically increase the probability of adaptive evolutionary change that would be highly unlikely to emerge in a single step.

The second meaning of *miracle* has to do with events that violate the laws of nature. As defined by the *Oxford English Dictionary* a miracle is

a marvelous event occurring within human experience, which cannot have been brought about by human power or by the operation of any natural agency, and must therefore be ascribed to the special intervention of the Deity or of some supernatural being; chiefly, an act (e.g. of healing) exhibiting control over the laws of nature, and serving as evidence that the agent is either divine or is specially favoured by God.

Abraham Lincoln coming back to life and walking the streets of Springfield, Illinois, would be such an event. So would the creation of all the earth's plants and animals in a single day. Such events are thought to be the work of God or some other powerful supernatural entity that has the power to circumvent the usual laws of nature.

The ever-skeptical Scottish philosopher David Hume, whose critique of induction was presented in chapter 6, also offered a critique of miracles, or rather, belief in miracles, in his *Enquiries Concerning Human Understanding* published in 1748. His reasoning on this subject involved two arguments. The first was that it was rational to believe that a miracle had occurred only if the evidence that it had occurred was stronger than the evidence that it had not. As Hume put it:

> The plain consequence is "That no testimony is sufficient to establish a miracle, unless the testimony be of such a kind, that its falsehood would be more miraculous, than the fact, which it endeavours to establish; and even in that case there is a mutual destruction of arguments, and the superior only gives us an assurance suitable to that degree of force, which remains, after deducting the inferior."[1]

To give an example, imagine that an old and trustworthy friend from Springfield told you that yesterday she saw the sixteenth president of the United States walking the streets of his hometown. Now, this would appear to be an event in flagrant violation of the laws of nature as we understand them. Hume would advise you to consider which would be more miraculous—Abraham Lincoln coming back to life, or your friend being either mistaken or deceitful—and accordingly believe the less miraculous. In this case, it would be rational to believe in Lincoln's resurrection only if it would be more of a miracle that your friend could have either been mistaken or deceitful in making her report.

The second part of Hume's argument consists of four reasons why in actual practice there could never be compelling evidence for a miracle, even though evidence for one could exist *in principle* that outweighed the evidence against it. Only one of these reasons will be mentioned here, that having to do with the unreliability of human reports of miraculous events:

> . . . there is not to be found, in all history, any miracle attested by a sufficient number of men, of such unquestioned good sense, education, and learning, as to secure us against all delusion in themselves; of such undoubted integrity, as to place them beyond all suspicion of any design to deceive others; of such credit and reputation in the eyes of mankind, as to have a great deal to lose in case of their being detected in any falsehood; and at the same time, attesting facts performed in such a public manner and in so celebrated a part of the world, as to render the detection unavoidable: All which circumstances are requisite to give us full assurance in the testimony of men.[2]

In other words, Hume insists that you simply cannot trust the testimony of your friend from Springfield, at least not when reporting a miracle, since it is always more likely that her testimony is unreliable. And such is the case not only for your friend but for any individual or group of individuals providing an account of any miraculous event.

It should be made clear that Hume does not discount the possibility that miracles may have in fact occurred, but only that it is never rational to believe that one has occurred. Neither does anything in his argument suggest that we might not be happier if we did believe in miracles, such as that the world and everything in it were made by a benevolent creator who continues to provide for our welfare. But a stubborn belief in miraculous accounts of events for which we have nonmiraculous accounts is inconsistent with the scientific enterprise that involves the continual search for the simplest and most parsimonious explanations of the goings-on of the universe.

It is of interest that Charles Darwin, certainly no philosopher, expressed very much the same conclusion as Hume in his autobiography where he recounts his gradual loss of Christian faith:

> By further reflecting that the clearest evidence would be requisite to make any sane man believe in the miracles by which Christianity is supported,—that the more we know of the fixed laws of nature the more incredible do miracles become,—that the men at that time were ignorant and credulous to a degree almost incomprehensible by us,—that the Gospels cannot be proved to have been written simultaneously with the events,—that they differ in many important details, far too important as it seemed to me to be admitted as the usual inaccuracies of eyewitnesses;—by such reflections as these, which I give not the least novelty or value, but as they influenced me, I gradually came to disbelieve in Christianity as a divine revelation. The fact that many false religions have spread over large portions of the earth like wild-fire had some weight with me. Beautiful as is the morality of the New Testament, it can hardly be denied that its perfection depends in part on the interpretation which we now put on metaphors and allegories.[3]

To the remote desert dweller witnessing the effects of refrigeration for the first time, seeing water turn into a solid may indeed appear to be a miracle, a remarkable violation of a very basic law of nature (that water is always fluid), which could only have been accomplished by a supernatural power. But science has a different view, since scientific progress would cease if it proposed and retained theories that could at any time be violated by miracles, instead of proposing theories and then rejecting them (and replacing them with better ones) when their predictions are in error, or when more powerful and parsimonious theories are generated.

This is why, in our scientific endeavor to make sense of the emergence of adapted complexity, we have no choice but to be rational and seek nonmiraculous explanations of the type that universal selection theory provides.

Notes

Chapter 1

1. Darwin (1859).
2. Campbell (1974b, p. 139).
3. Pinker & Bloom (1990, p. 709).

Chapter 2

1. Paley (1813, p. 539).
2. Lamarck (1809; translated by Burkhardt, 1977, p. 166).
3. Darwin (1859, p. 109).
4. Aristotle (1929, pp. 173, 175).
5. Paley (1813, p. 3).

6. The Latin verb from which the English *providence* derives can be translated as "to foresee," so that divine providence both provides and foresees, the latter being important to provide what will be useful at some future time.

7. The French encyclopedist Denis Diderot (1713–1784) presented much the same criticisms against the argument from design in his imaginary death-bed dialogue involving the blind English mathematician Nicholas Sanderson and a Reverend Holmes. Diderot creates these words for Sanderson:

> And even if the physical mechanism of animals is as perfect as you claim . . . what has that to do with a sovereignly intelligent being? If it is a matter of astonishment for you, then that may possibly be because you are in the habit of treating everything that is beyond your comprehension as a miracle.

> If nature presents us with a knot that is difficult to untie, then let us leave it as it is; let us not insist on cutting it there and then and on employing for the task the hand of a being who thereupon becomes a knot more difficult to untie than the first. Ask an Indian why the globe remains suspended in the air and he will reply that it is borne on the back of an elephant. And on what does the

elephant rest? On a tortoise. And the tortoise, who supports that? (reprinted in Bajema, 1983, pp. 50–51).

Dawkins made essentially the same argument (1986, p. 147).

8. Burkhardt (1977, p. 22).

9. Although Lamarck did not see God directly involved in the creation of current life forms, he referred to God as "the supreme author of all things" (quoted in Burkhardt, 1977, p. 185).

10. It should be mentioned that Lamarck was not primarily interested in accounting for the phenomenon of adaptation. Rather, he attempted to account for what he saw as the increasing complexity of organisms over evolutionary time. Nonetheless, he and many others did use his theory to explain the fit between organisms and their environments, as shown in his account of the giraffe's long neck quoted earlier (see Burkhardt, 1979, pp. 174–175).

11. Lamarck (1809; translated by Burkhardt, 1977, pp. 173–174).

12. Lamarck (1835; quoted in Løvtrup, 1987, p. 52).

13. Lamarck (1835; quoted in Løvtrup, 1987, p. 53).

14. Lamarck (1815–1822; quoted in Burkhardt, 1977, p. 166).

15. See Medvedev (1969) and Joravsky (1970) for interesting accounts of the life and times of Lysenko.

16. Indeed, the separation of germ and somatic cells is referred to as the central dogma in molecular biology. In slightly more technical terms, the central dogma states that although the DNA of genes directs the synthesis of proteins and thus has an important influence on the form of an organism and its constituent parts, the reverse is not true—proteins cannot instruct the DNA coded in the genes. Therefore, no acquired changes in an organism's structure can cause changes in the genes of germ cells that can be passed on to offspring and result in structures like those acquired by the parent.

17. See Ridley (1985, p. 32).

18. See Burkhardt (1977, p. 145).

19. A third problem with Lamarckian inheritance pointed out by Gregory Bateson (1979, pp. 151–153) is that it would ultimately make organisms less adaptable to changes in the environment.

20. Quoted in L. Huxley (1900, p. 170).

21. See, for example, Dawkins (1986) and Ridley (1985). To these should be added recent research documenting evolution occurring in the wild, the most extensive of which is the work of Peter and Rosemary Grant on Darwin's finches in the Galápagos Islands (see Weiner, 1994, for a delightful account of the Grants' work).

22. Ridley (1985).

23. Gould (1980, pp. 20–21).

24. Darwin (1859, p. 170).

25. See Dawkins (1986, chapter 3) for an explanation of the cumulative nature of biological evolution and how this differs from the one-step selection processes found in the inanimate world.

26. Wallace (reprinted in F. Darwin & Seward, 1903, p. 268).

27. I first heard the phrase "Darwin's hammer" used by William T. Powers in a criticism of certain aspects of Darwinian evolution.

28. Darwin (1859/1966, p. 134).

29. Kimura (1982).

30. Ho & Saunders (1984).

31. Dover (1982).

32. See Dawkins (1986, chapter 11) for an excellent discussion of how these challenges to natural selection as explanations of adaptive evolutionary change fall far short of their goal.

33. Sheler (1991, p. 59).

34. Eve & Dunn (1989).

35. Dennett (1984, p. 92, footnote).

36. Stein & Lipton (1989, p. 36).

37. Stein & Lipton (1989, p. 36).

38. Dawkins (1983).

Chapter 3

1. Paley (1802/1902, pp. 336–337).

2. Lamarck (translated by Burkhardt, 1977, pp. 170–171).

3. Darwin (1859, pp. 242–244).

4. Beach (1955).

5. Aquinas (1265–1273/1914, p. 460).

6. Richards (1987, pp. 22–23).

7. Paley (1813, p. 306).

8. Erasmus Darwin (quoted in Richards, 1987, p. 34).

9. This idea was to persist into the twentieth century in the developmental psychology of Jean Piaget.

10. Richards (1987, p. 94).

11. Richards (1987, p. 96).

12. Quoted in Richards (1987, p. 136).

13. Richards (1987, p. 136).

14. Darwin's difficulty in formulating a natural selection theory of instinctive behavior that could account for that of the neuter insects may have been an important factor in the long delay between his discovery of natural selection as a general

principle in evolution and his first publication of the theory 20 years later in *The Origin of Species* (see Richards, 1987, pp. 152–156).

15. Richards (1987, p. 145).

16. Darwin (1856–1858/1975, p. 370). Now it is known that many truly social (eusocial) insects are *haplodiploid*, meaning that females have a double set of genes (that is, are *diploid*, as are humans), but males develop only from unfertilized eggs, so that they have only one set of genes (are *haploid*). As a result, a female worker is more related to her sisters, sharing with them half of her mother's genes and all of her father's, than she would be to her own offspring. Thus it is in the genetic interest of a female worker to forgo having offspring of her own and devote her energies to caring and protecting those of her siblings (see Badcock, 1991, pp. 73–75). However, haplodiploidy is neither necessary nor sufficient for eusocial behavior since termites and naked mole rats are both eusocial and diploid, and not all haplodiploid insects are eusocial. Nonetheless, kin selection always underlies eusociality in insects or other animals, whether or not it is facilitated by haplodiploidy (see Sherman, Jarvis, & Braude, 1992, pp. 72–73).

17. See, for example, Axelrod (1984) and Hamilton (1964).

18. See Richards (1987, pp. 230–234) and Ghiselin (1969, pp. 203–206).

19. Gould (1980, p. 50).

20. Wallace (1867; quoted in Gould, 1980, p. 51).

21. Concerning their ability to sing, Wallace (1895; quoted in Gould, 1991a, p. 57) commented:

> The habits of savages give no indication of how this faculty could have been developed by natural selection, because it is never required or used by them. The singing of savages is a more or less monotonous howling. . . . This wonderful power . . . only comes into play among civilized people. It seems as if the organ had been prepared in anticipation of the future progress in man, since it contains latent capacities which are useless to him in his earlier condition.

22. Anthropomorphism refers to the ascribing of human qualities and traits to non-human animals or objects.

23. Burkhardt (1983, p. 433).

24. Lorenz (1981, p. 1).

25. Lorenz (1981, pp. 236–237).

26. James (1890, p. 7).

27. Burkhardt (1983, p. 437).

28. Lorenz (1981, p. 101; emphasis in original).

29. Burkhardt (1983, p. 436).

Chapter 4

1. Bibel (1988, p. 178).

2. Jerne (1967, p. 201).

3. Ontogeny, or ontogenesis, refers to the development and growth of an individual organism from embryo to adult.

4. Ehrlich (1900).

5. Tonegawa (1985, p. 105).

6. Breinl & Haurowitz (1930).

7. Pauling (1940). Pauling won the Nobel prize for chemistry in 1954 and the Nobel peace prize in 1963, the only person ever to win two unshared Nobel prizes.

8. Jerne (1955).

9. Burnet (1957).

10. Jerne (1951, 1975).

11. Ada & Nossal (1987, p. 52); Tonegawa (1985, pp. 104–105).

12. Janeway (1993, p. 75).

13. Tonegawa (1983, 1985).

14. Janeway (1993, p. 73).

15. Tonegawa (1985, p. 110).

16. Recent research suggests that the recombination of somatic genes may be involved in the development of the mouse brain (Matsuoka et al., 1991) and therefore may be involved in other mammalian organs as well.

17. Perelson & Oster (1979).

18. Farmer, Packard, & Perelson (1986, p. 190).

19. Marrack & Kappler (1993, p. 87).

20. It should be noted that recombination of immunoglobulin genes involved in the production of antibodies differs somewhat from the recombination of parental genes in sexual reproduction. In the former, nucleotides can be inserted and deleted at random from the recombined immunoglobulin gene. This adds an important additional source of diversity in the generation of antibodies (see Janeway, 1993, p. 75; Kallenbach et al., 1992).

Chapter 5

1. Changeux (1985, p. 248; first emphasis added).

2. Gazzaniga (1992, p. 50).

3. Not all dendrites serve to excite the attached neuron. Some are inhibitory in that they act to prevent the neuron from firing.

4. It has been said that "human beings have caused greater changes on earth in 10,000 years than all other living things in 3 billion years. This remarkable dominance is related to the development of the brain from the minute cerebrum of simple animals to the complex organ of about 1350 grams in man" (Sarnat & Netsky, 1981, p. 279).

5. Bullock (1977, p. 410).

6. See Gould (1991a, essay 9).

7. Gould & Vrba (1982, p. 13). See also Gould (1991b) for an introduction to the concept of exaptation.

8. Gazzaniga (1992, p. 35).

9. Changeux (1985, p. 206).

10. Eccles (1989, pp. 1, 4).

11. Changeux (1985, p. 206). Similar remarks have been made by Schatz (1992, p. 67) and Nobelist Manfred Eigen (1992, p. 47).

12. Changeux (1985, pp. 216–217).

13. Hamburger (1975).

14. Changeux (1985, p. 217).

15. Ramón y Cajal (quoted in Changeux, 1985, pp. 212–213).

16. Schatz (1992, p. 63).

17. Schatz (1992, pp. 66–67).

18. Gazzaniga (1992, p. 37).

19. Greenough & Black (1992).

20. Huttenlocher (1984, p. 490).

21. Werker & Tees (1984).

22. Curtiss (1977).

23. See Johnson & Newport (1989) and Newport (1994).

24. Hebb (1949, p. 65).

25. Eccles (1965; quoted in Rosenzweig et al., 1979).

26. Young (1964, p. 285).

27. Albus (1971; quoted in Rosenzweig et al., 1979).

28. Dawkins (1971).

29. Changeux (1985, p. 272).

30. Changeux (1985, p. 248).

31. Turner & Greenough (1985).

32. Black & Greenough (1986).

33. Black & Greenough (1986, p. 33). These experience-dependent changes in the mature brain are contrasted with the experience-expectant development of the maturing brain, the latter taking advantage of the great number of redundant connections already present in the postnatal brain.

34. Jones & Schallert (1992, 1994); Schallert & Jones (1993).

35. Calvin (1987, 1990).

36. Edelman (1987, 1988, 1989, 1992).

37. See Gazzaniga (1992, chapter 2). Gazzaniga's innatist construal of brain selectionism is critiqued at the end of chapter 15.

38. Changeux (1985, p. 272).

Chapter 6

1. Plato, Meno dialogue (1952, p. 183).

2. Locke (1690/1952, p. 121).

3. Lorenz (1941/1982, pp. 124–125).

4. The word *epistemology* is derived from the Greek word for knowledge, *episteme*.

5. Chomsky (1988b, p. 34).

6. Plato (1952, p. 179).

7. Plato (1952, p. 179).

8. See Petrie (1981, pp. 12–15).

9. Bereiter (1985, p. 203; emphasis added).

10. Plato (1952, p. 183).

11. Descartes's philosophy is often referred to as rationalist epistemology (from the Latin *ratio*, "reason"). Rationalist epistemology stresses the ability of the mind to come to knowledge about the world without the need for (or indeed, despite misleading) sensory experience. It thus stresses the role of reasoning and innate ideas that according to Descartes's preevolutionary philosophy, could be provided only by God.

12. The word *empiricism* has its roots in the Greek word *empeiria*, which was translated as *experientia* in Latin, and from which is derived the French and English word *experience*. In philosophy, empiricism is the view that we can obtain knowledge of the world through direct sensory experience of it. As such, it can be considered an instructionist epistemology.

13. Thomson (1964, p. 240).

14. See Musgrave (1993, chapter 8) for a more detailed discussion of Hume's irrationalism.

15. Russell (quoted in Popper, 1979, p. 1).

16. Walsh (1967, p. 306).

17. Although Kant did not make explicit use of divine providence in his epistemology, he wrote, "When we speak of the totality of nature, we must inevitably conclude that there is Divine regulation" (quoted in Lorenz, 1941/1982, p. 142).

18. Lorenz (1941/1982).

19. Lorenz (1941/1982, p. 127).

20. Lorenz (1941/1982, pp. 124–125).

21. Mayr (1966, pp. ix–x).

22. See Cziko & Campbell (1990) for nearly 1000 references related to evolutionary epistemology.

23. Munz (1993) and Plotkin (1994) provide additional arguments for a selectionist, Darwin-inspired epistemology.

Chapter 7

1. Skinner (1974, p. 17).

2. Skinner (1953, p. 430).

3. See Plotkin (1994, pp. 144–152) for a discussion of how learning is necessary to solve the "uncertain futures problem" that biological evolution alone cannot solve.

4. Pavlovian conditioning is also referred to by psychologists as classical or respondent conditioning.

5. See Boakes (1984, p. 121).

6. Pavlov (quoted in Boakes, 1984, p. 121).

7. Watson & Rayner (1920).

8. Watson (1917; quoted in Boakes, 1984, p. 220).

9. Thorndike (1911; quoted in Boakes, 1984, p. 74).

10. Thorndike (quoted in Boakes, 1984, p. 73).

11. Skinner (1948, 1971, 1974).

12. Brewer (1974).

13. The response studied was the galvanic skin response, a change in electrical resistance between two points on a person's skin.

14. Brewer (1974, p. 27). Brewer also questioned the belief that noncognitive, automatic, unconscious processes are involved in what appears to be the conditioning of children and nonhuman mammals.

15. Skinner (1957).

16. Plotkin (1987, p. 144).

17. Plotkin (1987, p. 144).

18. Thorndike (1911; quoted in Boakes, 1984, p. 75).

19. See in particular Skinner (1966).

20. Skinner (1974, p. 68).

21. Skinner (1971, pp. 130–131).

22. See Kohn (1993) for a review of how the application of the principles of operant conditioning as espoused by Skinner and other behaviorists have repeatedly failed in home, school, and work settings.

Chapter 8

1. Dewey (1896, p. 363).

2. James (1890, p. 7).

3. James (1890, p. 7).

4. Skinner (1974, p. 224).

5. Dewey (1896, p. 363).

6. Described in Tolman (1932, pp. 79–80) and in Boakes (1984, p. 232).

7. Tolman (1959, p. 100).

8. Tolman (1959, p. 103).

9. Peckham (1905, p. 123).

10. This description of a cruise control as an example of a control system was inspired by McClelland (1991).

11. Ashby (1952, 1956).

12. See Richardson (1991) for a historical account of the application of cybernetic and control system concepts to behavior, which includes Wiener, Ashby, and Powers, among others.

13. Powers (1973).

14. To show how control systems actually behave, Powers has created and made available two programs for IBM-compatible personal computers (*demo1* and *demo2*) to demonstrate both the phenomenon of control and the use of control systems as generative models of behavior. These programs are available on the Internet at http://www.ed.uiuc.edu/csg/csg.html and at gopher://gopher.ed.uiuc.edu following the path Higher Education Resources/ Professional societies and journals /Control Systems Group.

15. See Robertson & Powers (1990, pp. 67–80).

16. McPhail, Powers, & Tucker (1992).

17. Powers (1991b).

18. See Hershberger (1990).

19. See Powers (1973, p. 189).

20. Powers (1991a, p. 9).

21. Note that this describes a positive feedback loop.

22. See Staddon (1983, p. 241, figure 7.18).

23. Powers (1973, p. 179).

24. See Koshland (1980, pp. 14–15).

25. Brewer (1974).

26. Powers, Clark, & McFarland (1960).

27. See for example Pavlovski et al. (1990) and Bourbon (1990).

28. Ford, (1989); Goldstein (1989, 1990).

29. McClelland (1991, 1994); McPhail (1991); McPhail & Tucker (1990).

30. Gibbons (1990).

31. Plooij (1984).

32. Forssell (1993, 1994).

33. Petrie (1981); Cziko (1992).

34. That is, generative control system models have simulated behavior that typically correlates with individual human performance at values between .97 and .99, a level of precision that surpasses by far the explanatory and predictive power of the more traditional and established approaches to social and behavioral science (e.g., Bourbon, 1990; Marken, 1986, 1989, 1991).

Chapter 9

1. Köhler (1969, pp. 133–134).

2. See Gardner (1987) for an account of the history of the cognitive revolution.

3. Parts of the following account of Köhler's research are based on Boakes (1984, pp. 184–196). Koffka, Wertheimer, and Köhler were the leading proponents of what is called Gestalt psychology (*Gestalt* is a German word most often translated as "form"). Gestalt psychologists stress the importance of the integrity and wholeness of perception, and note that our perceptions are different from the sum of the parts of the stimuli that give rise to them. Thus ** ** ** is seen quite differently from *** ***, although both configurations are made up of six asterisks.

4. Köhler (1925, p. 17).

5. Köhler (1969, pp. 152, 153).

6. Köhler (1969, pp. 153, 154).

7. For example, you may have tried to calculate the length of line *l* using your knowledge of Pythagoras's theorem, which states that the square of the diagonal of a right triangle is equal to the sum of the square of the two bases.

8. Köhler (1969, p. 164).

9. Russell (1927; quoted in Boakes, 1984, p. 202).

10. Piaget was author or coauthor of more than 30 books.

11. Piaget's four major stages of human mental development are the sensorimotor stage, the preoperational stage, concrete operations, and formal operations.

12. See Vidal et al. (1983) and Vonèche (1985) for attempts to understand the reasons for Piaget's rejection of Darwinian theory. These include his religious beliefs (at one time he considered entering the ministry), his socialization into a Lamarckian form of biology, and his aversion to the wasteful picture of nature painted by the Darwinian concepts of blind chance and the struggle for survival.

13. Piaget (1976; quoted in Vidal et al., 1983, p. 87).

14. The difficulties encountered by students and scholars attempting to come to grips with Piaget's account of cognitive growth can be appreciated by pondering the following quotation:

> We still have to look for the reason why constructions required by the formation of reason become progressively necessary when each one begins by various trials that are partly episodic and that contain, until rather late, an important component of irrational thought (non-conservations, errors of reversibility, insufficient control over negations, and so on). The hypothesis naturally will be that this increasing necessity arises from autoregulation and has a counterpart with the increasing, parallel equilibration of cognitive structures. Necessity then proceeds from their "interlocking." (Piaget, quoted in Piattelli-Palmarini, 1980, p. 31)

15. Shortly after I wrote this paragraph, I was seated in a dentist's waiting room when a woman and her young daughter between one and two years of age sat down beside me. The girl had picked up a children's book that contained within its cardboard pages various types of surfaces to explore by touching. She touched one of these surfaces and said, "Wet!" Thereupon her mother immediately replied, "That's not wet, it's *sticky.*"

16. Chomsky (1988b, p. 34).

17. Fodor in Piattelli-Palmarini (1980, p. 149).

18. Piaget's Lamarckian tendencies are clearly revealed in his notion of the "phenocopy," a term he used to describe the inheritance of acquired characteristics that he believed to have observed in the freshwater snail *Limnea.* See Danchin (1980) for a refutation of Piaget's interpretation of the evolution of the *Limnea* and an alternative that is consistent with Darwinian theory.

19. Piattelli-Palmarini (1980).

20. Bain (1868, p. 596). See Roberts (1989) for a more up-to-date account of accidental scientific discoveries.

21. Bain (1868, p. 593).

22. Jevons (1874; quoted in Campbell, 1974a, p. 428).

23. Wright (1877/1971, pp. 115–116).

24. Although James claimed that his application of evolutionary principles to human thought was original, he was almost certainly influenced by discussions with Wright and knowledge of Wright's essays (see O'Hara, 1994).

25. James (1880, pp. 456, 457).

26. Souriau (1881; quoted in Campbell, 1974a, p. 429).

27. Baldwin (1889, p. 83).

28. Mach (1896; quoted in Campbell, 1974a, p. 427).

29. Poincaré (1913; quoted in Campbell, 1974a, pp. 427-428).

30. See Campbell (1974a).

31. The following four numbered paragraphs are taken from Campbell (1974a, p. 421).

32. The word *vicarious* refers to something being performed or experienced in a substitute fashion for another, in the same sense that a *vicar* is considered an earthly substitute and representative of Christ or God.

33. Campbell (1990, p. 9).

34. *Mnemonic* refers to memory. Thus, memory plays no role in nonmnemonic problem solving.

35. Campbell (1974a, p. 427).

36. Campbell (1974a, p. 472).

37. Campbell (1974a, p. 431).

38. Campbell (1974a, p. 432). Note that this account of imitation is quite consistent with the perceptual control theory account of behavior as described in chapter 8, with the criterion image serving as the reference level with the reorganization of control systems leading to the child's perception (of his song) eventually matching the reference level of the remembered song.

39. Campbell (1974a, p. 432).

40. Campbell (1974a, p. 432).

41. Campbell (1974a, p. 433).

42. Campbell (1974a, p. 434).

Chapter 10

1. Popper (1979, p. 261).

2. Kuhn (1970b, pp. 172–173).

3. Darwin recognized that natural selection would require a very long time for complex forms of life and new species to evolve. For this reason, he was troubled by Lord Kelvin's calculation (based on the temperature of the interior of the earth, the slowing of the earth's rotation by the action of tides, and the rate of decrease of the sun's heat) that the earth could not be more than 10 to 15 million years old, much too short to allow evolution to produce the diverse and complex forms of life existing on the earth. However, both radioactivity, which plays a major role in maintaining the earth's high interior temperatures, and nuclear fusion, which is the source of the sun's energy, were unknown phenomena in Kelvin's time. The earth is now considered to be about 4.5 billion (4500 million) years old, which is generally considered sufficient time for evolution to have brought forth the biosphere's current and past inhabitants. For an account of the conflict between Darwin and Kelvin concerning the age of the earth, see Burchfield (1975).

4. Much of the information in this section is taken from Reader (1988, chapter 3). This book provides fascinating case studies of 12 different cultural groups and describes the ways in which their behaviors and beliefs are adapted to their environments. These include the Pacific islanders of Yap, slash and burn farmers of New Guinea, alpine pastoralists in the Swiss Alps, potato growers in the Andes, hunter-gatherers of the Kalahari Desert, and city dwellers of Cleveland, among others.

5. The association of rice with fertility is not unknown in Western cultures, as revealed in the practice of throwing rice at newly married couples.

6. Reader (1988, pp. 71–72).

7. Reader (1988, p. 68).

8. Reader (1988, p. 68).

9. See Boyd & Richerson's (1985) concept of conformant transmission.

10. Srinivas (1952; quoted in Harris, 1966, p. 51).

11. See Badcock (1991, pp. 66–68).

12. Trivers (1971).

13. Axelrod (1984).

14. Campbell (1991, p. 107).

15. Campbell (1991, p. 98).

16. Campbell (1991, p. 99).

17. Campbell (1991, p. 99).

18. At least from a short- or medium-term economic perspective.

19. The sea otter places a flat rock on its chest and pounds clams and mussels against the rock to open them. Chimpanzees use long sticks to "fish" for termites. Two species of Darwin's finches on the Galápagos Islands trim cactus spines or leaf-stalks with their beaks and use these to extract grubs from the bark of dead tree branches (Weiner, 1994, p. 17)

20. See Basalla (1988, pp. 15–16).

21. Basalla (1988).

22. Basalla (1988, p. 45).

23. Basalla (1988, p. 136).

24. Basalla (1988, p. 137).

25. Mokyr (1990).

26. Mokyr (1990, p. 275).

27. Mokyr (1990, p. 276).

28. Vincenti (1990).

29. See also Bradshaw (1993a, 1993b).

30. Vincenti (1990, p. 247).

31. Vincenti (1990, pp. 247–248).

32. Petroski (1985, p. 45; emphasis added).

33. A distinction between experimental and analytical forms of vicarious trials has its merits. We will see, however, in chapter 13 how the increasing use of computer simulations in science and engineering has blurred this distinction.

34. In addition to wind tunnel tests, the Wright brothers made use of kites to test wing shapes, miniature propellers to find ways of providing maximum thrust, and gliders to test means of flight control (see Bradshaw, 1993a, 1993b).

35. Vincenti (1990, p. 250).

36. It could similarly be argued that Einstein's formulation of relativity theory owed relatively little to technology, since Einstein relied primarily on his knowledge of physics, mathematics, and his own thought experiments to arrive at his theory.

37. Bacon's method is still very much with us today, particularly in the social and behavioral sciences, where attempts are made to tease apart variables to determine which one is the actual cause of a certain phenomenon (e.g., is it the home environment or school environment that is responsible for success in school?).

38. Bacon (quoted in Hesse, 1964, p. 144).

39. It is not the case, as some critics have concluded, that Popper was a "naive falsificationist" who believed that a single disconfirmation of a theory entails its rejection. Instead, refutations of a theory may require many repeated disconfirming studies since in any one experiment equipment or methodological problems can lead to erroneous disconfirming results.

40. Popper (1979, p. 261).

41. See Popper (1964).

42. Although Popper puts forth important philosophical and logical arguments for a selectionist theory of science, he does not provide a picture of how scientists go about their day-to-day business interacting—both cooperating and competing—with other scientists. American biologist and philosopher David Hull spent many years investigating how scientists and scientific communities work, and he shares his findings in an interesting book (1988b).

43. See Schilpp (1974) for a collection of 33 critiques of Popper's philosophy, followed by Popper's replies.

44. Contemporary philosophers rejecting providential (foundational) and instructionist (empiricist, inductive) views of science include Campbell (1990), Feyerabend (1975), Giere (1988), Hanson (1958), Hull (1988a, 1988b), Kuhn (1970a, 1970b, 1991), Laudan (1984), Polanyi (1958), Quine (1960), Root-Bernstein (1989), Toulmin (1972), and van Fraassen (1980). Among these, Campbell, Giere, Hull, Kuhn, Root-Bernstein, and Toulmin make explicit use of an evolutionary or selectionist perspective to understand the development of science.

45. See Hull (1988a, pp. 28ff.)

46. Dawkins (1989, p. 192) coined the term *meme* (rhymes with *cream*) in 1976 and it has enjoyed growing popularity ever since. It is meant to represent a unit of imitation and is a shortened form of the Greek word *mimeme*, which also resembles memory and the French word *même* ("same").

47. Dawkins (1989, p. 192).

48. For example, Boyd & Richerson (1985) and Cavalli-Sforza & Feldman (1981). Indeed, the title of the latters' book is *Cultural Transmission and Evolution*.

49. However, even this apparent transmission of genetic information involves selection mechanisms. The DNA molecule makes copies of itself by splitting lengthwise along its double helix. Each strand then selects complementary nucleotide bases

from the soup of various molecules present in the cell's nucleus. It is through this process of blind molecular variation in the form of random molecular shuffling and selective retention that each single strand becomes another double-stranded DNA molecule that is almost always identical to the original parent molecule. The few errors that do arise constitute an important source of genetic variation on which natural selection can operate (see Li & Graur, 1991).

Chapter 11

1. Darwin (1874, pp. 88–89).

2. See Berk (1994) for an account of how the private, self-directed speech of the child gradually turns into the silent thoughts of the adult.

3. Mentioning these advantages of spoken language is by no means intended to imply that sign language is not a powerful and expressive system of communication. No hearing communities have been known to use sign language instead of spoken language, however.

4. Bickerton (1990, p. 106).

5. Pinker & Bloom (1990, p. 711).

6. Darwin (1859/1966, p. 191).

7. It is of interest to note that the human infant also shares the high larynx of non-human mammals, making it possible simultaneously to drink through the mouth and breathe through the nose. This ability is soon lost and linguistic ability gained as the larynx drops, permitting the infant to make all the human speech sounds.

8. Lieberman (1991, p. 56).

9. Letters in square brackets are symbols for language sounds (phonemes) as used in the International Phonetic Alphabet.

10. Lieberman (1991, p. 65).

11. Lieberman (1984) proposes:

> ... that the extinction of Neanderthal hominids was due to the competition of modern human beings who were better adapted for speech and language. The synergetic effect of rapid data transmission through the medium of encoded speech and the cognitive power of the large hominid brain probably yielded the full human linguistic system. The rapid changes in human culture that occurred shortly after the replacement of the Neanderthals could be the result of a difference in the way in which humans thought. Though it is impossible to prove that human language and thought were the causative agents, the replacement of the Neanderthal population—adapted for strength and agility—by a population that was inferior save for enhanced speech abilities is consistent with this hypothesis. (p. 329)

12. Kimura (1979; quoted in Lieberman, 1991, p. 79).

13. See Lieberman (1991, p. 80).

14. Goodall (1986; quoted in Lieberman, 1991, p. 52).

15. Lieberman (1991, p. 109).

16. Lieberman (1991, pp. 80, 81).

17. Chomsky (1988a, p. 23).

18. Gould & Lewontin (1979).

19. Gould & Lewontin (1979, p. 285).

20. The following is adapted from Lieberman (1991, pp. 8, 9).

21. Pinker & Bloom (1990, p. 726). This paper provides an excellent discussion of issues involved in understanding language as the product of Darwinian natural selection. Of particular interest are the 31 critical commentaries that follow the article and Pinker and Bloom's subsequent responses.

22. Pinker & Bloom (1990, p. 721).

23. Reader (1988, p. 143).

24. Konner (1982, p. 171).

25. Calvin (1990, p. 207) discusses the ideas of Nicholas Humphrey and Richard Dawkins as aired in a BBC radio program.

26. The argument that language is necessary for a complete human sense of self and consciousness was made by Helen Keller, who became deaf and blind shortly after birth and learned a language based on touch at the age of eight. As she recounts: "When I learned the meaning of 'I' and 'me' and found that I was something, I began to think. Then consciousness first existed for me" (Keller, 1904, p. 145).

27. Skinner (1957, p. 81).

28. Skinner (1957, pp. 1, 2).

29. Skinner (1957, p. 164).

30. Skinner (1957, p. 58).

31. Skinner (1957, p. 206).

32. Chomsky's (1959) review of Skinner's book is considered to be an important event in the start of what has become known as the cognitive revolution in psychology.

33. Chomsky (1964, p. 558).

34. Bickerton (1990, pp. 57, 58).

35. On this characteristic English constrasts strongly with languages such as Latin, Navajo, and Walpiri, which have quite free word orders.

36. Fodor (1980, p. 143).

37. Fodor (1980, p. 149).

38. Munz's (1993, p. 154) concept of organisms as "embodied theories" is useful here.

39. Bickhard (1991, pp. 16–17). See also Bickhard & Terveen (1995, pp. 25ff.) for additional critique of cognitive innatism.

40. Note that this is not the case for the genes that determine antibody production, as discussed in chapter 4.

41. Again, except for that part of the genome underlying the production of antibodies.

42. Quine (1960).

43. See Macnamara (1972, p. 3).

44. Campbell (1973).

45. Pinker (1994, p. 154).

46. See Gleitman (1994) and Markman (1994) for additional constraints that children bring to the task of learning word meanings.

47. The tentative and fallible nature of our learning of word meanings was made clear to me several years ago when I realized that the word *befriend* made no sense in the context of a newspaper article I was reading. Checking the meaning of this word in my dictionary, I was quite surprised to learn that it means "to make a friend" whereas for more than 30 years of my life I had understood it as meaning "to lose a friend," somewhat analogous, I suppose, to the way that *behead* involves losing a head.

48. These four sentences are taken from Pinker (1989, pp. 19ff.).

49. A study conducted by Brown & Hanlon (1970) indicated that parents do not provide corrective information concerning the grammatical errors made by their children. This study has been widely cited, especially by those who make innatist arguments for language acquisition. However, more recent studies provide evidence that children do have access to information from adults concerning the grammaticality of their utterances, information that would make it even easier for them to reject incorrect hypotheses about the language being learned. These studies found that parents respond differentially to the ungrammatical utterances of their children by often repeating verbatim well-formed sentences in contrast to repeating with changes, or requesting clarification for, sentences containing errors (Bohannon & Stanowicz, 1990; Demetras, Post, & Snow, 1986; Hirsch-Pasek, Treiman, & Schneiderman, 1984; Penner, 1987; see also Gordon, 1990; and Bohannon, MacWhinney, & Snow, 1990 for contrasting views of this research and its importance). Although it has not yet been demonstrated convincingly that children actually use such information in learning language, the availability of such negative evidence has the potential of making language acquisition easier for the child without relying on innate linguistic knowledge. But then again, see Marcus (1993) for arguments against the role of adult feedback in language acquisition.

50. Pinker (1989, p. 255; comments added in brackets).

51. Pinker (1989, p. 255).

52. Pinker (1989, p. 290).

53. MacWhinney (1987, p. 287).

54. MacWhinney (1987, p. 292).

55. MacWhinney (1989, p. 65).

56. See Skinner (1966, 1981).

57. Pinker (1989, pp. 166, 167). I would like to see Pinker's use of quotation marks around "confirmation factor" as a recognition that hypotheses can never be confirmed. And of course the environment does not actually tell anyone which hypotheses to keep, but rather provides information for the learner (or scientist) concerning which ones should be rejected or revised.

58. Pinker & Bloom (1990).

59. See Clark & Roberts (1993) for an interesting application of a cumulative selectionist learning process (called genetic algorithms and described later in this book in chapter 14) to language acquisition and language change based on Chomsky's principles-and-parameters approach to linguistic theory.

60. Georgia Green brought these examples to my attention.

61. Bransford & Johnson (1972, p. 722).

62. Wells (1986, pp. 216, 217).

63. See Tannen (1990) for an interesting account of language-based misunderstandings that arise between men and women.

Chapter 12

1. Augustine (389/1938, pp. 47–48).

2. Comenius (1623/1896, p. 441).

3. Perkinson (1993, p. 34).

4. Augustine (389/1938, p. 48).

5. Augustine (389/1938, p. 46).

6. Perkinson (1984, p. 15).

7. A trip north of the Arctic Circle during the end of December (or south of the Antarctic Circle at the end of June) will clearly reveal that the belief that the sun rises every morning is mistaken.

8. Perkinson (1984, p. 35).

9. Popper (1963, p. 45).

10. Piaget (1972; quoted in Perkinson, 1984, pp. 71–72).

11. Perkinson (1984, p. 93).

12. Perkinson (1984, p. 165). (Quotations from this and the previous three paragraphs.)

13. Montessori (1967, pp. 246–247).

14. Perkinson (1984).

15. Perkinson (1984, p. 180).

16. Perkinson (1984, p. 190). It should also be mentioned that a selectionist view of education also has important implications for educational research, implications that reject the current standard practice of attempting to improve education by investigating the relationships between independent variables (related to aspects of

the transmission process) and dependent variables (measures of the success or fidelity of the transmission process). See Cziko (1989, 1992) for discussions of these and related issues.

17. In this case, the crucial controlled variable is actually the carbon dioxide content of the blood. However, since we will assume that the student already knows that to breathe she must keep her mouth and nose above water, we will consider the latter to be the controlled variable.

18. Recall from chapter 10 how the Wright brothers' breaking down the problem of flight into components led more quickly to a solution of the overall problem.

19. See Farnham-Diggory (1992, pp. 173–175).

20. A good example of scaffolding for a physical skill is holding a bicycle for a child learning to ride (or providing training wheels). In this way the child can learn to pedal and steer without worrying about balancing. As the child develops better control of balance, the instructor gradually withdraws his support (or moves the training wheels higher, eventually removing them altogether).

21. An example of these techniques for teaching the complex cognitive skill of reading comprehension can be found in the research of Palincsar and Brown (1984) on reciprocal teaching. They were successful in teaching reading comprehension skills to children who were having particular difficulty, by demonstrating and then having the children imitate concrete activities involving certain subgoals of reading comprehension. These included reading a text and then summarizing, questioning, clarifying, and predicting its contents.

22. Powers (1973, p. 223).

23. They are reciprocals of each other.

Chapter 13

1. Goldberg (1986; quoted in Levy, 1992, p. 153).

2. But the effects of evolution can be seen if one carefully studies the right species in the right location for a long enough time, as in the Grants' 20-year study of the evolution of Darwin's finches on the Galápagos Islands (see Weiner, 1994).

3. Dawkins (1986).

4. A Macintosh version of Dawkins's program is available from the publisher W. W. Norton in New York.

5. Dawkins (1986, pp. 59–60).

6. Much of the following account of Ray's work is based on Levy (1992, pp. 216–230).

7. Quoted in Levy (1992, p. 221).

8. See Dawkins (1986, chapter 7).

9. Readers with access to either an IBM-compatible computer or a Unix workstation or mainframe computer can also observe the processes of evolution using programs available by anonymous FTP at tierra.slhs.udel.edu or life.slhs.udel.edu.

10. See Levy (1992) for a more detailed account of the work of pioneers in the field of artificial life.

11. Bell (quoted in Levy, 1992, p. 320).

12. Fogel, Owens, & Walsh (1966, p. 113).

13. One particularly surprising outcome of the work done by Holland and his students on genetic algorithms was the importance of crossover and the relative insignificance of mutation in the evolutionary process. It was found that the new recombinations provided by the crossover of genes in sexual reproduction allowed the genetic algorithm to find building blocks during the evolutionary process so that fitness could better be accumulated from one generation to the next. In contrast, mutation (the occasional random changing of a 0 to 1 or vice versa) turned out to be relatively unimportant, serving primarily as a way of resurrecting old algorithms that had already been discarded but whose descendants would nonetheless provide useful solutions.

14. A nontechnical account of genetic algorithms can be found in Holland (1992). A more comprehensive, technical treatment is provided by Goldberg (1989).

15. See Goldberg (1994) for more information on these and other real-world applications of genetic algorithms.

16. Evolver, developed by Axcellis in Seattle, is designed to work in conjunction with Microsoft Excel.

17. Koza's follow-up book *Genetic Programming II*, was published in 1994 together with another accompanying videotape.

18. Koza (1992, pp. 257–258).

19. For more information on these and other applications of genetic programming, see Kinnear (1994) and appendix F of Koza (1994). Singleton (1994) provides information on how genetic programming can be done with the widely used C++ programming language.

20. The International Conference on Genetic Algorithms had its first meeting in 1985 and has met on every odd-numbered year since. The Workshop on the Foundations of Genetic Algorithms meets in even-numbered years. Other regular meetings on related topics include Parallel Problem Solving from Nature, the IEEE International Conference on Evolutionary Computing, Artificial Life Workshops, and the European Conference on Artificial Life.

Three young journals that publish related research are *Artificial Life*, *Adaptive Behavior*, and *Evolutionary Computation*, all published by the MIT Press.

Readers with access to the Internet can learn more about genetic algorithms by subscribing (for free) to the genetic algorithm (ga-list-request@aic.nrl.navy.mil) and genetic programming (genetic-programming-request@cs.stanford.edu) electronic mailing lists. A repository of programs and papers on genetic programming is available on the Internet through anonymous FTP from the site ftp.cc.utexas.edu in the pub/genetic-programming directory.

21. Levy (1992, p. 179; bracketed material in original).

22. Vincenti (1990, p. 248).

23. See Larijani (1994) for an introduction to virtual reality and many additional examples of its use and potential.

24. See Gibbs (1994).

25. The field of neural networks overlaps with the study of parallel distributed processes (PDP), both of which are referred to as connectionist approaches to artificial intelligence.

26. See Caudill & Butler (1990) for a nonmathematical introduction to neural networks (including the ones described here) and what they can be made to do.

27. Caudill & Butler (1990, pp. 194–195).

28. Two other neural network architectures that learn by instruction are known are perceptrons and adalines.

29. The term stochastic is often applied to selectionist neural networks, referring to the random variations used to develop the networks.

30. Barto (1994, p. 202) uses the term supervised learning to refer to tasks where a teacher or supervisor not only knows the right answer but also how the network should change its internal organization in order to reduce its error. This is therefore an instructionist approach to learning. In contrast, reinforcement learning employs a critic who provides the system information about the degree of error only and "does not itself indicate how the learning system should change its behavior to improve performance; it is not directed information. . . . [The system] has to probe the environment—perform some form of exploration—to obtain information about how to change its behavior" (pp. 203, 218). Reinforcement learning is selectionist in that the learning system must be active in exploring possible solutions using the information provided by the critic to reject some but select and cumulatively modify others.

31. Barto, Sutton, & Brouwer (1981).

32. Barto & Sutton (1981).

33. See Caudill & Butler (1990, pp. 105–109).

34. Caudill & Butler (1990, p. 208).

35. Edelman (1993); Friston et al. (1994); Sporns & Edelman (1993).

36. See Edelman (1989, pp. 58–63; 1992, pp. 91–94); Reeke & Sporns (1990). Kinesthesia refers to sensory information on the position of body parts; for example, lets you know where your hand is when you are not looking at it.

37. Edelman et al. (1992). See also Levy (1994) for a brief description of Darwin IV as well as an interesting account of Edelman's work, personality, and ambitions.

38. See Albrecht, Reeves, & Steele (1993); Koza (1992, pp. 513–525). Another computational technique used to solve problems like those encountered in neural networks is referred to as simulated annealing. This is a process by which a solution is found to a complex problem (such as determining the structure of a molecule based on data provided by magnetic resonance imaging) by taking a tentative solution and effectively shaking it up to see if a better solution can be found. Since this shaking introduces a source of blind variation in an attempt to find and select a

better solution, the evolutionary parallel is quite obvious. See Davis (1987) for a collection of chapters providing an introduction to simulated annealing and its applications.

Chapter 14

1. From Darwin's 1858 letter to American biologist Asa Gray. Reprinted in Bajema (1983, pp. 191–192).

2. Much of the information presented in this section on plant breeding was obtained from Stoskopf (1993).

3. Darwin (1859/1966, p. 30).

4. Stoskopf (1993, p. 5).

5. See Stoskopf (1993, p. 439).

6. See Murrell & Roberts (1989) for an introduction to genetic engineering from which much of the following information on the subject was obtained. The special report on medicine and health published by the *Wall Street Journal* on May 20, 1994, also provides a useful collection of articles on the methods, products, promises, and problems of genetic engineering.

7. See Capecchi (1994).

8. See Salmond (1989, pp. 56–61).

9. In his delightful book on accidental scientific discoveries, Roberts (1989) provides many other examples of serendipitous drug discoveries.

10. Silverman (1992, especially chapter 2) describes many of these techniques for modifying lead compounds in the attempt to find better drugs.

11. Bugg, Carson, & Montgomery (1993, p. 94).

12. Much of the information presented here on Spiegelman and directed molecular evolution is taken from Joyce (1992).

13. Kelly (1994, p. 301).

14. Flam (1994).

15. Quoted in Kelly (1994, p. 302).

16. See Kelly (1994) for an interesting account of the increasing use and influence of artificial forms of evolution.

Chapter 15

1. Jerne (1967, p. 204).

2. Changeux (1985, pp. 279–281).

3. Monod (1971, pp. 112, 113). Monod (1910–1976) shared the Nobel prize for physiology or medicine in 1965 for his work on messenger RNA and operator genes.

4. Darwin (1860/1952, p. 239).

5. At a deeper level, such a theory could, of course, be selectionist if it recognized that the antibody-specific information in the genome had been achieved through biological evolution.

6. Wilson (1975, 1978).

7. Changeux has also noticed the reoccurring sequence of instructionist (his "instructive") theories preceding selectionist (his "selective") ones as indicated in the epigraph at the beginning of this chapter.

8. Gunther Stent remarked of Edelman's writing: "I consider myself not too dumb. I am a professor of molecular biology and chairman of the neurobiology section of the National Academy of Sciences, so I should understand it. But I don't" (quoted in Johnson, 1992, p. 22).

9. Since switching from immunology to the neurosciences, Edelman has been a prolific writer, publishing books in 1987, 1988 (two), 1989, and 1992.

10. Vincenti (1990, p. 246).

11. Thagard (1988, p. 103).

12. Powers, who has provided what is arguably the only working model of purposeful human behavior, does not hesitate to invoke blind variation to explain how individuals are able to develop new control systems to control new aspects of their environment that they could not control before. "This is what I assume to be the basic principle of reorganization, which I could not put any better than Campbell did. Act at random, and select future actions on the basis of the consequences" (Powers, 1989, p. 288).

13. Campbell (1977, p. 506).

14. See Stein & Lipton (1989) for a comparison of such heuristic constraints in the development of science to preadaptations (exaptations) in biological evolution.

15. Gamble (1983, p. 358).

16. Thagard (1988, p. 107).

17. Thagard (1988, p. 108).

18. Hull (1988b, p. 476).

19. Campbell (1977, p. 506).

20. Gould claimed that ". . . cultural evolution is direct and Lamarckian in form: The achievements of one generation are passed by education and publication directly to descendants" (1991a, p. 65).

21. Dawkins (1982) stated that "complex and elaborate adaptive fits can be achieved by instruction, as in the learning of a particular human language" (p. 173). Chapter 11 of this book argues otherwise.

22. Hull (1988b, p. 453).

23. It could be said, however, that since no task is ever exactly the same on different occasions, some variation and selection is always required since old knowledge must be continually adapted to new situations.

24. Gamble (1983, p. 358).

25. See, for example, Bradie (1986).

26. The discussion here assumes that Pavlovian conditioning is an instruction-ist process, but there are some reasons to question such an interpretation. Let us examine some of these briefly.

Of all the sensations taking place during the presentation of the first neutral then conditional stimulus, only one or a few may turn out to be reliable indicators of the impending unconditional stimulus. In the example just given, many things may happen immediately before the dark acidic liquid is placed in the dog's mouth—footsteps are heard; a researcher appears; liquid is poured from a large jar into a small beaker; a small amount is sucked up into a pipette; the researcher comes very close to the dog; the pipette is placed in the dog's mouth; and the acid is finally released. Which of these many stimuli can the dog use as a reliable indicator of the impending acid? There may well be some selection going on here, with the dog "jumping to conclusions" concerning the conditional stimulus and ultimately rejecting those conclusions that are not reliable. Selectionism may also be involved in the "blooming and pruning" of neurons that accompany such learning as discussed in chapter 5.

27. See Maynard Smith (1978).

28. See Popper (1966) for a discussion of the necessity of an open society for the continued improvement of government policies.

29. Gamble (1983, pp. 359, 360).

30. Piattelli-Palmarini (1986, 1989).

31. Piattelli-Palmarini (1989, p. 17).

32. Piattelli-Palmarini (1986, p. 127; emphasis in original).

33. See Janeway (1993).

34. Capecchi (1994, p. 52).

35. See, for example, Fodor (1975, pp. 79–97).

36. These two examples are similar to ones suggested to me by Paul Bloom.

37. Gazzaniga (1992, p. 2).

38. Gazzaniga (1992, p. 134).

39. Tononi (1994, p. 298).

40. Kant (quoted in Weiner, 1994, p. 3).

Chapter 16

1. Popper (1979, p. 261).

2. Dawkins (1983).

3. Campbell (1974a).

4. Campbell (1974b, p. 147).

5. This observation of the apparent tautological nature of biological evolution as survival of the fittest has been made many times. For some responses, see Maynard Smith (1969), Stebbins (1977), and Alexander (1980).

6. Campbell (1974a, p. 421).

7. For philosophers, an analytic statement is one that, like a tautology, is true by definition and necessity such as "a triangle is a closed figure having three straight sides," and a synthetic statement is true or false according to some condition of the world that it describes such as "all birds have feathers."

8. Gamble (1983, p. 359).

9. Gamble (1983, p. 362).

10. Mosaic, for Macintosh, MS-DOS Windows, and X-Windows environments can be obtained from the National Center for Supercomputing Applications at the University of Illinois at Urbana-Champaign through FTP from ncsa.uiuc.edu.

11. If antibody production depended solely on the recombination of the variable B lymphocyte genes followed by selection, this would be single-step generative selection. But the continued hypermutation of antibodies and continued selection make antibody production a more powerful cumulative generative selection process (see chapter 4).

12. If there already is some knowledge concerning the solution, this simply means that some of the values of x need not be tested or that other values should be tested first. But this does not change in any substantial way the nature of the problem, since a range of x values still must be investigated.

13. See Goldberg (1989, chapters 1 & 2) for a highly readable account of how cumulative blind variation and selective retention can discover building blocks to find solutions in complex problem spaces.

14. Plotkin (1994, p. 84).

15. Plotkin (1994, p. 139).

16. Dawkins (1986, pp. 317–318).

17. This, however, has not dissuaded some creationists from embracing saltationism, since if adaptive macromutations were the rule in evolution, this, they argue, could only be the result of a providential creator.

18. See, for example, Eldredge (1985), and Gould (1980, chapter 17).

19. See Dawkins (1986, chapter 9) for additional discussion about how the theory of punctuated equilibrium is entirely consistent with the Darwinian view of evolution.

20. Cairns, Overbaugh, & Miller (1988).

21. Cairns, Overbaugh, & Miller (1988, p. 145).

22. MacPhee (1993).

23. The view that evolution could involve a type of control process by which greater variability is produced in response to environmental stress was to my knowledge first proposed by Powers (1989b, pp. 124–127), whose seminal work on

applying control systems theory to understanding the behavior of living organisms was introduced in chapter 8.

24. However, Gould (1993, pp. 109–120) believes that Darwin got this sequence backward.

25. Piattelli-Palmarini (1989, p. 6).

26. Fernald (1992, p. 395).

27. Margulis (quoted in Kelly, 1994, p. 365).

28. Kauffman (1993).

29. Kauffman (1993, p. 644).

30. This point was brought to my attention by Henry J. Perkinson.

Appendix

1. Hume (1748/1952, p. 491).

2. Hume (1748/1952, p. 491).

3. Darwin (1887/1959, pp. 86).

References

Ada, G. L., & Nossal, G. (1987). The clonal-selection theory. *Scientific American, 257*(2), 50–57.

Albrecht, R. F., Reeves, C. R., & Steele, N. C. (Eds.). (1993). *Artifical neural nets and genetic algorithms: Proceedings of the international conference in Innsbruck, Austria.* Vienna: Springer-Verlag.

Albus, J. S. (1971). A theory of cerebellar function. *Mathematical Biosciences, 10,* 25–61.

Alexander, R. D. (1980). *Darwinism and human affairs.* London: Pitman.

Anglin, J. (1993). Vocabulary development: A morphological analysis. *Monographs of the Society for Research in Child Development* (Ser. No. 238), *58*(10).

Aquinas, T. (1914). *The "Summa theologica" of St. Thomas Aquinas* (Part II). London: W. & T. Washbourne. (Original work written 1265–1273)

Aristotle. (1929). *The physics* (P. H. Wicksteed & F. M. Cornford, Trans.). London: Heinemann.

Ashby, W. R. (1952). *Design for a brain.* New York: Wiley.

Ashby, W. R. (1956). *Introduction to cybernetics.* New York: Wiley.

Augustine, S. A. (1938). *Concerning the teacher* and *On the immortality of the soul* (G. G. Leckie, Trans.). New York: D. Appleton-Century. (Original work completed 389)

Axelrod, R. M. (1984). *The evolution of cooperation.* New York: Basic Books.

Badcock, C. (1991). *Evolution and individual behavior: An introduction to sociobiology.* Cambridge, MA: Basil Blackwell.

Bain, A. (1868). *The senses and the intellect* (3rd ed.). London: Longman, Green.

Bajema, C. J. (Ed.). (1983). *Natural selection theory.* Stroudsburg, PA: Hutchinson Ross.

Baldwin, J. M. (1889). On selective thinking. *Psychological Review, 5*(1), 1–24.

Barto, A. G. (1994). Some learning tasks from a control perspective. In L. Nadel & D. Stein (Eds.), *1990 lectures in complex systems* (Santa Fe Institute Studies in the Sciences of Complexity, Lect. Vol. III, pp. 195–223). Reading, MA: Addison-Wesley.

Barto, A. G., & Sutton, R. S. (1981). Landmark learning. An illustration of associative search. *Biological Cybernetics, 42*, 1–8.

Barto, A. G., Sutton, R. S., & Anderson, C. W. (1983). Neuronlike adaptive elements that can solve difficult learning control problems. *IEEE Transactions on Systems, Man, and Cybernetics, 13*(5), 834–846.

Barto, A. G., Sutton, R. S., & Brouwer, P. S. (1981). Associative search network: A reinforcement learning associative memory. *Biological Cybernetics, 40*, 201–211.

Basalla, G. (1988). *The evolution of technology*. Cambridge: Cambridge University Press.

Bateson, G. (1979). *Mind and nature*. New York: Dutton.

Beach, F. A. (1955). The descent of instinct. *Psychological Review, 62*, 401–410.

Bereiter, C. (1985). Toward a solution of the learning paradox. *Review of Educational Research, 55*(2), 201–226.

Berk, L. E. (1994). Why children talk to themselves. *Scientific American, 271*(5), 78–83.

Bibel, D. J. (1988). *Milestones in immunology: A historical exploration*. Madison, WI: Science Tech.

Bickerton, D. (1984). The language bioprogram hypothesis. *Behavioral and Brain Sciences, 7*, 173–212.

Bickerton, D. (1990). *Language and species*. Chicago: University of Chicago Press.

Bickhard, M. H. (1991). The import of Fodor's anti-constructivist argument. In L. P. Steffe (Ed.), *Epistemological foundations of mathematical experiences* (pp. 14–25). New York: Springer-Verlag.

Bickhard, M. H., & Terveen, L. (1995). *Foundational issues in artificial intelligence and cognitive science: Impasse and solution*. Amsterdam: Elsevier.

Black, J. E., & Greenough, W. T. (1986). Induction of pattern in neural structure by experience: Implications for cognitive development. In M. E. Lamb, A. L. Brown, & B. Rogoff (Eds.), *Advances in developmental psychology* (Vol. 4, pp. 1–50). Hillsdale, NJ: Erlbaum.

Boakes, R. (1984). *From Darwin to behaviourism: Psychology and the minds of animals*. Cambridge: Cambridge University Press.

Bohannon III, J. N., MacWhinney, B., & Snow, C. (1990). No negative evidence revisited: Beyond learnability or who has to prove what to whom. *Developmental Psychology, 26*(2), 221–226.

Bohannon III, J. N., & Stanowicz, L. (1990). The issue of negative evidence: Adult responses to children's language errors. *Developmental Psychology, 24*(5), 684–689.

Bourbon, W. T. (1990). Invitation to the dance: Explaining variance when control systems interact. *American Behavioral Scientist, 34*(1), 95–105.

Boyd, R., & Richerson, P. J. (1985). *Culture and the evolutionary process*. Chicago: University of Chicago Press.

Bradie, M. (1986). Assessing evolutionary epistemology. *Biology and Philosophy*, *1*(4), 401–459.

Bradshaw, G. L. (1993a). Why did the Wright brothers get there first? Part 1. *Chemtech*, *23*(6), 8–13.

Bradshaw, G. L. (1993b). Why did the Wright brothers get there first? Part 2. *Chemtech*, *23*(7), 16–22.

Bransford, J. D., & Johnson, M. K. (1972). Contextual prerequisites for understanding: Some investigations of comprehension and recall. *Journal of Verbal Learning and Verbal Behavior*, *11*, 717–726.

Breinl, F., & Haurowitz, F. (1930). Chemische Untersüchung des Präzipitates aus Hämoglobin und Anti-Hämoglobin-Serum und Bemerkungen über die Nature der Antikörper. *Zeitschrift der Physiologischer Chemie*, *192*, 45.

Brewer, W. F. (1974). There is no convincing evidence for operant or classical conditioning in adult humans. In W. B. Weimer & D. S. Palermo (Eds.), *Cognition and symbolic processes* (pp. 1–42). Hillsdale, NJ: Erlbaum.

Brown, R., & Hanlon, C. (1970). Derivational complexity and the order of acquisition in child speech. In R. Brown (Ed.), *Psycholinguistics* (pp. 155–207). New York: Free Press.

Bugg, C. E., Carson, W. M., & Montgomery, J. A. (1993). Drugs by design. *Scientific American*, *269*(6), 92–98.

Bullock, T. H. (1977). *Introduction to nervous systems*. San Francisco: Freeman.

Burchfield, J. D. (1975). *Lord Kelvin and the age of the earth*. New York: Science History Publications.

Burkhardt, R. W., Jr. (1977). *The spirit of the system: Lamarck and evolutionary biology*. Cambridge: Harvard University Press.

Burkhardt, R. W., Jr. (1983). The development of an evolutionary ethology. In D. S. Bendall (Ed.), *Evolution from molecules to men* (pp. 429–443). Cambridge: Cambridge University Press.

Burnet, F. M. (1957). A modification of Jerne's theory of antibody production using the concept of clonal selection. *Australian Journal of Science*, *20*, 67–68.

Cairns, J., Overbaugh, J., & Miller, S. (1988). The origin of mutants. *Nature*, *335*, 142–145.

Calvin, W. H. (1987). The brain as a Darwin machine. *Nature*, *330*, 33–34.

Calvin, W. H. (1990). *The ascent of mind: Ice age climates and the evolution of intelligence*. New York: Bantam.

Campbell, D. T. (1973). Ostensive instances and entitativity in language learning. In W. Gray & N. D. Rizzo (Eds.), *Unity through diversity: A Festschrift for Ludwig von Bertalanffy* (pp. 1043–1057). New York: Gordon & Breach.

Campbell, D. T. (1974a). Evolutionary epistemology. In P. A. Schilpp (Ed.), *The philosophy of Karl Popper* (Vol. 1, pp. 413–463). La Salle, IL: Open Court.

Campbell, D. T. (1974b). Unjustified variation and selective retention in scientific discovery. In F. J. Ayala & T. Dobzhansky (Eds.), *Studies in the philosophy of biology* (pp. 139–161). Berkeley: University of California Press.

Campbell, D. T. (1977). Comment on "The natural selection model of conceptual evolution." *Philosophy of Science, 44,* 502–507.

Campbell, D. T. (1988a). A general "selection theory," as implemented in biological evolution and in social belief-transmission-with-modification in science [Commentary on preceding target article by D. L. Hull]. *Biology and Philosophy, 3,* 171–177.

Campbell, D. T. (1988b). Popper and selection theory. *Social Epistemology, 2*(4), 371–377.

Campbell, D. T. (1990). Epistemological roles for selection theory. In N. Rescher (Ed.), *Evolution, cognition, and realism* (pp. 1–19). Lanham, MD: University Press of America.

Campbell, D. T. (1991). A naturalistic theory of archaic moral orders. *Zygon, 26*(1), 91–114.

Capecchi, M. R. (1994). Targeted gene replacement. *Scientific American, 270*(3), 52–59.

Carver, C. S., & Scheier, M. F. (1981). *Attention and self-regulation: A control theory approach to human behavior.* New York: Springer-Verlag.

Caudill, M., & Butler, C. (1990). *Naturally intelligent systems.* Cambridge: MIT Press.

Caudill, M., & Butler, C. (1992). *Understanding neural networks: Vol. 1. Basic explorations.* Cambridge: MIT Press.

Cavalli-Sforza, L. L., & Feldman, M. W. (1981). *Cultural transmission and evolution: A quantitative approach.* Princeton, NJ: Princeton University Press.

Changeux, J.-P. (1985). *Neuronal man: The biology of mind.* New York: Oxford University Press.

Chomsky, N. (1959). A review of B. F. Skinner's *Verbal behavior. Language, 35*(1), 26–58.

Chomsky, N. (1964). A review of B. F. Skinner's *Verbal behavior.* In J. A. Fodor & J. J. Katz (Eds.), *The structure of language: Readings in the philosophy of language* (pp. 547–578). Englewood Cliffs, NJ: Prentice-Hall.

Chomsky, N. (1988a). *Noam Chomsky on the generative enterprise: A discussion with Riny Huybregts and Henk van Riemsdijk.* Dordrecht, The Netherlands: Foris Publications.

Chomsky, N. (1988b). *Language and problems of language: The Managua lectures.* Cambridge: MIT Press.

Churchland, P. M. (1990). *Matter and consciousness* (rev. ed.). Cambridge: MIT Press.

Clark, R., & Roberts, I. (1993). A computational model of language learnability and language change. *Linguistic Inquiry, 24*(2), 299–345.

Comenius, J. A. (1896). *The great didactic.* London: Adam & Charles Black. (Original work completed 1623)

Curtiss, S. (1977). *Genie: A psycholinguistic study of a modern-day "wild child."* New York: Academic Press.

Cziko, G. A. (1989). Unpredictability and indeterminism in human behavior: Arguments and implications for educational research. *Educational Researcher, 18*(3), 17–25.

Cziko, G. A. (1992). Purposeful behavior as the control of perception: Implications for educational research. *Educational Researcher, 21*(9), 10–18.

Cziko, G. A., & Campbell, D. T. (1990). Comprehensive evolutionary epistemology bibliography. *Journal of Social and Biological Structures, 13*(1), 41–81.

Danchin, A. (1980). A critical note on the use of the term "phenocopy." In M. Piattelli-Palmarini (Ed.), *Language and learning: The debate between Jean Piaget and Noam Chomsky* (pp. 356–360). Cambridge: Harvard University Press.

Darwin, C. (1859). *On the origin of species by means of natural selection.* London: John Murray. (Facsimile edition published in 1966 with an introduction by E. Mayr, Cambridge: Harvard University Press)

Darwin, C. (1874) *The descent of man and selection in relation to sex* (2nd ed.). New York: Crowell.

Darwin, C. (1952). *The origin of species by means of natural selection.* In R. M. Hutchins (Ed.), *Great books of the Western world* (Vol. 49, pp. 1–251). Chicago: Encyclopædia Britannica. (Original work published 1860)

Darwin, C. (1959). *The autobiography of Charles Darwin.* New York: Harcourt, Brace. (Original work published 1887)

Darwin, C. (1966). *The origin of species by means of natural selection.* (A facsimile of the first edition originally published in 1859, with an introduction by Ernst Mayr.) Cambridge: Harvard University Press.

Darwin, C. (1975). *Charles Darwin's natural selection, being the second part of his big species book written from 1856–1858.* (P. Barrett, Ed.; 2 vols.). Cambridge: Cambridge University Press.

Darwin, F., & Seward, A. C. (Eds.). (1903). *More letters of Charles Darwin* (Vol. 1). London: Murray.

Davis, L. (Ed.). (1987). *Genetic algorithms and simulated annealing.* London: Pitman.

Dawkins, R. (1971). Selective neurone death as a possible memory mechanism. *Nature, 229,* 118–119.

Dawkins, R. (1982). *The extended phenotype: The gene as the unit of natural selection.* San Francisco: Freeman.

Dawkins, R. (1983). Universal Darwinism. In D. S. Bendall (Ed.), *Evolution from molecules to men* (pp. 403–425). Cambridge: Cambridge University Press.

Dawkins, R. (1986). *The blind watchmaker.* New York: Norton.

Dawkins, R. (1989). *The selfish gene* (rev. ed.). Oxford: Oxford University Press. (Original work published 1976)

Demetras, M., Post, K., & Snow, C. (1986). Feedback to first language learners: The role of repetitions and clarification questions. *Journal of Child Language, 13,* 275–292.

Dennett, D. (1984). *Elbow room: The varieties of free will worth wanting.* Cambridge: MIT Press.

Dewey, J. (1896). The reflex arc concept in psychology. *Psychological Review, 3*(4), 357–370.

Dover, G. A. (1982). Molecular drive: A cohesive mode of species evolution. *Nature, 299,* 111–117.

Eccles, J. C. (1989). *The evolution of the brain: Creation of the self.* London: Routledge.

Edelman, G. M. (1987). *Neural Darwinism: The theory of neuronal group selection.* New York: Basic Books.

Edelman, G. M. (1988). *Topobiology: An introduction to molecular embryology.* New York: Basic Books.

Edelman, G. M. (1989). *Remembering the present: A biological theory of consciousness.* New York: Basic Books.

Edelman, G. M. (1992). *Bright air, brilliant fire: On the matter of the mind.* New York: Basic Books.

Edelman, G. M. (1993). Selection and reentrant signaling in higher brain function. *Neuron, 10*(2), 115–125.

Edelman, G. M., Reeke, G. N., Gall, W. E., Tononi, G., Williams, D., & Sporns, O. (1992). Synthetic neural modeling applied to a real-world artifact. *Proceedings of the National Academy of Sciences of the United States of America, 89*(15), 7267–7271.

Ehrlich, P. (1900). On immunity: With special reference to cell life. *Proceedings of the Royal Society of London, 66*(432), 424–448.

Eigen, M. (1992). *Steps toward life: A perspective on evolution.* Oxford: Oxford University Press.

Eldredge, N. (1985). *Time frames: The rethinking of Darwinian evolution and the theory of punctuated equilibria.* New York: Simon & Schuster.

Eve, R. A., & Dunn, D. (1989). High school biology teachers and pseudoscientific belief: Passing it on? *Skeptical Inquirer, 13,* 260–263.

Fairchild, D. (1930). *Exploring for plants.* New York: Macmillan.

Farmer, J. D., Packard, N. H., & Perelson, A. S. (1986). The immune system, adaptation and machine learning. *Physica, 22D,* 187–204.

Farnham-Diggory, S. (1992). *Cognitive processes in education* (2nd ed.). New York: HarperCollins.

Fernald, A. (1992). Human maternal vocalizations to infants as biologically relevant signals: An evolutionary perspective. In J. H. Barkow, L. Cosmides, & J. Tooby (Eds.), *The adapted mind: Evolutionary psychology and the generation of culture* (pp. 391–428). New York: Oxford University Press.

Feyerabend, P. K. (1975). *Against method: Outline of an anarchistic theory of knowledge.* London: NLB.

Fisher, R. A. (1930). *The genetical theory of natural selection.* Oxford: Clarendon Press.

Flam, F. (1994). Co-opting a blind watchmaker. *Science, 265,* 1032–1033.

Fodor, J. A. (1975). *The language of thought.* New York: Crowell

Fodor, J. A. (1980). Fixation of belief and concept acquisition. In M. Piattelli-Palmarini (Ed.), *Language and learning: The debate between Jean Piaget and Noam Chomsky* (pp. 143–149). Cambridge: Harvard University Press.

Fogel, L. J., Owens, A. J., & Walsh, M. J. (1966). *Artificial intelligence through simulated evolution.* New York: Wiley.

Ford, E. E. (1989). *Freedom from stress.* Scottsdale, AZ: Brandt.

Forssell, D. C. (1993). Perceptual control: A new management insight. *Engineering Management Journal, 5*(4), 17–25.

Forssell, D. C. (1994). Perceptual control management: Insight for problem solving. *Engineering Management Journal, 6*(4), 31–39.

Friston, K. J., Tononi, G., Reeke, G. N., Sporns, O., & Edelman, G. M. (1994). Value-dependent selection in the brain: Simulation in a synthetic neural model. *Neuroscience, 59*(2), 229–243.

Gamble, T. J. (1983). The natural selection model of knowledge generation: Campbell's dictum and its critics. *Cognition and Brain Theory, 6*(3), 353–363.

Gardner, H. (1987). *The mind's new science: A history of the cognitive revolution* (paperback ed.). New York: Basic Books.

Gazzaniga, M. S. (1992). *Nature's mind: The biological roots of thinking, emotions, sexuality, language, and intelligence.* New York: Basic Books.

Ghiselin, M. T. (1969). *The triumph of the Darwinian method.* Berkeley: University of California Press.

Gibbons, H. (1990). *The death of Jeffrey Stapleton.* Concord, NH: Franklin Pierce Law Center.

Gibbs, W. W. (1994). Virtual reality check. *Scientific American, 271*(6), 40, 42.

Giere, R. N. (1988). *Explaining science: A cognitive approach.* Chicago: University of Chicago Press.

Gleitman, L. (1994). The structural sources of verb meanings. In P. Bloom (Ed.), *Language acquisition: Core readings* (pp. 174–221). Cambridge: MIT Press.

Goldberg, D. E. (1989). *Genetic algorithms in search, optimization, and machine learning.* Reading, MA: Addison-Wesley.

Goldberg, D. E. (1994). Genetic and evolutionary algorithms come of age. *Communications of the Association for Computing Machinery, 37*(3), 113–119.

Goldstein, D. M. (1989). Control theory applied to stress management. In W. A. Hershberger (Ed.), *Volitional action: Conation and control* (pp. 481–491). Amsterdam: North-Holland.

Goldstein, D. M. (1990). Clinical applications of control theory. *American Behavioral Scientist, 34*(1), 110–116.

Gordon, P. (1990). Learnability and feedback. *Developmental Psychology, 26*(2), 217–220.

Gould, S. J. (1980). *The panda's thumb: More reflections in natural history.* New York: Norton.

Gould, S. J. (1991a). *Bully for brontosaurus: Reflections in natural history.* New York: Norton.

Gould, S. J. (1991b). Exaptation: A crucial tool for an evolutionary psychology. *Journal of Social Issues, 47*(3), 43–65.

Gould, S. J. (1993). *Eight little piggies: Reflections in natural history.* New York: Norton.

Gould, S. J., & Lewontin, R. C. (1979). The spandrels of San Marco and the Panglossian paradigm: A critique of the adaptationist programme. *Proceedings of the Royal Society of London, 285,* 281–288.

Gould, S. J., & Vrba, E. S. (1982). Exaptation—A missing term in the science of form. *Paleobiology, 8,* 4–15.

Greenough, W. T., & Black, J. E. (1992). Induction of brain structure by experience. In M. Gunnar & C. Nelson (Eds.), *Developmental behavioral neurosciences. Vol. 24. Minnesota symposia on child development* (pp. 155–200). Hillsdale, NJ: Erlbaum.

Hamburger, V. (1975). Cell death in the development of the lateral motor column of the chick embryo. *Journal of Comparative Neurology, 160,* 535–546.

Hamilton, W. D. (1964). The genetical evolution of social behaviour (I and II). *Journal of Theoretical Biology, 7,* 1–16, 17–52.

Hanson, N. R. (1958). *Patterns of discovery.* Cambridge: Cambridge University Press.

Harris, M. (1966). The cultural ecology of India's sacred cattle. *Current Anthropology, 7*(1), 51–66.

Harris, M. (1989). *Our kind.* Grand Rapids, MI: Harper & Row.

Hebb, D. O. (1949). *The organization of behavior: A neuropsychological theory.* New York: Wiley.

Hershberger, W. A. (1990). Control theory and learning theory. *American Behavioral Scientist, 34*(1), 55–66.

Hesse, M. B. (1964). Francis Bacon. In D. J. O'Connor (Ed.), *Critical history of Western philosophy* (pp. 141–152). London: Free Press of Glencoe/Collier-Macmillan.

Hirsch-Pasek, L., Treiman, R., & Schneiderman, M. (1984). Brown and Hanlon revisited: Mother's sensitivity to ungrammatical forms. *Journal of Child Language, 11*, 81–88.

Ho, M.-W., & Saunders, P. (1984). *Beyond neo-Darwinism*. London: Academic Press.

Holland, J. H. (1992). Genetic algorithms. *Scientific American, 267*(1), 66–72.

Hull, D. L. (1988a). Interactors versus vehicles. In H. C. Plotkin (Ed.), *The role of behavior in evolution* (pp. 19–50). Cambridge: MIT Press.

Hull, D. L. (1988b). *Science as progress: An evolutionary account of the social and conceptual development of science*. Chicago: University of Chicago Press.

Hume, D. (1952). An inquiry concerning human understanding. In R. M. Hutchins (Ed.), *Great books of the Western world* (Vol. 35, pp. 451–509). Chicago: Encyclopædia Britannica. (Original work published 1748)

Huttenlocher, P. R. (1984). Synapse elimination and plasticity in developing human cerebral cortex. *American Journal of Mental Deficiency, 88*(5), 488–496.

Huxley, L. (1900). *The life and letters of Thomas Henry Huxley* (Vol. 1). London: Macmillan.

James, W. (1880, October). Great men, great thoughts, and the environment. *Atlantic Monthly, 46*(276), 441–459.

James, W. (1890). *The principles of psychology* (2 vols.). New York: Holt.

Janeway, C. A., Jr. (1993). How the immune system recognizes invaders. *Scientific American, 269*(3), 73–79.

Jerne, N. K. (1951). A study of avidity. *Acta Pathologica Microbiologica Scandinavica Supplement, 87*, 1–183.

Jerne, N. K. (1955). The natural-selection theory of antibody formation. *Proceedings of the National Academy of Sciences of the United States of America, 41*(11), 849–857.

Jerne, N. K. (1967). Antibodies and learning: Selection versus instruction. In G. C. Quarton, T. Melnechuk, & F. O. Schmitt (Eds.), *The neurosciences: A study program* (pp. 200–205). New York: Rockefeller University Press.

Jerne, N. K. (1975). Structural analogies between the immune system and the nervous system. In I. R. Miller (Ed.), *Stability and origin of biological information* (pp. 201–204). New York: Wiley.

Johnson, G. (1992, April 19). Evolution between the ears. *New York Times Book Review*, p. 22.

Johnson, J., & Newport, E. (1989). Critical period effects in second language learning: The influence of maturational state on the acquisition of English as a second language. *Cognitive Psychology, 21*, 60–99.

Jones, T. A., & Schallert, T. (1992). Overgrowth and pruning of dendrites in adult rats recovering from neocortical damage. *Brain Research, 581*, 156–160.

Jones, T. A., & Schallert, T. (1994). Use-dependent growth of pyramidal neurons after neocortical damage. *Journal of Neuroscience, 14*, 2140–2152.

Joravsky, D. (1970). *The Lysenko affair.* Cambridge: Harvard University Press.

Joyce, G. F. (1992). Directed molecular evolution. *Scientific American, 267*(12), 90–97.

Kallenbach, S., Doyen, N., Fanton d'Andon, M., & Rougeon, F. (1992). Three lymphoid-specific factors account for all junctional diversity characteristic of somatic assembly of T-cell receptor and immunoglobulin genes. *Proceedings of the National Academy of Sciences of the United States of America, 89*, 2799–2803.

Kauffman, S. A. (1993). *The origins of order: Self-organization and selection in evolution.* New York: Oxford University Press.

Keller, H. (1904). *The world I live in.* London: Hodder & Stoughton.

Kelly, K. (1994). *Out of control: The rise of neo-biological civilization.* Reading, MA: Addison-Wesley.

Kimura, M. (1982). *The neutral theory of molecular evolution.* Cambridge: Cambridge University Press.

Kinnear, K. E., Jr. (Ed.). (1994). *Advances in genetic programming.* Cambridge: MIT Press.

Köhler, W. (1925). *The mentality of the apes* (2nd rev. ed., E. Winter, Trans.). New York: Harcourt, Brace.

Köhler, W. (1969). *The task of Gestalt psychology.* Princeton, NJ: Princeton University Press.

Kohn, A. (1993). *Punished by rewards: The trouble with gold stars, incentive plans, A's, praise and other bribes.* Boston: Houghton Mifflin.

Konner, M. (1982). *The tangled wing: Biological constraints on the human spirit.* New York: Holt, Rinehart & Winston.

Koshland, D. E., Jr. (1980). *Bacterial chemotaxis as a model behavioral system.* New York: Raven Press.

Koza, J. R. (1992). *Genetic programming: On the programming of computers by means of natural selection.* Cambridge: MIT Press.

Koza, J. R. (1994). *Genetic programming II: Automatic discovery of reusable programs.* Cambridge: MIT Press.

Kuhn, T. S. (1970a). Reflections on my critics. In I. Lakatos & A. Musgrave (Eds.), *Criticism and the growth of knowledge* (pp. 231–278). Cambridge: Cambridge University Press.

Kuhn, T. S. (1970b). *The structure of scientific revolutions* (2nd ed., enlarged). Chicago: University of Chicago Press.

Kuhn, T. S. (1991). The road since structure. *PSA 1990, 2*, 3–13. (East Lansing, MI: Philosophy of Science Association)

Larijani, L. C. (1994). *The virtual reality primer.* New York: McGraw-Hill.

Laudan, L. (1984). *Science and values: The aims of science and their role in scientific debate.* Berkeley: University of California Press.

Levy, S. (1992). *Artificial life: The quest for a new creation.* New York: Pantheon Books.

Levy, S. (1994, May 2). Dr. Edelman's brain. *New Yorker*, pp. 62–73.

Li, W.-H., & Graur, D. (1991). *Fundamentals of molecular evolution.* Sunderland, MA: Sinauer.

Lieberman, P. (1984). *The biology and evolution of language.* Cambridge: Harvard University Press.

Lieberman, P. (1991). *Uniquely human: The evolution of speech, thought, and selfless behavior.* Cambridge: Harvard University Press.

Locke, J. (1952). An essay concerning human understanding. In R. M. Hutchins (Ed.), *Great books of the Western world* (Vol. 35, pp. 85–395). Chicago: Encyclopædia Britannica. (Original work published 1690)

Lorenz, K. (1981). *Foundations of ethology.* New York: Simon & Schuster.

Lorenz, K. (1982). Kant's doctrine of the a priori in the light of contemporary biology. In H. C. Plotkin (Ed.), *Learning, development, and culture: Essays in evolutionary epistemology* (pp. 121–143). Chichester: Wiley. (Reprinted from *General Systems*, 1962, 7, 23–35. Original work published as Kants Lehre vom apriorischen im Lichte gegenwärtiger Philosophie. *Blätter für Deutsche Philosophie*, 1941, *15*, 94–125.)

Løvtrup, S. (1987). *Darwinism: The refutation of a myth.* London: Croom Helm.

McClelland, K. (1991, August). *Perceptual control and sociological theory.* Paper presented at the annual meeting of the Control Systems Group, Durango, CO.

McClelland, K. (1994). Perceptual control and social power. *Sociological Perspectives, 37*(4), 461–496.

McClure, F. A. (1986). *The bamboos: A fresh perspective.* Cambridge: Harvard University Press.

Macnamara, J. (1972). Cognitive basis of language learning. *Psychological Review, 79*(1), 1–13.

McPhail, C. (1991). *The myth of the madding crowd.* Greenwood, NY: Aldine/de Gruyter.

McPhail, C., Powers, W. T., & Tucker, C. T. (1992). Simulating individual and collective action in temporary gatherings. *Social Science Computer Review, 10*(1), 1–28.

McPhail, C., & Tucker, C. T. (1990). Purposive collective action. *American Behavioral Scientist, 34*(1), 91–94.

MacPhee, D. G. (1993). Directed evolution reconsidered. *American Scientist, 81*, 554–561.

MacWhinney, B. (1987). The competition model. In B. MacWhinney (Ed.), *Mechanisms of language acquisition* (pp. 249–308). Hillsdale, NJ: Erlbaum.

MacWhinney, B. (1989). Competition and teachability. In M. L. Rice & R. L. Schiefelbusch (Eds.), *The teachability of language* (pp. 63–104). Baltimore: Brookes.

Marcus, G. F. (1993). Negative evidence in language acquisition. *Cognition, 46,* 53–85.

Marken, R. S. (1986). Perceptual organization of behavior. *Journal of Experimental Psychology: Human Perception and Performance, 12,* 67–76.

Marken, R. S. (1989). Levels of intention in behavior. In Wayne A. Hershberger (Ed.), *Volitional action: Conation and control* (pp. 299–314). Amsterdam: North-Holland.

Marken, R. S. (1991). Degrees of freedom in behavior. *Psychological Science, 2,* 92–100.

Markman, E. M. (1994). Constraints children place on word meanings. In P. Bloom (Ed.), *Language acquisition: Core readings* (pp. 154–173). Cambridge: MIT Press.

Marrack, P., & Kappler, J. W. (1993). How the immune system recognizes the body. *Scientific American, 269*(3), 81–89.

Matsuoka, M., Nagawa, F., Okazaki, K., Kingsbury, L., Yoshida, K., Müller, U., Larue, D. T., Winer, J. A., & Sakano, H. (1991). Detection of somatic DNA recombination in the transgenic mouse brain. *Science, 254,* 81–86.

Maynard Smith, J. (1969). The status of neo-Darwinism. In C. H. Waddington (Ed.), *Towards a theoretical biology. Vol. 2. Sketches* (pp. 82–89). Edinburgh: Edinburgh University Press.

Maynard Smith, J. (1978). *The evolution of sex.* Cambridge: Cambridge University Press.

Mayr, E. (1966). Introduction to the facsimile edition of Charles Darwin's first edition (1859) of *The origin of species by means of natural selection* (pp. i–x). Cambridge: Harvard University Press.

Medvedev, Z. A. (1969). *The rise and fall of T. D. Lysenko* (L. G. Lawrence, Trans.). New York: Columbia University Press.

Mokyr, J. (1990). *The lever of riches: Technological creativity and economic progress.* New York: Oxford University Press.

Monod, J. (1971). *Chance and necessity: An essay on the natural philosophy of modern biology.* New York: Knopf.

Montessori, M. (1967). *The absorbent mind.* New York: Holt, Rinehart & Winston.

Munz, P. (1993). *Philosophical Darwinism: On the origin of knowledge by means of natural selection.* London: Routledge.

Murrell, J. C., & Roberts, L. M. (Eds.). (1989). *Understanding genetic engineering.* New York: Halsted Press (Wiley).

Musgrave, A. (1993). *Common sense, science and scepticism: A historical introduction to the theory of knowledge.* Cambridge: Cambridge University Press.

Nagy, W., & Herman, P. A. (1987). Breadth and depth of vocabulary knowledge: Implications for acquisition and instruction. In M. McKeown & M. Curtis (Eds.), *The nature of vocabulary acquisition* (pp. 19–59). Hillsdale, NJ: Erlbaum.

Newport, E. (1994). Maturational constraints on language learning. In P. Bloom (Ed.), *Language acquisition: Core readings* (pp. 543–560). Cambridge: MIT Press.

O'Connor, D. J. Locke. In D. J. O'Connor (Ed.), *A critical history of Western philosophy* (pp. 204–219). London: Free Press of Glencoe/Collier-Macmillan.

O'Hara, R. J. (1994, March-April). Chauncey Wright: Brief life of an "indolent genius": 1830–1875. *Harvard Magazine, 96*(4), 42.

Paley, W. (1813). *Natural theology* (14th ed.). London: Hamilton.

Paley, W. (1902). *Natural theology*. New York: American Tract Society. (Original work published 1802)

Palincsar, A. S., & Brown, A. L. (1984). Reciprocal teaching of comprehension-fostering and comprehension-monitoring activities. *Cognition and Instruction, 1*(2), 117–175.

Pauling, L. (1940). A theory of the structure and process of formation of antibodies. *Journal of the American Chemical Society, 62*, 2643–2657.

Pavlovski, R. P., Barron, G., & Hogue, M. A. (1990). Reorganizations of a control theory. *American Behavioral Scientist, 34*(1), 32–54.

Peckham, G. W. (1905). *Wasps, social and solitary*. Boston: Houghton Mifflin.

Penner, S. (1987). Parental responses to grammatical and ungrammatical child utterances. *Child Development, 58*, 376–384.

Perelson, A. S., & Oster, G. F. (1979). Theoretical studies of clonal selection: Minimal antibody repertoire size and reliability of self-non-self discrimination. *Journal of Theoretical Biology, 81*, 645–670.

Perkinson, H. J. (1980). *Since Socrates: Studies in the history of Western educational thought.* New York: Longman.

Perkinson, H. J. (1984). *Learning from our mistakes: A reinterpretation of twentieth-century educational theory*. Westport, CT: Greenwood Press.

Perkinson, H. J. (1993). *Teachers without goals/Students without purposes*. New York: McGraw-Hill.

Petrie, H. (1981). *The dilemma of enquiry and learning*. Chicago: University of Chicago Press.

Petroski, H. (1985). *To engineer is human: The role of failure in successful design*. New York: Martin's Press.

Piattelli-Palmarini, M. (Ed.). (1980). *Language and learning: The debate between Jean Piaget and Noam Chomsky*. Cambridge: Harvard University Press.

Piattelli-Palmarini, M. (1986). The rise of selective theories: A case study and some lessons from immunology. In W. Demopoulos & A. Marros (Eds.), *Language learning and concept acquisition* (pp. 117–130). Norwood, NJ: Ablex.

Piattelli-Palmarini, M. (1989). Evolution, selection and cognition: From "learning" to parameter setting in biology and in the study of language. *Cognition, 31*, 1–44.

Pinker, S. (1989). *Learnability and cognition: The acquisition of argument structure*. Cambridge: MIT Press.

Pinker, S. (1994). *The language instinct*. New York: Morrow.

Pinker, S., & Bloom, P. (1990). Natural language and natural selection. *Brain and Behavioral Sciences, 13*, 707–784.

Plato. (1952). The dialogues of Plato. In *Great books of the Western world* (Vol. 7, pp. 1–251). Chicago: Encyclopædia Britannica

Plooij, F. X. (1984). *The behavioral development of free-living chimpanzee babies and infants*. Norwood, NJ: Ablex.

Plotkin, H. (1987). The evolutionary analogy in Skinner's writings. In S. Modgil & C. Modgil (Eds.), *B. F. Skinner: Consensus and controversy* (pp. 139–149). New York: Falmer Press.

Plotkin, H. (1994). *Darwin machines and the nature of knowledge*. Cambridge: Harvard University Press.

Polanyi, M. (1958). *Personal knowledge*. London: Routledge & Kegan Paul.

Popper, K. R. (1963). *Conjectures and refutations*. London: Routledge & Kegan Paul.

Popper, K. R. (1964). *The poverty of historicism*. New York: Harper & Row.

Popper, K. R. (1966). *The open society and its enemies* (5th ed.). Princeton, NJ: Princeton University Press.

Popper, K. R. (1979). *Objective knowledge: An evolutionary approach* (rev. ed.). Oxford: Clarendon Press.

Powers, W. T. (1973). *Behavior: The control of perception*. Greenwood, NY: Aldine/de Gruyter.

Powers, W. T. (1989a). An outline of control theory. In *Living control systems: Selected papers of William T. Powers* (pp. 253–293). Gravel Switch, KY: Control Systems Group.

Powers, W. T. (1989b). *Living control systems: Selected papers of William T. Powers*. Gravel Switch, KY: Control Systems Group.

Powers, W. T. (1990, August). Personal communication.

Powers, W. T. (1991a, February). Skinner's mistake. *Control Systems Group Newsletter*, 7–14.

Powers, W. T. (1991b). Evolution and purpose. *Control Systems Group Network*, July 4, csg-l@vmd.cso.uiuc.edu.

Powers, W. T., Clark, R. K., & McFarland, R. L. (1960). A general feedback theory of human behavior (Parts 1 & 2). *Perceptual and Motor Skills, 11*, 71–88, 309–323. (Reprinted in Powers, W. T. (1989). *Living control systems* (pp. 1–45). Gravel Switch, KY: Control Systems Group.)

Quine, W. V. (1951). Two dogmas of empiricism. *Philosophical Review, 60*, 20–43. (Reprinted in Quine, W. V. (1980). *From a logical point of view* (pp. 20–46). New York: Harper Torchbooks.)

Quine, W. V. (1960). *Word and object.* New York: Wiley.

Reader, J. (1988). *Man on earth.* Austin: University of Texas Press.

Reeke, G. N., Jr., & Sporns, O. (1990). Selectionist models of perceptual and motor systems and implications for functionalist theories of brain function. In S. Forrest (Ed.), *Emergent computation* (pp. 347–364). Cambridge: MIT Press.

Richards, R. J. (1987). *Darwin and the emergence of evolutionary theories of mind and behavior.* Chicago: University of Chicago Press.

Richardson, G. P. (1991). *Feedback thought in social science and systems theory.* Philadelphia: University of Pennsylvania Press.

Ridley, M. (1985). *The problems of evolution.* Oxford: Oxford University Press.

Roberts, R. M. (1989). *Serendipity: Accidental discoveries in science.* New York: Wiley.

Robertson, R. J., & Powers, W. T. (1990). *Introduction to modern psychology: The control-system view.* Gravel Switch, KY: Control Systems Group.

Root-Bernstein, R. S. (1989). *Discovering.* Cambridge: Harvard University Press.

Rosenzweig, M. R., Møllgaard, K., Diamond, M. C., & Bennett, E. L. (1979). Negative as well as positive synaptic changes may store memory. *Psychological Review, 79*(1), 93–96.

Salmond, G. (1989). Gene cloning in *Escherichia coli.* In J. C. Murrell & L. M. Roberts (Eds.), *Understanding genetic engineering.* New York: Halsted Press (Wiley).

Sarnat, H. B., & Netsky, M. G. (1981). *The evolution of the nervous system* (2nd ed.). New York: Oxford University Press.

Schallert, T. & Jones, T. A. (1993) "Exuberant" neuronal growth after brain damage in adult rats: The essential role of behavioral experience. *Journal of Neurotransplantation and Plasticity, 4*(3), 193–198.

Schatz, C. J. (1992). The developing brain. *Scientific American, 267*(3), 60–67.

Schilpp, P. A. (Ed.). (1974). *The philosophy of Karl Popper* (2 vols.). La Salle, IL: Open Court.

Sheler, J. L. (1991, December 23). The creation. *U.S. News & World Report,* pp. 57–62, 64.

Sherman, P. W., Jarvis, J. U. M., & Braude, S. H. (1992). Naked mole rats. *Scientific American, 267*(2), 72–78.

Silverman, R. B. (1992). *The organic chemistry of drug design and drug action.* San Diego, CA: Academic Press.

Singleton, A. (1994, February). Genetic programming with C++. *Byte,* pp. 171–176.

Skinner, B. F. (1948). *Walden two.* London: Collier-Macmillan.

Skinner, B. F. (1953). *Science and human behavior*. New York: Free Press.

Skinner, B. F. (1957). *Verbal behavior*. New York: Appleton-Century-Crofts.

Skinner, B. F. (1966). The phylogeny and ontogeny of behavior. *Science, 153,* 1205–1213.

Skinner, B. F. (1971). *Beyond freedom and dignity*. New York: Knopf.

Skinner, B. F. (1974). *About behaviorism*. New York: Knopf.

Skinner, B. F. (1981). Selection by consequences. *Science, 213,* 501–504.

Sporns, O., & Edelman, G. M. (1993). Solving the Bernstein problem: A proposal for the development of coordinated movement by selection. *Child Development, 64*(4), 960–981.

Staddon, J. E. R. (1983). *Adaptive behavior and learning*. Cambridge: Cambridge University Press.

Stebbins, G. L. (1977). In defense of evolution: Tautology or theory? *American Naturalist, 111,* 386–390.

Stein, E., & Lipton, P. (1989). Where guesses come from: Evolutionary epistemology and the anomaly of guided variation. *Biology and Philosophy, 4,* 33–56.

Stoskopf, N. C. (1993). *Plant breeding: Theory and practice*. Boulder, CO: Westview Press.

Tannen, D. (1990). *You just don't understand: Women and men in conversation*. New York: Morrow.

Thagard, P. (1988). *Computational philosophy of science*. Cambridge: MIT Press.

Thomson, J. F. (1964). Berkeley. In D. J. O'Connor (Ed.), *A critical history of Western philosophy* (pp. 236–252). London: Free Press of Glencoe/Collier-Macmillan.

Tolman, E. C. (1932). *Purposive behavior in animals and men*. New York: Century.

Tolman, E. C. (1959). Principles of purposive behaviorism. In S. Koch (Ed.), *Psychology: A study of a science* (Vol. 2, pp. 92–157). New York: McGraw-Hill.

Tonegawa, S. (1983). Somatic generation of antibody diversity. *Nature, 302,* 575–581.

Tonegawa, S. (1985). The molecules of the immune system. *Scientific American, 253*(4), 104–113.

Tononi, G. (1994). Book review of Michael Gazzaniga's *Nature's mind. American Scientist, 82,* 289.

Toulmin, S. (1972). *Human understanding* (Vol. 1). Princeton, NJ: Princeton University Press.

Tributsch, H. (1982). *How life learned to live: Adaptation in nature*. Cambridge: MIT Press.

Trivers, R. L. (1971). The evolution of reciprocal altruism. *Quarterly Review of Biology, 46*(4), 35–57.

Turner, A. M., & Greenough, W. T. (1985). Differential rearing effects on rat visual cortex synapses. I. Synaptic and neuronal density and synapses per neuron. *Brain Research*, *329*, 195–203.

van Fraassen, B. C. (1980). *The scientific image*. Oxford: Clarendon Press.

Vidal, F., Buscaglia, M., & Vonèche, J. J. (1983). Darwinism and developmental psychology. *Journal of the History of the Behavioral Sciences*, *19*, 81–94.

Villegas, M. (1990). *Tropical bamboo*. New York: Rizzoli.

Vincenti, W. G. (1990). *What engineers know and how they know it: Analytical studies from aeronautical history*. Baltimore: Johns Hopkins University Press.

Vonèche, J. (1985). Genetic epistemology in the context of evolutionary epistemology. In J. C. Pitt (Ed.), *Change and progress in modern science* (pp. 199–232). Dordrecht, The Netherlands: Reidel.

Walsh, W. H. (1967). Kant, Immanuel. In P. Edwards (Ed.). *Encyclopedia of philosophy* (Vol. 4, pp. 305–324). New York: Macmillan.

Watson, J. B., & Rayner, R. (1920). Conditioned emotional reactions. *Journal of Experimental Psychology*, *3*, 1–14.

Weiner, J. (1994). *The beak of the finch: A story of evolution in our time*. New York: Knopf.

Wells, G. (1986). *The meaning makers: Children learning language and using language to learn*. Portsmouth, NH: Heinemann.

Werker, J. F., & Tees, R. C. (1984). Cross-language speech perception: Evidence for perceptual reorganization during the first year of life. *Infant Behavior and Development*, *7*, 49–63.

White, T., Graves, M., & Slater, W. (1990). Growth of reading vocabulary in diverse elementary schools: Decoding and word meaning. *Journal of Educational Psychology*, *82*, 281–290.

Wiener, N. (1948). *Cybernetics: Or control and communication in the animal and the machine*. Cambridge: MIT Press.

Wilson, E. O. (1975). *Sociobiology: The new synthesis*. Cambridge: Harvard University Press.

Wilson, E. O. (1978). *On human nature*. Cambridge: Harvard University Press.

Wright, C. (1971). *Philosophical discussions*. New York: Burt Franklin. (Original work published 1877)

Youatt, W. (1834). *Cattle: Their breeds, management, and disease*. London: Library of Useful Knowledge.

Young, J. Z. (1964). *A model of the brain*. Oxford: Clarendon.

Index